Emergence and Modularity in Life Sciences

Lars H. Wegner · Ulrich Lüttge
Editors

Emergence and Modularity in Life Sciences

Springer

Editors
Lars H. Wegner
Karlsruher Institut für Technologie
Eggenstein-Leopoldshafen, Germany

Ulrich Lüttge
FB Biologie
TU Darmstadt
Darmstadt, Germany

ISBN 978-3-030-06127-2 ISBN 978-3-030-06128-9 (eBook)
https://doi.org/10.1007/978-3-030-06128-9

Library of Congress Control Number: 2018966398

This Springer imprint is published by the registered company Springer Nature Switzerland AG
The registered company address is: Gewerbestrasse 11, 6330 Cham, Switzerland

Preface

Natural sciences tend to follow reductionist concepts leading to the identification of smaller and smaller entities, which we may call modules. Specialization has the inherent tendency of being driven too far so that the specialist in the end knows more and more about less and less until knowing all about nothing (Lorenz 1977, p. 51). Emergence of new unitary systems by integration and self-organization is juxtaposed to modularity (Lüttge 2012). Empirically we do need to describe and know the modules well. However, as already noted by Aristotle (384–322 BC), the whole is more than the sum of its parts, and reductionist approaches tend to neglect features of complex systems that result from non-reducible emergence. While the reference to Aristotle demonstrates that the phenomenon itself has been recognized very early, we are still in need of rigorous scientific concepts systematically describing how new features are generated in complex systems by the integration of modules.

The Aristotelian view is often quoted as the essence of emergence. Lorenz (1977, pp. 47–49) characterizes emergence as a sudden "fulguration" with lightning of innovation. For Karl Popper as a rule emergence appears to happen by a change of function of something that has already existed (Popper 2013[1]). Jan C. Schmidt recognizes in emergence the answer to the question if there is anything new under the sun. In this vein, from the perspective of the philosophy of science with the introductory chapter (Chapter "Is There Anything New Under the Sun?"), he lays the ground to this book on the relations of emergence and modularity where in an unstable world self-organization and emergence are twins. It remains a paradox how entities and processes with inherent instabilities integrate to form remarkably robust complex systems—in fact, instabilities appear to be the very ontological core of self-organization, allowing emergence and restricting methodological reductionism.

[1]P. 19; "In der Regel scheint Emergenz sich durch eine Funktionsänderung von etwas zu ereignen, das schon existierte".

In Part I of our book on the basics, Marc-Thorsten Hütt (Chapter "Modular Organization and Emergence in Systems Biology") then moves on to explore how modularity and emergence give birth to systems in systems biology and systems medicine. Systems biology, using a bottom-up approach, describes interactions ("links" or "edges") of elements ("nodes") forming a network that represents a complex system. In this terminology, "modules" are subgroups of elements that are densely interconnected. In this view, "emergence" results from the fact that patterns of preferred states are established and sustained in these networks, and provide systems with stable, new features. Hence, emergence corresponds to the spontaneous onset of a collective ("self-organized") state under variation of an intrinsic (local) parameter. For living systems, which are always open systems not in a thermodynamic equilibrium, but rather in a dynamic pseudo-state state (Chapter "Is There Anything New Under the Sun?"), questions of stability and sustainability are paramount. Fabio R. Scarano (Chapter "The Emergence of Sustainability") fathoms the emergence of sustainability in general terms where he asks for glue concepts to bind together social, economic and environmental dimensions of sustainability, such as policy, moral value and science emerging as a new utopia. Sustainment of the biosphere or Gaia sensu Lovelock (1979) highlights the dilemma of failure or success of sustainability which might determine the fate of humankind on the planet. This aspect is revisited and extended in the last two chapters of this book.

In Part II, we deal with modularity and emergence of systems of life, although we note that naturally emergence also occurs in the inorganic world. Robert B. Laughlin (2005) considers that in scaling up all physical laws become emergent. Emergence is day-to-day experience in engineering and technology where out of unstructured raw materials modules are created and their integration leads to the construction of sophisticated emergent systems serving various purposes and applications. Some aspects of that come up in Chapter "Emergence in Biomimetic Materials Systems" on biomimetics.

Before that Lars H. Wegner (Chapter "Modularity *Versus* Emergence: How to Cope with Complexity in Whole-Plant Physiology?") introduces modular features of biological systems, which result from autonomously acting components, and emergent features emanating from their interaction. Both juxtaposition and interrelation are considered in a synthesis of the two concepts. Degrees of modularity and emergence are assessed in a quantitative way; formalism is suggested to characterize, e.g., the contribution of single proteins to a complex, scalable trait (taken as a given phenomenon that cannot be reduced to the activity of a certain set of proteins). This approach is top-down, in contrast to the mathematical framework offered by systems biology (compare Chapter "Modular Organization and Emergence in Systems Biology"; note also the somewhat different understanding of "modularity" in both approaches). This theoretical assessment is illustrated by examples from systemic plant physiology, i.e. specifically by functional integration of the proteins involved in self-organization of photosynthetic metabolism and in stomatal regulation in networks at higher organizational levels.

The chapter on biomimetics by Thomas and Olga Speck (Chapter "Emergence in Biomimetic Materials Systems") looks at the emergent characteristics in the knowledge transfer from biological role models to biomimetic materials systems. We learn something in general about emergence in technological innovation and, more specifically, how biomimetics combine the disciplines of biology and technology for solving technical problems through abstraction, transfer and application of insights derived from biological models. Actual examples are described and depicted. Four types of emergence can be distinguished. Combining individual elements with specific physical properties causes emergence (1) of an additional "complex" physical property of the same class or (2) of another class. Societal values emerge and are belonging to the biomimetic product which (3) can be attributed to the biological role models or (4) cannot be attributed to them. It is evaluated if and how biomimetic products may contribute to bioinspired sustainability, here defined as the trade-off between the fulfilment of specific functions and the respective resource demand.

After this interlude on technical applications of emergence from biology, specific complex biological systems at increasing levels of scale are discussed in Chapters "Roots of Complexity in the Self-referential Genetic Code"–"Ecology: Ecosystems and Biodiversity", namely the genome, brains, whole plants and ecosystems. In Chapter "Roots of Complexity in the Self-referential Genetic Code" Romeu Cardoso Guimarães elucidates molecular details contributing to an evaluation of the emergence of the genetic code in pre-Darwinian evolution in a self-organized process. A self-referential model is presented involving feedback between proto-transfer-RNA (tRNA) and the peptides produced by it, and where the self-stimulation capacity (positive feedback) is explaining the formation of the genetic code based on protein synthesis directed by dimers of proto-tRNA. A concise overview of embryonic development of animals in the chapter by Paul Gottlob Layer (Chapter "Brains Emerging: On Modularity and Self-organisation of Neural Development In Vivo and In Vitro") shows that the field of developmental biology is governed by the genesis of emergent properties. Embryonic development is not only determined by a set of specific genes but also by space–time conditions of its developing modules which shape the environment for the successive developmental step. The biological emergence is based on gene–protein networks and declines exclusive gene centrism. The development of brains originates from the formation of cellular spheres (blastulae) and their transformation into planar epithelia and the neural tube from which also the eye is derived. Tissue culture experiments show that individual cells can also self-organize to spheroids in vitro demonstrating the enormous plasticity of the developmental process (which nonetheless seems to be "directed" in a way that remains elusive yet). Some readers may miss a separate chapter here dealing with the human mind, self-consciousness and how they emerged from matter. However, summarizing the abundant recent literature on this topic adequately would fill a book of its own. Hence, this topic is not covered in depth here, but some aspects are discussed in Chapter "Brains Emerging: On Modularity and Self-organisation of Neural Development In Vivo and In Vitro" and further below in Chapter "Emergence and Sustainment of Humankind on Earth: The Categorical Imperative".

While animals possess a central organ of computing input of signals and information and output of reactions, such as the brain, this is lacking in plants. They have often been considered merely modular organisms. However, the chapter by Ulrich Lüttge (Chapter "Plants: Unitary Organisms Emerging From Integration and Self-organization of Modules") shows that they emerge as unitary organisms by integration of their organs (roots, stems and leaves) and other modules on the finer scaling of tissues, cells and molecules. Integration is organized during development. There is signalling carrying information accepted by signal receptors and signal transduction as instruction for coordinated reactions in the whole plant, e.g. source–sink relations, induction of flowering, orientation in space under gravity and light, and responses to environmental cues. Plants are unitary organisms without a neural system. The first time when a systemic approach was used in biology probably was in 1935 when Sir Arthur Tansley coined the term "ecosystem" comprehending nature as a living quasi-organism (Tansley 1935). Systems are intrinsically emergent as elucidated for ecosystems in the chapter by Erwin Beck (Chapter "Ecology: Ecosystems and Biodiversity"). Ecosystems are structured by abiotic and biotic components and function via their interactions. They are emerging from coordination of their components, viz. species and populations and functional modules. Two types of experimental ecological approaches can be used to characterize emergent ecosystem functions, an analytical one by artificially imposing environmental changes in natural ecosystems and a synthetic one by using artificial species compositions in an otherwise natural environment. Ecological models are built up from that, allowing simulations of heuristic potential which are also relevant for predictions and for developing strategies for human land use under the challenges of vanishing resources (see also further below, Chapter "Emergence and Sustainment of Humankind on Earth: The Categorical Imperative") and global change.

For visualizing ultimate integration in emergence in Part III, it is necessary to consider the complex hierarchical organization of biological systems as in Chapter "Emergent Properties and Stability in Hierarchical Biosystems: There Is no Privileged Level of Causation" by Gustavo M. Souza and Marcelo N. Do Amaral. Each level of organization consists of multiple elements of the next lower level. This implies the occurrence of bottom-up and top-down events. The central gene dogma of molecular biology is that in a bottom-up or upward causation all is determined by the information contained in the DNA. However, transmission of information is not one way, upper levels can regulate lower levels including genes at the bottom, biological functionality is multilevel, and there is no privileged direction of causality. A system hierarchical view in biological thinking reveals that in a true dynamic feedback circuit in close relation to bottom-up (upward) causation there is also top-down (downward) causation. With the last two chapters, we eventually arrive at the top, i.e. at the highest level of organization with holobionts (or, rather, holobiont-like systems, see below) such as the entire biosphere and Gaia. All life is emergent, and this includes humankind. Therefore, Ulrich Lüttge and Fabio R. Scarano (Chapter "Emergence and Sustainment of Humankind on

Earth: The Categorical Imperative") then specifically address the sustainment of life in view of massive extinction waves with submergence of the old and emergence of the new in the history of life on Earth. Mastering the challenge of sustainability in the contemporary sixth and now anthropogenic extinction wave decides about plausible futures in the new utopia. The basic definition of holobiont is that it is a host organism (plant or animal) in interaction with all associated microorganisms as an entity for selection in evolution. We can extend it from microorganisms to organisms in general. A holobiont is the network of interactions which as a whole is subject to natural selection rather than the individual organism per se, and this network comprises information which can persist beyond the lifespan of the individual organisms participating in it. With an evolutionary motivation generalizing the concept, the holobiont-like system is introduced by Wolfgang zu Castell, Ulrich Lüttge and Rainer Matyssek (Chapter "Gaia—A Holobiont-like System Emerging From Interaction"). This shifts the focus from carrier of information in the structure of genes and other components to the process of the transmission of "infotypes" (which is the total information stored in a system needed for survival in its environment). Performance of Gaia or the entire biosphere as a holobiont-like system is built on multilevel interactions similar to the outcome of the considerations of hierarchy in Chapter "Emergent Properties and Stability in Hierarchical Biosystems: There Is no Privileged Level of Causation". Gaia is optimizing interaction. With the notion of self-organization and self-stabilization of Gaia the question of sustainment of the biosphere and life on Earth in future evolution is implicated, and this brings us back to Chapter "The Emergence of Sustainability".

Eggenstein-Leopoldshafen, Germany — Lars H. Wegner
Darmstadt, Germany — Ulrich Lüttge

References

Laughlin RB (2005) A different universe—reinventing physics from the bottom down. Basic Books, New York

Lorenz K (1977) Die Rückseite des Spiegels. Versuch einer Naturgeschichte menschlichen Erkennens. Deutscher Taschenbuch Verlag, München

Lovelock JE (1979) Gaia: a new look at life on Earth. Oxford University Press, Oxford

Lüttge U (2012) Modularity and emergence: biology's challenge in understanding life. Plant Biol 14: 865–871

Popper KR (2013) Eine Neuinterpretation des Darwinismus. Die erste Medawar-Vorlesung 1986. Aufklärung und Kritik 1:7–20

Tansley AG (1935) The use and abuse of vegetational concepts and terms. Ecol 16:284–307

Acknowledgements

We are grateful to all authors for their willingness to contribute to this book, making it a success by investing their profound expertise and their valuable time. We would also like to thank Dr. Christina Eckey from the publisher Springer Life Sciences for her encouragement and help during the initiation of this book project and her constant support and her patience during the process of its compilation. Thanks are due to Dr. Francisco Cánovas for giving us advice during the review process.

Contents

Part III Ultimate Integration

Part I
The Basics

Is There Anything New Under the Sun?

Instability as the Core of Emergence

Jan Cornelius Schmidt

Abstract This chapter aims to give substance to a contemporary understanding of emergence from the perspective of the philosophy of science. Looking at progress in the natural sciences, it draws on the concept of self-organization in order to provide a characterization of emergence. From the 1960s onward, theories of self-organization (including complex systems theory, nonlinear dynamics, chaos theory, synergetics, dissipative structures, fractal geometry, and autopoiesis theory) explicitly addressed emergent behavior. Referring to those theories, this chapter enquires into the ontological core as well as into the methodological and epistemological characteristics of self-organization—and, hence, of emergence. It is asked whether there is unity in the diversity of self-organizing phenomena in nature. It will be shown that instabilities constitute the ontological core of self-organization and are, therefore, central to any semantically meaningful understanding of emergence. Besides this ontological condition for the possibility of emergence (instability), the chapter also reveals that there are three further ontological characteristics of emergence, namely, (1) novelty, (2) processuality/temporality, (3) internality ("self"). In addition, related methodological and epistemological characteristics encompass: (4) limits in reproducibility/repeatability, (5) obstacles to predictability and (6) deficits in testability and reductive describability/explainability. In sum, since instabilities play a crucial role in nature, they are essential to any present-day concept of emergence.

Keywords Progress in physical sciences · Philosophy of science · Emergence · Self-organization · Unity in diversity · Instability · New view of nature

J. C. Schmidt (✉)
Philosophy of Science and Technology, Department of Social Sciences, Darmstadt University of Applied Sciences, Haardtring 100, D-64295 Darmstadt, Germany
e-mail: jan.schmidt@h-da.de

L. H. Wegner and U. Lüttge (eds.), *Emergence and Modularity in Life Sciences*,
https://doi.org/10.1007/978-3-030-06128-9_1

Introduction

Is there anything new under the sun?[1] If there is, how and to what extent can we grasp, describe and explain it?—These questions constitute the very center of the debate on the concept of emergence. Despite the historically late appearance of the term "emergence"—it was coined by George H. Lewes in his *Problems of Life and Mind* in 1875 (Lewes 1875, p. 412), made popular in the first decades of the twentieth century by the scholars of British Emergentism,[2] and has experienced a renaissance in the debate on neuro-/cognitive sciences in the last 25 years—the content of this term has a fairly long history.

Since ancient Greek times the appearance of novelty, the creation of newness, and the "fulguration" of new systemic properties have been ongoing topics of philosophical and scientific inquiry. At the end of the eighteenth century Immanuel Kant, in his *Critique of Judgment,* speaks of "self-organizing beings"—a notion that shares much in content with "emergence". Similar ideas can be found in the works of Johann Wolfgang von Goethe or Friedrich Schelling. Particularly Schelling embraces Kant's idea of "self-organizing beings" and envisions a new science—"speculative physics"—to conceptualize a processual worldview: *natura naturans* as active, creative, acting nature. Atleast since Charles Darwin's *Origins of Species* in 1859 and since the advancement of biology and the establishment of an evolutionary view of nature it has become scientifically evident that there is newness under the sun.

The current philosophical discourse on the concept of emergence is inherently intertwined with the notion and concept of self-organization—even in the domain of what has become popular under the umbrella term neuro-philosophy. While "emergence" appears to be, more or less, a philosophic or reflective term, "self-organization" is used much more commonly within the sciences—and also among the public. Both terms, emergence and self-organization, can, in fact, be considered as twin sisters.[3] From the 1960s, theories of self-organization, such as complex systems theory, nonlinear dynamics, chaos theory, synergetics, dissipative structures, fractal geometry, catastrophe theory, hyper-cycle theory, and autopoiesis explicitly addressed emergent behavior, including self-structuring, pattern formation, growth processes, information generation and, also, the arrow of time. These new approaches were enabled by the rapid advancement of computer technology and the new ability to numerically handle nonlinear equations that govern emergent processes. Self-organization theories were tremendously influential on scientific development in the second half of the twentieth century. New and highly relevant research fields such as brain/neuro/cognitive sciences, artificial intelligence, robotics, and autonomous agents have been established. Even traditional disciplines like physics are widening

[1] Well known in the Latin statement: Nihil sub sove novum (colloquial translation at Ecclesiastes 1:9).

[2] For instance: Conway Lloyd Morgan, Samuel Alexander, Roy W. Sellars, and William McDougall.

[3] The concepts are often used interchangeably, although "self-organization" is broader in content than "emergence".

their methodological focus. This all gives substance to the diagnosis of a structural-scientific paradigm shift within the architecture of sciences.

Induced by epochal advancements in self-organization theories since the 1960s, a new debate on emergence and the related "isms"—emergentism, reductionism and holism—began.[4] Many of the scholars involved in the debate subscribe to—at least some of—the following characteristics that any sufficiently rich concept of emergence has to encompass.[5]

First, *novelty of systemic properties, entities, structures, patterns, or functionalities.* Emergence refers to the relationship between a system's properties and their microstructure. Emergent properties are considered systemic properties: ones that only a system, but no single part of the system, possesses. They are novel with regard to what exists at the level of the microstructure. Many classic concepts of emergence focus on mental properties or mind as the one and only emergent property of interest. Others lean more toward physical or biological entities, structures, patterns, or functionalities.[6]

Second, *processuality, temporality, time-dependency, or evolution.* Emergence is not a static concept; it is highly dynamic. When we speak of emergence, we are referring to a process of evolving—and we consider how things come into being. Processuality is, indeed, a central element that is inherently linked with novelty.

Third, *internality* or *synchronic determination.* The emergent properties or entities existing or coming into existence depend only on the system's microstructure; they consist solely of natural, physical, and/or biological parts. In the notion of self-organization, the term "self" underlines that the properties classified as emergent are instantiated by the system itself—and not by its environment. No supernatural external actor is necessary to change the state of the system from the outside. This represents, at least to some extent, a physical monism or a minimal naturalist viewpoint.

Further characteristics, as set forth as follows, are epistemological or methodological in nature. Fourth, in contrast with other branches of sciences that do not deal with emergent phenomena, we experience problems of *reproducibility.* Emergent properties are often hard to reproduce in detail. Fifth, some aspects of *unanticipability* and *unpredictability* are related to the emergent phenomena. Sixth, certain *limits in testability* of theories and the explanatory *irreducibility* of knowledge, theories, models, or description length are associated with emergence. Emergence places limits on reductionist approaches, notably on types of reductionist explanations.

It follows from the above that the notion of "emergence" is interlaced with ontological (1–3) and with epistemological and methodological (4–6) dimensions. Concepts of emergence may differ since some of them put strong emphasis on novelty (1) but

---

[4] This vagueness of the term has challenged philosophers such as Ansgar Beckermann, Hans Flohr, Jaegwon Kim and Manfred Stöckler to provide clarification.

[5] The main points of the list that follows can also be found in Stephan (2007) and in more detail in Schmidt (2015). However, Stephan (2007) distinguishes between different types of emergence. As we are here less interested in semantic details and more concerned with providing a conceptual framework, we will not make further differentiations at this point.

[6] The properties might also include aspects of information, dispositions, or risks.

are less stringent with respect to internality (3) and may even advocate a non-specific understanding of physical monism. Besides novelty, other positions mainly consider irreducibility (6) and the lack of (mechanical or nomological) explanations but are reluctant to employ characteristics such as processuality (2). In sum, strong concepts of emergence require all six characteristics; weaker versions claim that only a few of them are relevant.

In this chapter, I aim to give further substance to a present-day concept of emergence by providing some background reflection on the above characteristics. I will begin by underlining that a current concept of emergence needs to relate, as already briefly stated, to the advances in self-organization theory or in the mostly synonymously understood concept of complex systems theory[7] interlaced with the crucial role of instabilities. Self-organization theory is a young and fascinating field of scientific inquiry that spans many established disciplines (cp. Mainzer 1996; Schmidt 2008a, b, 2015). Since the advent of self-organization theory and its advancement in the realm of physical sciences in the 1960s, further evidence for the existence and prevalence of emergence in nature has arisen. These developments in the physical sciences lend weight to findings already presented in disciplines such as biology, in particular in evolutionary and development biology—and give them a further, more profound, foundation.[8]

The core element and the source of nature's capacity[9] to enable emergence are instabilities—this is the main thesis put forward in this chapter. The same holds for macro-phenomena, say, in classic statistical thermodynamics such as entropy and the arrow of time, where the collision of particles causes instabilities at the micro-level.[10] Generally speaking, although instabilities call some implicit metaphysical-methodological convictions and well-established underlying prerequisites of physical sciences into question—such as predictability and reproducibility, and especially: reductionism—, they are by no means viewed in just a negative way today. On the contrary, instabilities are highly valued; we diagnose a "positivization" of instabilities insofar as they constitute the central element of emergence and self-organization. Without instability, there is no change, no emergence, no complexity, no evolution. In the face of the obvious phenomena around us, instabilities cannot be considered as exceptions within a stable world. Rather, it is the other way around: stability is only local, it is restricted to certain spatial and temporal domains—and it is mostly based on instability. Instability is the root of many familiar phenomena in our day-to-day experience—for example, living organisms and humans as well as water freezing to ice in a refrigerator.[11]

[7]Including Nonlinear Dynamics, Chaos Theory, Synergetics, Dissipative Structures, Fractal Geometry, and Catastrophe Theory.

[8]In fact, we can observe a convergence of some sub-disciplines of physics, informatics, biology, chemistry, and medicine.

[9]I wish to stress here at the beginning of the chapter that the same holds for social or technical complex systems.

[10]For instance, as in a billiard system, which has become a paradigm in this regard.

[11]Instability has, therefore, to be regarded as an empirical fact of our life-world and beyond—not just as a contingent convention.

Discovering and acknowledging instabilities as the core of emergence impel a reconsideration of the metaphysical view that undergirded the traditional, mistaken focus on stable, unchanging properties for a long time in the history of science. The outline of this chapter is as follows: in section "Instabilities as the Ontological Core of Emergence: Novelty, Processuality, Internality", I show the history leading to the acknowledgement of instabilities and provide a characterization of emergence. In section "Types of Instabilities", I discuss the nature of instabilities in more detail and distinguish between three kinds of instability. In section "Limits of Physical Sciences?—Methodological and Epistemological Issues", I focus on methodological and epistemological problems and challenges induced by instabilities; the limitations of traditional concepts of modern sciences will be discussed. These issues are also central to providing an understanding of what constitutes "emergence". In section "Science in an Unstable World", I show some methodological aspects of how present-day physical science manages to cope, at least to some degree, with instabilities.

Instabilities as the Ontological Core of Emergence: Novelty, Processuality, Internality

Nothing New Under the Sun and the Disregard of Instability

For a long time, physical scientists disregarded instabilities—and, in doing so, they overlooked emergent phenomena. "Nature" was defined as nature by stability. Stability metaphysics dominated our history. Since ancient Greek times, it was not doubted that instabilities allude to a lack of knowledge. Nature, stability, and knowledge were mainly used as synonyms. Plato's cosmos was structured by a Demiurge according to the idea of simple mathematical laws, intrinsic harmony, time-invariant order, and universal stability. Even if nature's harmony, simplicity and stability seemed obscure at first glance, there was no doubt about the existence of such characteristics behind the apparent complexity, guaranteeing the timeless and infinite existence of nature as a whole. Plato's foundation of theoretical science was so influential that, from the beginning of modern science in the early seventeenth century, mathematical invariants—from Platonic bodies to Newton's laws and Einstein's equations—were considered the core of nature.

In modern times, stability metaphysics can be found in the works of outstanding physicists such as Newton and Einstein. For instance, in his *Opticks* Newton did not trust his own nonlinear equations for three- and n-body systems which potentially can exhibit unstable solutions (Newton 1770). He required God's frequent supernatural intervention to stabilize the planetary system. In the same vein, Einstein (1917) introduced ad hoc, and without any empirical evidence or physical justification, the cosmological constant in the framework of general relativity to guarantee a static and stable cosmos, "Einstein's cosmos". Both examples, from Newton and Einstein illustrate that metaphysical convictions—*what nature is* or *has to be*!—can be incredibly strong, even if they contradict what is known about nature at the time. In other words, modern science, on the whole, did not scrutinize whether nature is stable or not. Science operated on the implicit stability assumption: all mathematical laws, models, and theories have to be stable in order to represent a stable world. Indeed, framing nature as nature insofar as it is stable was a successful strategy to advance a specific physical knowledge.[12]

A more in-depth look shows that the neglect and ignorance of instabilities in the history of sciences are remarkable because physical scientists (and later, biologists) have always been aware of the existence of unstable processes in nature as well as in models. The problem has been broadly identified ever since the development of hydrodynamics in the nineteenth century. The basic equations in this field include the Navier–Stokes equations describing unstable phenomena such as turbulent flows. In addition, Poincaré's description of the solar system and Einstein's general relativity theory reveal several types of instabilities. And even before then, instabilities were discussed in Newton's theory of the moon, and in Maxwell's classic mechanics: "matter and motion". Maxwell was a particularly prominent precursor in identifying instabilities. According to Maxwell, in physics "there is [... a stability] maxim which asserts 'That like causes produce like effects'. [But,] [t]his is only true when small variations in the initial circumstances produce only small variations in the final state of the system [...]. [T]here are other cases in which a small initial variation may produce a very great change in the final state of the system, as when the displacement of the 'points' causes a railway train to run into another instead of keeping its proper course" (Maxwell 1991, p. 13f). Maxwell was the first to explicitly identify the relevance of sensitive dependence on initial conditions—which is the nucleus of dynamical instability. Nevertheless, the prevalence of instabilities was not obvious in the nineteenth century. Instabilities were not broadly acknowledged at that time. Instead, they were regarded as exceptions within a stable world; instability was located just on the fringes of stability.

Acknowledging Instabilities

Conceiving instabilities as existing just on the fringes of stability—this view essentially changed during the 1960s. In that decade, we experienced the beginning of an interdisciplinary or, more specifically, a structural-scientific revolution caused by the rapid advancement of microelectronics and the development of computer technology. Lorenz (1963) discovered instabilities in weather forecast, Haken (1977) developed his synergetics, and Prigogine and Glansdorff (1971) formulated the nonlinear thermodynamics far-from-equilibrium of dissipative structures. In the early 1970s, Ruelle and Takens (1971) presented an influential paper on solutions of the hydro-mechanical Navier–Stokes equations and coined the term "chaos" metaphor-

[12] In this regard, stability metaphysics precedes scientific methodology.

ically for a certain class of solutions to partial differential equations. In 1975, Li and Yorke (1975) published an epoch-making mathematical article on one-dimensional difference equations, highlighting "chaos" explicitly in the title and providing a mathematically clear definition: "Period Three implies Chaos". During the 1970s, Mandelbrot (1991) elaborated his "fractal geometry", and Thom (1975) presented the "catastrophe theory". Many ideas were developed simultaneously. It was an exciting time of change in how nature and sciences were viewed, inducing a *structural-scientific* paradigm shift. The main overarching achievement was that instability was acknowledged as a basic feature of nature, and the relation of stability and instability was reversed. No longer was instability viewed as a minor subset of stability. On the contrary, stability became one tiny island in an ocean of instability: "order at the edge of chaos". Instability—not stability—is the prevalent case in nature; such an insight in the very fundament of nature should not be regarded as simply a negative message of destruction.[13]

In general, instabilities reflect situations in which a system is on a razor's edge: criticalities, flip points, thresholds, watersheds, sharp wedges. They generate butterfly effects or sensitive dependencies, bifurcations, points of structural changes, and phase transitions. The list of examples is extensive: the emergence and onset of a chemical oscillation, the roll dynamics of a fluid in heat transfer, an enzyme kinetic reaction, a gear chattering, or turbulence of a flow. A fluid becomes viscous, ice crystallization emerges, a phase transition from the fluid to a gas phase takes place, a solid state becomes super-fluid, laser light issues, a water tap begins to drip, a bridge crashes down, an earthquake or a tsunami arises, a thermal conduction comes to rest and a convection sets in, e.g., Bénard instability. New patterns and structures appear. Instabilities are the necessary condition for novelty—and for dynamical stability on a higher level, for metabolism and for living organisms.[14]

Today, ex post and thanks to the advancement of physical sciences, we can identify a "dogma of stability" that has implicitly determined the selection or construction not only of the models, laws, and theories but also of the objects of study. "We shall question the conventional wisdom that stability is an essential property for models of physical systems [...]. The logic which supports the *stability dogma* is faulty" (Guckenheimer and Holmes 1983, p. 259). The stability assumption is itself unstable! Our world is essentially a world of emergence, of dynamics, change, temporality, and evolution. The mere empirical fact that emergence and, more generally, self-organization is possible in the world we live in provides evidence that instabilities are not a mere convention but actually exist in the world. From an ontological and epistemological perspective, this point gives substance to a position that could be called *minimal instability realism*. In this vein, the physicist Langer (1980) underlines the role of "instabilities for any kind of pattern formation". According to Ebeling and Feistel

[13]As was seen in the 19th century in the debate on the heat death of the universe.

[14]Moreover, the various definitions and meanings of complexity—even if they do not refer to the genesis and evolution of a new pattern such as those in more geometric definitions of "complexity" via "dimensions"—refer directly or indirectly to instabilities (Atmanspacher et al. 1992; Wackerbauer et al. 1994).

(1990, p. 46), "self-organization is always induced by instability of the 'old' structure via small fluctuations". "This is why studying instability is of major importance." Nicolis and Prigogine (1977, p. 3f) argue that "instabilities" are the "necessary condition for self-organization". And Krohn and Küppers (1992, p. 3) emphasize that "instabilities are the driving force for systems evolution and development." In addition to novelty, some scientists in this field refer to the conditions of life or the "arrow of time". Traditional modern (mathematical) science was "unable to tackle and to resolve the antagonism between reversibility and irreversibility. To accomplish this, it was necessary to introduce a new concept: the concept of instability" (Ebeling and Feistel, 1990, p. 197).

Novelty, Processuality, and Internality

Paraphrasing Kant, instability has to be considered the condition for the possibility of self-organization and emergence. Although such a characterization draws on a very fundamental ontological element, it is not very specific. What can be deduced from this crucial fact? Now, since the notion of "self-organization" refers to the formation of new patterns, the emergence of new order, the creation of new structures, or the becoming of novel entities, it has at least two necessary dimensions: novel order on the one hand, and the process of creating novelty on the other. In other words, (1) novelty and (2) processuality (or temporality)—enabled by instabilities—constitute the very core of the notion of "self-organization". These two characteristics are in line with the works of two pioneers in the research on self-organizing phenomena—with Hermann Haken's *synergetics* and Ilya Prigogine's dissipative structures. Whereas Haken mainly focusses on new order, structures, and patterns (Haken 1977),[15] Prigogine advances a processual view of nature and draws upon the aspect of "becoming" (Prigogine 1980). In addition to the two characteristics of "novelty" and "processuality", the notion of "self-organization" points to (3) an internal source of the emergence of new order; no external force or actor is necessary. This internality of pattern formation is apparent in the prefix "self" of the word "self-organization".

Besides physical systems, biological systems or living organisms are particularly prominent examples to illustrate the overall enabling capacity of instability—and the three associated characteristics: (1) novelty, (2) processuality, and (3) internality. Living organisms are, in general, open systems through which a continuous flow of material, energy and information takes place. In fact, they are not stable in the sense that they attain a steady-state equilibrium. Such a steady-state would mean that they are dead; stability as understood in the physical sciences entails such a situation. However, living organisms do not follow the laws of equilibrium thermodynamics but the principle of nonlinear thermodynamics of irreversible processes far from equilibrium. Being open systems, they manage to maintain their overall structure for a while—this is why we speak of dynamic equilibria or pseudo-steady states.

---

[15]The term "organization" in "self-organization" emphasizes such characteristics.

A closer look at the models of the underlying mathematical structure—which are based on nonlinear thermodynamics (cf. Prigogine and Glansdorff 1971; Nicolis and Prigogine 1977)—would reveal that, in order to maintain the dynamical stability of the whole living organism on the meso-level, instabilities are a prerequisite for enabling the material, energy and information flow through the organism on the micro-level. Therefore, stability and instability are not disjunct; stability in complex systems such as in (and of) living organisms has to be regarded as a derivative of instability. Certainly, this insight reverses the view about the relationship between stability and instability.

In sum, instabilities are not solely the common denominator of all theories in the realm of self-organization.[16] In addition and more fundamentally, instabilities can be regarded as the ontological[17] core of the three central characteristics of emergence, namely of (1) novelty, (2) processuality/temporality, and (3) internality. In other words, instabilities are ontologically constitutive for any concept of emergence. I will now turn to providing a more detailed explication of what instability means—also for the purpose of showing that the other three characteristics of emergence (4–6) mentioned at the beginning of this article hold true. These latter characteristics are not ontological, but epistemological and methodological in nature. Now, at least three different, but cognate types of instability can be distinguished.

Types of Instabilities

Watersheds—and the Static Instability

Static instabilities and, synonymously, watersheds are ubiquitous. The related phenomena are well known from our everyday experience. For instance, a pen standing on its tip will eventually fall down to one side or another. Similarly, a ball on a roof ridge will roll down to one side or the other. The same occurs with a mass of a physical oscillator at the unstable point of maximum potential energy. In a pinball machine, the ball hits sharp wedges and jumps to the right or left. The quincunx, or "Galton's apparatus", consists of rows of alternating pins that are the source of instabilities. A tiny ball that falls on a pin passes it either on the right or on the left—this occurs with the ball several times on several rows. Ultimately, we observe a random-looking Gaussian distribution of all tiny balls at the bottom. That is why

[16]The priority of instability can also be shown for: symmetry breaking, phase transitions, time's arrow/irreversibility, deterministic chance, randomness, chaos, turbulence, catastrophes, fractals, information loss or gain. In general, however, the same argument holds with respect to the essential role of nonlinearity in these phenomena. But nonlinearity per se does not challenge traditional physical methodology. Many systems are nonlinear but stable, for example, simple two-dimensional planetary systems. They are, like linear systems, without chaos and turbulence, without self-organization and symmetry breaking. *Instability, and not nonlinearity by itself, makes the difference.*

[17]Or, more precisely, we could speak of the "onto-nomological core" of emergence (cf. Schmidt 2015).

instabilities once served as an (implicit) underlying idea for the development of classic probability theory. For Jacob Bernoulli, instabilities in coin tossing—and the resulting binary sequence of 0 (pitch) and 1 (toss)—are paradigmatic for producing randomness. Instabilities generate nomological randomness or, to put it in other words, deterministic chance.

The central idea of static instability traces back to Henri Poincaré at the end of the nineteenth century, when he inquired into issues of stability and instability within the planetary system and, more specifically, within the three-body system sun-earth-moon (Poincaré 1892): "A very small cause which escapes our notice determines a considerable effect that we cannot fail to see, and then we say that the effect is due to chance" (Poincaré 1914, p. 56f).[18] At points of static instability, a sensitive dependence on initial conditions occurs. Alternative trajectories from two nearby initial points separate and will never again become neighbors. In the course of the system's time evolution, one raindrop falling near the watershed may be transported to the Mediterranean Sea, while its neighbor goes to the North Sea. Between the two raindrops, there is a basin boundary or watershed—in other words, points of static instability. It deserves to be mentioned here that the fact that small differences in the initial conditions can cause large effects in the future is not only characteristic of what has been called deterministic chaos, but is—as we have seen—much more common.[19]

Overall, time evolution is of minor relevance for cases of static instability. After a certain time, the initial point has left the neighborhood of the watershed; the watershed no longer has any impact on time evolution. Instability, then, is not to be considered an epistemological or methodological challenge.

Chaos—and the Dynamical Instability

The relevance and impact of dynamical instability, or chaos, is more far-reaching. According to Friedrich Nietzsche (1930, p. 127), the "underlying character of the world is chaos". For Nietzsche, chaos is not associated with lawlessness and disorder but with an unforeseeable dynamics or a complex time evolution. His view of chaos differs from the messy *Tohuwabohu* of the Judeo-Christian tradition. That is why Nietzsche can be regarded as a precursor of our recent understanding of "chaos". Following Nietzsche, Martin Heidegger (1986, p. 87) advocates a positive under-

[18]Compare Maxwell's (1991, p. 13) remark above, forty years earlier.

[19]A special case of static instability is one that shows sensitive dependence of *all* points in the state space. A prominent example is compound interest. Tiny differences in seed capital grow exponentially into enormous differences years later. The linear differential equation $dx/dt = ax$ yields a solution that can be determined analytically: $x(t) = x(0) \cdot e^{at}$, with a constant Lyapunov coefficient: $a > 0$. Trajectories starting from a neighborhood reveal a static continuous divergence. In an open-ended state space, they need never converge to each other again. However, a closed solution does not exist for this linear differential equation. If the initial conditions are known, a calculation for all times is feasible and a prediction is possible.

standing of chaos: chaos means "dynamics, change, transition—whereas regularity seems to be hidden: we do not know the underlying law at first glance."[20] Clearly, Nietzsche and Heidegger did not define "chaos" in scientific terms. It was left up to the mathematicians of the 1970s to provide a distinct definition.

Chaos is essentially characterized by dynamical instability. The time evolution of a chaotic system is continuously on a razor's edge. Nearly all points in the state space exhibit the property of sensitivity: the trajectories separate exponentially by time evolution. Because chaotic attractors are bounded in the state space, we find exponential separation merely as a mean value. For short timescales, two trajectories may converge before they diverge again, and so on—an interplay of divergence and convergence among trajectories occurs. However, dynamically unstable (chaotic) trajectories do not show an exact recurrence. Henri Poincaré, the precursor of modern nonlinear dynamics and self-organization theory, studied the phenomenon of recurrence more than one hundred years ago.[21] As Poincaré observed, the dynamically unstable trajectory of a chaotic attractor resembles a ball of wool. It covers the attractor densely. Nearly all points are reached approximately by the trajectory after a certain time, for instance in the case of a chaotic double pendulum.

Chaos—and its core: dynamical instability—is associated with random-like behavior generated by deterministic models in a spatial-continuous state space in which every point is a watershed or divergence point: a random *pheno*type, in spite of a nomological *geno*type. Although Poincaré and later Birkhoff (1927) were aware of the possibility of random-like behavior, they did not have computer resources and numerical simulation tools to deal approximately with dynamically unstable systems. In the 1960s, Ed Lorenz, who discovered chaos in atmospheric dynamics through computer simulations, coined the term "butterfly effect" to illustrate its real impact: for dynamically unstable systems a tiny event such as a butterfly's wing flap in South America can cause a thunderstorm in the USA.[22] The butterfly effect is frequently associated with "weak causality" in order to underline that dynamical instability can

[20]Translation from German (JCS).

[21]Today, we utilize Poincaré's fundamental ideas to construct a specific map in order to obtain insights into the time evolution of a dynamical system—the so-called "Poincaré map".

[22]The term "butterfly effect", central to the notion of "chaos", was coined by Lorenz (1963, 1989). In addition, and in more detail, more than 15 competing notions and definitions of "chaos" exist (cp. Brown and Chua 1996, 1998). One of the most relevant chaos definitions nowadays has been suggested by Devaney (1987). Devaney requires a system's dynamics to exhibit three features in order to be called chaotic: (a) sensitive dependence on initial conditions (unpredictability); (b) topological transitivity (non-decomposability, wholeness); and (c) denseness of periodic points (elements of regularity). Banks et al. (1992) show from a mathematical perspective for one-dimensional maps that sensitivity (a) follows from properties (b) and (c). Thus, sensitivity does not seem to be a fundamental requirement for providing an adequate understanding of dynamical instability. From an empirical perspective, Devaney's and Banks' definitions—focusing on (b) and (c)—are regarded as interesting but hard to determine. Unlike mathematicians, physicists mostly advocate an understanding of chaos based on sensitivity coefficients. A classical sensitivity coefficient is Lyapunov's exponent that measures the average divergence of trajectories ("chaos coefficient"). For physics and its methodological need to deal with experimental uncertainties, a positive Lyapunov exponent is one major characteristic of dynamical instability.

still be considered part of a causal nexus of nature though not in the traditional sense of "strong causality".

Criticalities and Bifurcations—the Structural Instability

Although structural instability is central to nature—to enable biological and cosmological evolution, and to secure an overall stability of organisms—, it has not received the attention it deserves. Structural instability is associated with small changes in a system's structure. This type of instability occurs even in simple physical systems, for example in the heating of water, more specifically in the Rayleigh–Bénard convection where a fluid layer is confined between two thermally conducting plates and is heated from below, creating a vertical temperature gradient. For lower temperature gradients, and hence for smaller density gradients, an upward conductive heat transfer through horizontally uniform layers takes place. If the temperature gradient is sufficiently large, a point of structural instability or, synonymously, a bifurcation point or criticality occurs: The hot fluid rises, leading to a circular convective flow in three-dimensional cells, which results in enhanced transport of heat between the two plates.

In contrast to what we have discussed in the case of dynamical instability, structural aspects are far more fundamental. Structural instability does not solely refer to initial points or a particular trajectory, but to the whole dynamical system, represented by the mathematical structure of the model itself, namely the law or equation. In general, structural instability is expressed as follows: if one alters the structure of a model (equation, law) slightly, the overall dynamics changes qualitatively. The equivalence class of the system or model is not preserved; the disturbed and undisturbed models show different topological characteristics, for instance, different types of orbits or, at least, different periodicities.[23] Such a case occurs, for instance, in the onset of a chemical oscillation or of fluid roll dynamics under heat transfer, the emergence of laser light, or a water tap beginning to drip.

The first exact definition of structural stability and instability traces back to Andronov and Pontryagin (1937). They coined the term "robustness", or "coarse-grainedness", for structural stability. Birkhoff (1927) is another forerunner of the concept of structural instability/structural stability. In his influential book *Dynamical Systems,* he defines and distinguishes between different types of stability.[24] Some of Birkhoff's definitions paved the way and gave rise to what would later be known as structural instability. When *dynamical systems theory* was further developed, Andronov's and Pontryagin's term "robustness" was replaced by "structural

[23] Depending on the initial model class, altering a model can be performed by varying the model's equations, the system parameters or exponents, by adding mathematical functions or by modifying the state space.

[24] Such as trigonometric stability, stability of the first order, permanent stability, semi-permanent stability, unilateral stability, regions of stability, stability in the sense of Poisson (cp. Birkhoff 1927).

stability" with no change in content. Today, structural stability, with its converse of structural instability, is a major part of bifurcation theory (cp. Wiggins 1988; Ruelle 1989).

Structurally unstable phenomena in nature are not just exceptions in an ocean of stability; rather, the reverse is the case. According to Devaney (1987, p. 53), structural instability is most prevalent, even in classical physics. "Most dynamical systems that arise [even] in classical mechanics are not structurally stable." Given its prevalence, structural instability deepens and extends our discussion. Since Plato's time, theories (models, laws, equations) have been considered the nucleus of nature as well as the goal of science. In contrast, initial conditions—most important in cases of static and dynamical instability—were seen as merely contingent factors or an outer appendix. Since structural instability does not refer to initial conditions but to theories, it can be regarded as the most central (non-contingent) type of instability.

Limits of Physical Sciences?—Methodological and Epistemological Issues

It is undebatable that methods constitute what we call science: Methodology is indispensable for characterizing science. Although the debate is still ongoing without a final consensus, the center of scientific methodology can be considered as consisting of four interlaced elements that also highlight distinguishable objectives and different understandings of science: (1) reproducibility/repeatability, (2) predictability, (3) testability, and (4) describability/explainability/reducibility. Janich (1997, p. 62), for instance, identifies the "striking technical, prognostic and explanatory success" of physical sciences. Janich's list of three characteristics might be complemented by a further element relating to the testability of theories in experiments and, thus, falsifiability or verifiability.[25]

Instabilities—and emergent phenomena insofar as instabilities constitute their ontological core—pose challenges to the above mentioned four well-established methodological elements of physical science and reveal several problems, obstacles, and limitations. The implicit anticipation of some of the issues also underlies the "stability dogma" that we have already discussed. In order to show the pervasiveness of instability-based issues for physical sciences in general, we will look at the challenges relating to all four elements.

[25] It is important to stress, as Janich does convincingly, the experimental/engineering aspects of physics: physics aims to produce experimentally reproducible phenomena, and it is based on experimentation (cp. Hacking 1983).

Limits with Respect to Reproducibility and Experimentation

The core of modern science, which was introduced by Galileo and Bacon, is the method-based experiment. By means of intentional interventions in nature, we constitute, construct, and control empirical facts of nature. These facts are mostly not given but have to be produced or created (cp. Hacking 1983). According to Bacon and Kant, experiments serve as "interrogation techniques" to "force nature to reveal her secrets". Experiments—and not passive observation with unaided senses—guarantee empirical evidence, objectivity or, at least, intersubjectivity: anybody performing a particular experiment in any place and at any time will obtain the same result—if she or he acts in accordance with the norms of experimentation. Classical-modern science is based on the methodological assumption of person, location, and time independency.

Traditionally, scientists presuppose that they can, at least in principle, control the experimental setup, including the relevant boundary conditions. In other words, we assume the validity of the *principle of isolation*, namely that we can sufficiently isolate the system from its environment. For example, we do not expect a bumble-bee on the double star Sirius or a falling leaf in another country to have an overall influence on our experiment. Otherwise a particular phenomenon under consideration would not be reproducible. In its core, the distinguishing character of an experiment is, therefore, reproducibility. Scientists have, for a long time, not doubted that reproduction is in principle feasible, even if it may not yet have been accomplished with regard to a specific phenomenon. Mittelstraß (1998, p. 107) believes that "reproducibility [...] is the major scientific norm". "The requirement of reproducibility is an indispensable criterion to define 'science' as science: in fact, it is the necessary condition for the possibility of science and scientific knowledge!" Batterman (2002, p. 57) stresses that "any experiment in which a phenomenon is considered to be manifest must be repeatable". Böhme and van den Daele (1977, p. 189) argue that the "methodological concept of science is the regular fact that includes the condition under which it can be reproduced by anybody any time". Hund (1972, p. 274) states that physical science is the "realm of reproducible phenomena". And Pauli (1961, p. 94) emphasizes that—in spite of astrophysics and cosmology—"exceptional and extraordinary events are far beyond the scope of physics; these events cannot be grasped by experiments".

If we were to subscribe to these statements and restrict science to reproducible phenomena, we would find ourselves in a dead end: science and instability would be like fire and water. But this is not the case. Admittedly, the challenges induced by instabilities are severe. Instabilities convey unobservable small effects to empirically accessible scales. That is why they bridge different spatial domains, e.g., microcosm, mesocosm, and macrocosm, and induce problems regarding experimentation. The devil is in the detail. Because of thermodynamic and quantum mechanical effects, initial and boundary conditions cannot be measured exactly or be controlled in detail. Making reference to the three-body problem and its instability, Poincaré (1914, p. 56) emphasizes that "even if it were the case that the natural laws no longer had any secret for us, we could still only know the initial situation approximately". Inherent inexactness causes severe problems for physical sciences. Repeatability is limited; the intentional reproduction of events is difficult to achieve with unstable objects; the dynamics of unstable objects cannot be controlled by the experimenter. However, the lack of control is not just a pragmatic or epistemic boundary that could be overcome by improved methods and more advanced technology. It is part of nature since it is inherent to objects and not just a challenge to knowledge: it is ontological rather than epistemological or methodological.

Obstacles to Predictability and Calculability

Anticipation of the future and the ability to predict outcomes have been one of the most important qualifiers of what constitutes science. Among those who embrace predictability are instrumentalists and pragmatists, and sometimes others such as realists and transcendental epistemologists. Carl Friedrich von Weizsäcker (1974, p. 122) maintains that "determining and predicting the future" is the "major essence of exact science". In line with Weizsäcker, Drieschner (2002, p. 90) emphasizes: "We find 'prediction' as the key term for understanding and defining physics". Einstein, Podolsky and Rosen (1935, p. 777f) regard predictability as the sole possibility for ascertaining whether any entity actually exists. "If, without in any way disturbing a system, we can predict (at least in principle) with certainty (i.e., with probability equal to unity) the value of a physical quantity, then there exists an element of physical reality corresponding to this physical quantity". According to Pietschmann (1996, p. 166), "the power and accuracy of prediction indicates whether to accept or to refute a theory under consideration". Salmon's (1989, p. 119) main criterion for a good explanation is the success of predictions: Explaining means to "show that the event to be explained was to be expected." According to Psillos (1999, p. xix), scientific realists regard "predictively successful scientific theories as well-confirmed and approximately true of the world." It is well known that Auguste Comte once said: "Savior pour prévoir" (knowing in order to foresee) (cp. Comte 2006).

Instabilities, however, challenge the prediction-oriented understanding of physical science. Bunge (1987, p. 188) perceives the "immense mathematical difficulties with nonlinear laws and, in addition, with unstable solutions."[26] A differential equation—such as Newton's or Einstein's laws of motion—provides a general corpus for a law of nature, but not a specific solution or a particular trajectory. The specific solution is not given with the differential equation itself; it has to be computed in order to enable predictions. Linear differential equations pose no trouble; most of them can be integrated analytically. In contrast, nonlinear differential equations often possess no analytic solutions. Newton, Euler, Laplace, Poincaré, and others were frustrated by this fact. Even if we were ever to succeed in finding a nonlinear theory of every-

[26]Bunge believes that instabilities and nonlinearity make it too difficult to assign determinism to unstable dynamics.

thing—the equation that governs the universe—, we likely would not be able to predict specific events in the far future. According to Maxwell (1873, p. 440), "it is manifest that the existence of unstable conditions renders impossible the prediction of future events, if our knowledge of the present state is only approximate, and not accurate". In some cases, numerical algorithms can help to integrate nonlinear differential equations and handle unstable solutions.[27] But in other, most prevalent cases of dynamical instability, unstable chaotic orbits cannot be approximated by numerical solutions. They are effectively non-computable. Devaney (1987, p. 53) stresses that "most dynamical systems that arise in classical mechanics are not stable. [...] These systems cannot be approximated" employing the classic approach of asymptotic convergence of mathematical functions. No numerical solution is accurate enough to determine an unstable trajectory; for unstable systems, these prediction tasks would require an accuracy that is de facto impossible (cf. Schmidt 2003).

There are two kinds of obstacles involved here. A *first* numerical obstacle is due to the impossibility of digitalizing real numbers. Digital computers only calculate and store results to finite precision, so there are rounding errors at every calculation step. In an unstable system, these errors will grow exponentially, and so the model's trajectory (when initialized from a particular state) will quickly differ significantly from the evolution of the (real exact) system. A *second* numerical obstacle arises because the computer is not a formal logical system or Turing machine, but a real physical machine. A physical machine is subject to physical, and especially thermodynamic, relativistic, and quantum mechanical limitations. Of practical relevance are mainly thermodynamic limitations: calculations, information processing, and annihilation require thermodynamic energy and produce entropy. Depending on the thermodynamic costs one is prepared to pay, different numerical limitations exist with regard to the quality of approximation. General limitations are due to the maximum energy that is available in our cosmos for performing a calculation. The limitation of predictions is, therefore, built into the structure of the world—it is not an epistemological artifact of the way we are, nor is it just a methodological or pragmatic limitation that can be overcome. Consequently, the limitation is not "human-dependent" in any subjective sense. When dealing with unstable systems, "you [might] know the right equations but they're just not helpful", as one of the founders of fractal geometry, Mitchell Feigenbaum, argues.[28]

In other words, instabilities drive a wedge between theories (models, equations, laws) and predictability—with consequences for a deterministic worldview.[29]

---

[27]In the case of *static instability*, if the watershed is not fractal or the initial states are not located within the neighborhood of the watershed, numerical integration is feasible. Small perturbations do not make any difference in the final result. In other cases, two initial states differing by imperceptible amounts may evolve into two considerably different states. Then, if any observation error of the present state is made, an acceptable prediction of the state (to within only small tolerances) in the distant future may well be impossible. Similar conclusions hold for *structural instability*.

[28]Feigenbaum cited in Gleick (1987, p. 174f). The differential equation of a deterministic system is effectively worthless since reliable predictions would require exact information.

[29]In particular, they drive a wedge between (deterministic) laws and (prediction-relevant, single) trajectories.

Despite Hume's skepticism, traditionally successful predictions of deterministic theories have been considered as the main argument in favor of a deterministic worldview. Whether the presence of reduced predictability might still provide a convincing argument for determinism remains open to debate.

Boundaries of Testability and Confirmability

Besides focusing on reproducibility and predictability, another well-established position in philosophy argues that science, insofar as it is science, has to meet the criterion of testability: both realists and empiricists—falsificationists and verificationists alike—consider empirical testability as being methodologically essential to science. Testability ensures the gaining of objectivity, truth, or at least inter-subjective evidence.

Among other advocates of such a view, Mach (1988, p. 465) strongly claims: "In domains where neither confirmation nor refutation does exist, we do not meet scientific standards: in fact, there is no science!" For Duhem (1991, p. 22) the "main criterion for scientific evidence" is for a law or theory to be in "accordance with experience". On a similar note, Carnap (1928) requires that "a statement to be meaningful" has to be based on empirical verification; "testability and meaning" turn out to be synonyms. "Verifiability"—later Carnap preferred "confirmability"—is the major criterion to demarcate science from metaphysics. Popper (1934), from another perspective, believes in the principle of refutability. According to Popper, the growth of scientific knowledge is based on risky conjectures that (normatively: have to!) allow refutations. Central to performing refutations are the so-called *experimenta crucis* that are capable of revealing the falsehood of a proposition or a theory. Such crucial, decision-making experiments are indispensable to qualify a theory as a scientific one—and they also show the asymmetry between the methodologically justified, because logically valid, refutation on the one hand and the unsound, because logically invalid, verification on the other hand.

The requirement of testability as a criterion to qualify science is, therefore, broadly shared among critical rationalists, hypothetical realists, and logical empiricists.[30] Any type of empirical test, for both Carnap and Popper, is based on a constant relation between the mathematical-theoretical side and the empirical-experimental side. However, when dynamical or structural instabilities are present, this relation is no longer given. The behavior of a single trajectory or a few orbits deduced from the mathematical model "cannot be compared with experiment, since any orbit is effectively non-correlated with any other and numerical round-off or experimental precision will make every orbit distinct" (Abarbanel et al. 1993, p. 1334). Similarly, Rueger and Sharp (1996, p. 103) stress: "If we test a theory in this way we will not find a precise quantitative fit, and this is to be expected if the theory is true of the

[30]It is still an ongoing debate whether testability is to be understood as a normative requirement or merely as a description of what in fact happens in science.

system". Theory and experiment are separated into two disjunct worlds, as Harrell and Glymour (2002) also maintain.

There are two related reasons for the issues of testability. Both the instability on the object's side and the instability on the model's side contribute to limiting testability: Because of instability, it is impossible to reproduce the object's behavior, and it is hard to make predictions. One of the two problems would be hard enough to cope with, but they emerge simultaneously. We are entering the realm of double uncertainty. For any unstable model that refers to an unstable object, Guckenheimer and Holmes (1983, p. 256) emphasize that "details of the dynamics, which do not persist in perturbations, may not correspond to testable [...] properties". A classical-modern test, based on quantification and the assumption of a quantitatively constant relation, is not possible.

Limitations of Describability and Reductive Explainability

René Descartes is a key figure in the early modern age. With others such as Bacon or Galilei, he coined and shaped the concept of modern science. The objective of modern science is to "trace back vague, clouded and dark propositions step-by-step to simple and evident propositions [or laws]" (Descartes 1979, p. 16/379). Later, in the middle of the twentieth century, Hempel and Oppenheim proposed the deductive-nomological model to describe (and to *evaluate*!) scientific explanations. This approach has become known as the covering-law model: a phenomenon is deemed explained by "subsuming it under general laws, i.e., by showing that it occurred in accordance with these laws, in virtue of the realization of certain specified conditions" (Hempel 1965, p. 246).[31] To put it in normative terms: experimental phenomena have to be described by the shortcut of mathematical structure; data needs to be compressed by models or laws. This claim comes close to the view that has become known as Ockham's razor. In this rationalist tradition, the aim of science is to achieve a minimal, simple, non-redundant description of the world. In other words, the challenge for physical sciences is to find laws or models as the major syntactic elements of any theory. The elimination of redundancy and the downsizing of description length are regarded as the necessary (first) condition for explanatory success. According to Hertz (1963, p. xxv), "all scientists agree upon the main task: Tracing the phenomena of nature back to simple laws". This position is sometimes called explanatory reductionism or, more strongly, inter-theoretical reductionism.[32]

Instabilities challenge reductive explanations. To some extent, this is not new. Early perceptions of these challenges date back to David Hume in the eighteenth

[31] In this line, Woodward (2000, p. 197) has very recently claimed: "Explanation in the special sciences involves subsumption under general laws."

[32] In the history of the philosophy of science, there has been a strong focus on these challenges, in particular on ontological, epistemological, and methodological types of reductionism. The unification of three of the four fundamental theories in physics shows the success of explanatory reductionism.

century. Hume anticipated problems of instability in his *Inquiry Concerning Human Understanding*—although he did not use the term "instability": "It is only on the discovery of extraordinary phenomena, such as earthquakes [...], that they [=scientists] find themselves at a loss to assign a proper cause, and to explain the manner in which the effect is produced by it. It is usual for men [=scientists], in such difficulties, to have recourse to some invisible intelligent principle as the immediate cause of that event which surprises them" (Hume 1990, p. 69). Today, scientists regard earthquakes as a paradigm of unstable natural processes. Likewise, Hume considered earthquakes to be unstable, irregular real phenomena of nature.[33] In addition to Hume, one hundred years later Maxwell (1991, p. 13) pointed out that "in so far as the weather may be due to an unlimited assemblage of local instabilities it may not be amenable to a finite scheme of law at all". That is to say, "only in so far as stability subsists [...] principles of natural laws can be formulated". Only a few scientists in Maxwell's time were aware of this challenge. Nowadays, the challenge provoked by instabilities to any reductive explanation is broadly acknowledged and deeply discussed within the context of self-organization theory and nonlinear dynamics. "Nonlinear dynamical systems theory [...] studies properties of physical behavior that are inaccessible to micro reductive analytical techniques", as Kellert (1993, p. 115) emphasizes. Where instabilities are present, we can concur with Battermann (2002) in stressing that the "devil is in the detail": Instabilities limit "approximate reasoning" and "reductive explainability."[34] In sum, unstable processes cannot be reductively condensed or compressed to a simple law.

The fact that there is an effective incompressibility of data sequences or phenomena generated by unstable processes is well known from information theory[35] and from chaos theory.[36] The key notion here is informational incompressibility, which is linked to essential unpredictability. According to von Neumann's ideas on self-organization, emergence, and complexity, a complex process—induced by underlying instabilities—is defined as one for which the simplest model is the realization of the process itself. The only way to determine the future of the system is to run it; there are no shortcuts and no compact laws at all. Insofar as instability underlies complexity, the simplest model of the unstable process is the process itself. Dynamical instability could also be regarded as the core of deterministic random processes and various types of non-white noise. Although these random processes might be generated by a deterministic law, it is impossible with traditional statistical tools to find and, further, to reconstruct the law-like structure from the data

[33]By referring to earthquakes Hume also supports his arguments against anti-naturalist positions since phenomena that appear irregular can be generated by regular laws.

[34]This point is, in fact, the nucleus of the classic deductive-nomological account of explanation (cp. Hooker 2004).

[35]For an in-depth discussion see, most prominently: Chaitin (1971, 1987, 2001).

[36]For example, see Crutchfield et al. (1986), Devaney (1987), Peitgen et al. (1992), and Jackson (1989).

sequences or to obtain a simple law or algorithm.[37] In this vein, James Crutchfield et al. (1986, p. 56) stress that "the hope that physics could be complete with an increasingly detailed understanding of fundamental physical forces and constituents is unfounded. The interaction of components on one scale can lead to complex global behavior on a larger scale that in general cannot be deduced from knowledge of the individual components".

Thus, instabilities restrict the elimination of redundancies. Unstable processes cannot be reduced to laws governing the microscopic level of atoms or molecules. Nietzsche (1930, p. 127) might have gone too far ahead in stating, "There are no laws of nature!", but he seems to be right in stressing that there are limits to finding laws. An unstable world is only partly accessible and knowable. Our non-knowledge regarding the behavior of unstable systems has thus nothing to do with a temporary insufficiency of our knowledge. It has everything to do with unstable characteristics of nature itself.

Critique of Standard Concepts of the Philosophy of Science

If we acknowledge the existence and prevalence of instabilities in nature, we are not only faced with the four methodological issues that are central to any semantically meaningful concept of emergence for the physical sciences themselves.[38] In addition, traditional concepts in philosophy of science are called into question. Challenging these concepts does not mean completely rejecting them, but a need for rethinking them is indicated.

Instabilities, *first* of all—as we have shown—reveal limits of *reproducibility*. But experiments and the reproduction of phenomena constitute a central point of reference for many concepts in philosophy of science, such as *experimentalism*, *methodological constructivism*, and, to some extent, *pragmatism*. One of the precursors of modern experimentalism was Francis Bacon. He advocated an intervention and action-oriented notion of science: forced by experiments, nature should unveil her nomological secrets. Kant later supported Bacon and argued in favor of an experimentation-oriented constructivist's point of view by which laws are imposed: nature is experimentally constructed and empirically conceptualized by human reason. To the extent that intervention and action-oriented concepts of science rely on strong, quantitatively oriented views of reproducibility, such concepts are called into question by instabilities. Arguably, instabilities do not only challenge experimentation but in a general sense also human action in an unstable environment. However,

---

[37] Of course, nonlinear techniques such as nonlinear data analysis, phase space reconstruction, surrogate data analysis, and other tools provide some options to find deterministic structures and separate them from white noise.

[38] See the six characteristics described in the first part of this article, in particular the methodological and epistemological characteristics 4–6, such as non-reproducibility/non-repeatability (4), unpredictability (5), and irreducibility (6).

besides from philosophers of sciences, notably action and planning theorists have not yet perceived this challenge or tackled the problems involved.

Second, where instabilities exist *predictability* is effectively reduced. Many philosophers who advocate an *instrumentalist* view of science uphold the predictive power of science. The success of predictions is the core argument to support their position. In addition to instrumentalism, other concepts also build their arguments on predictive success. For instance, in the tradition of Kant, *transcendental epistemologists* regard prediction as a key element of science and see it as a consequence of the necessary condition for the possibility of acquiring knowledge—and, therefore, of the world's structure: the possibility of knowledge is transcendentally guaranteed by a stable, predictable, deterministic world. According to this concept, prediction means learning from the past for the future.[39] For *realists* and *empiricists,* predictions play an important and necessary, but secondary role in their understanding of science. They regard predictability only as a means, whereas for *instrumentalists* the objective and summit of science is prediction itself. According to instrumentalists, predictive success is deemed to be an indispensable criterion of justification and the basis for deciding about the truth or refutation of propositions or theories. In fact, instabilities challenge all concepts referring to predictive success.

Third, *testability* is called into question. Most *realists* and, from a different angle, many *empiricists* draw their arguments from the (presupposed) empirical testability of theoretically derived propositions. In other words, realists' and empiricists' concepts are based on stronger claims than the above-mentioned reproducibility and predictability, but to advocate their concepts both predictability and reproducibility are necessary. Realists and empiricists claim that the theoretical side (predictability) and the experimental side (reproducibility) must be joined together: knowledge is qualified as *scientific* knowledge insofar as theories refer to empirical propositions and can be tested. In this sense, theories approximate the empirical laws or propositions in order to provide empirical evidence for the theories. Contrary to this presupposition, instabilities call the existence of such a constant relation between theory and experiment into question. According to Rueger and Sharp (1996, p. 94), "theories of nonlinear dynamics appear to generate a peculiar […] problem for the realist […]: in dealing with chaotic systems improved input for the theory will in general not lead to better predictions about the features relevant for confirmation."

Finally, *describability and explainability* are challenged by instabilities. *Conventionalists* and scientific *rationalists* are concerned with the descriptive and explanatory power of science. They draw their main arguments from condensed descriptions and the economy of thought. The Hempel-Oppenheim scheme is the preferred type of what characterizes a satisfactory explanation. In fact, instabilities question all concepts that refer to condensed descriptions. An unstable time evolution can hardly be represented by a simple law.

It remains an open question whether, and to what extent, instabilities—and the four issues induced by them: limits of reproducibility, predictability, testability, and

[39]Prediction and time's arrow are twins. And time is a necessary condition for the possibility of scientific experience.

describability/reductive explainability—severely challenge the traditional concepts of philosophy of science or not. We cannot go into further details here, but the need for further clarification is obvious. Some lines of thought toward a new epistemological concept have been drafted by Mitchell (2002). She labels the new concept developed in the realm of a biological approach to complex self-organizing systems "integrative pluralism" since it faces the plurality and variety of different natural phenomena from an integrative or synthetic perspective.

Let us now move on to discuss how self-organization theory, including complex systems theory, manages to deal with instabilities.

Science in an Unstable World

History of Science: Perceiving Instability, Imposing Stability

Self-organization phenomena and emergent properties pose severe challenges to sciences, in particular to physical sciences—however, this does not make science entirely impossible.[40] Throughout the history of science, physical scientists seem to have always had an inkling of the methodological problems provoked by instabilities. In addition to considerations of methodology, the restriction of physical sciences to stable systems was also supported by the underlying stability metaphysics common during the historical development of the sciences. Stability metaphysics was undoubtedly a methodologically beneficial approach from a historical perspective; it is evident that the stability assumption was crucial for the progress and explanatory success of physical sciences. All this underpins that there was at least some rationale behind the implicit restriction of physical science to stability.

It took historically a long time from the first discovery of instabilities to their broader acknowledgment.[41] In the early twentieth century, the challenge turned out to be pressing: how to deal with instabilities? A classic possibility was to renew what had proven a successful implicit assumption throughout the history of science. So the climax of the methodological challenges posed by instabilities, as well as the very last attempt to refute them and restrict physical sciences to stable systems, can be found in the works of Pierre Duhem and Alexandr Andronov. In *Aim and Structure of Physical Theory* (1908; cp. Duhem 1991) Duhem pointed to Poincaré's and Hadamard's work on unstable systems.[42] Even though, and probably because, Duhem was aware of instabilities, he believed physical sciences, and physics in particular, had to rely on stability. He restricted physical methodology—notably the deductive-nomological structure of explanation—explicitly to the requirement of stability. According to

---

[40]For a brief history from a different point of view, see: Holmes (2005). In general, these challenges—and the related problems—can also be observed in the nineteenth century in the realm of hydrodynamics (cf. Darrigol 2006; Swinney and Gollub 1981).

[41]An analysis of the historical development can be found in Aubin and Dalmedico (2002).

[42]For an introduction to Duhem's philosophy of instability see, Schmidt (2017).

Duhem's methodology, deductions are theory-based predictions designed to reveal empirically accessible consequences of a physical theory (cp. Duhem 1991, p. 55f). In order to be "useful for physics", deductions have to take experimental uncertainties into account. Duhem's bundle concept, a key element in his hypothetico-deductive approach, comes into play here (Schmidt 2017). Specific types of non-diverging bundles reflect the experimental uncertainty of facts and guarantee the application of well-established mathematical tools, such as quantitative error estimation, error theory, statistical analysis, hypothesis testing theory, and approximation theory. In contrast, a diverging bundle corresponds to an unstable trajectory, and for this trajectory "mathematical deduction is and always will remain useless to the physicist; [... it] will not permit us to predict what [will ...] happen." (Duhem 1991, p. 139)[43] According to Duhem (1991, p. 143), when faced with the methodological issues of making deductions that will satisfy the objectives of science, the scientist is normatively "bound to impose [...] rigorous conditions" on bundles: the requirement for non-divergence or, which is equivalent, for approximability. "To be useful to the physicist, it must still be proved that the second proposition remains approximately exact when the first is only approximately true." That is, deductions have to be restricted to stable trajectories in order to enable predictions and guarantee the soundness of conclusions—this is a strong normative requirement, equivalent to non-divergence. Stability is a necessary condition for approximation: deductions—if they are considered "physically or scientifically relevant"—*must* preserve the likeness of the neighborhoods within the bundles; the second bundle must be structurally similar to the first to guarantee coping with experimental uncertainties. Scientific relevance is equated with stability. The historian Jules Vuillemin (1991, p. xxviii) concludes that "Duhem limits his reflections to extrapolations concerning the stability of systems." Unstable phenomena—and, thus, phenomena of emergence—are not regarded as being located in a branch of physical sciences.[44]

In the 1930s, the Russian physicist Alexandr Andronov concurred with Duhem in restricting physical sciences to stable objects. On the one hand, Andronov was a pioneer in the awareness of instability. On the other hand, he believed, like Duhem, that physics was threatened by instabilities and, based on such considerations, was a particular proponent of (structural) stability: if stability cannot be taken for granted, it has to be imposed. As a consequence he formulated a stability requirement which Guckenheimer and Holmes (1983, p. 256f) later called "stability dogma". According to Andronov, stability has to be normatively imposed in physical methodology. He asked: "Which properties must a dynamical system [=model] possess in order to cope with a physical problem?" (Andronow et al. 1965/1969, p. 403f): the "system must be

[43]"Deduction is sovereign mistress, and everything has to be ordered by the rules she imposes" (ibid., 266). A "deduction is of no use to the physicist so long as it is limited to asserting that a given *rigorously true* proposition has for its consequence the *rigorous accuracy* of some other proposition" (Duhem 1991, p. 143).

[44]Further, Prigogine stresses that Duhem was pioneering in his reflection on instabilities, but–according to Prigogine–Duhem went too far in his general critique concerning the ultimate uselessness of "unstable deductions". Duhem assessed instabilities in a negative sense only, as a threat to classical-modern physical methodology.

structurally stable". The "stability dogma" was laid down as a strong methodological norm; it served also as a selection criterion for "physically relevant" objects.

In short, Duhem and Andronov are threshold figures indicating an age of transition. Whenever dogmatization occurs, it can be perceived as a signal of a crisis: what has until now been an unquestionable implicit belief turns out to be questionable and becomes a matter of dispute. What was formerly implicit is made explicit. To counteract the issues raised and to refute the crisis, the very first attempt is always to *re*-introduce dogmatically what seems to have been lost. This dogmatization has also been observed and described by Thomas S. Kuhn and Imre Lakatos from different angles.

The Qualitative: Complexity Characteristics at the Center of Testability

Since the 1960s, the rejection of the methodological challenges raised by instabilities is no longer convincing. In fact, we can observe a *Gestalt* switch in the methodology of physical sciences and in scientists' understanding of nature. Some of the most pressing problems posed by instabilities have been resolved thanks to the advancement of complex systems and self-organization theory over the past 30 years.[45] The very first step was to get rid of the stability dogma. After having disregarded and damned instability over hundreds of years, the stability dogma was replaced by an approach employing more qualitative features relating to those properties of a theory and of an experiment which are relevant for the specific situation in question—namely, specific characteristics often referred to as *complexity characteristics*. "The definition of physical relevance will clearly depend upon the specific problem. This is quite different from the original statement [...] of the stability dogma" (Guckenheimer and Holmes 1983, p. 259).

It is widely acknowledged throughout the methodological debate on instability that weaker requirements are sufficient for a sound methodology of science and, also, that they are necessary to grasp unstable objects. The weaker requirements are based on qualitative rather than on quantitative properties.[46] For instance, Benoit Mandelbrot (1991) calls fractal geometry a "qualitative theory of different kinds of similarities." And Thom (1975) stresses that his catastrophe theory is a "qualitative science of morphogenesis." Hirsch (1984, p. 11) maintains that "dynamics is used more as a source of *qualitative* insight than making *quantitative* predictions." All of these works have a common background: the "qualitative theory of differential equations" which can be traced back to Birkhoff (1927). Some ideas also originate from the work of Poincaré (1892) and in the mathematical theory of topology and

[45] However, a proper methodology of the sort that works for stable objects is, and will always remain, impossible.

[46] The traditional quantitative, metrically oriented stability dogma has been replaced by weaker, more qualitative topological characteristics.

differential topology. Thus, according to the advocates of instabilities, the qualitative aspect plays a major role in the post-stability methodology.

Another cognate point deserves to be mentioned. Instabilities provoke a kind of underdetermination of the characteristics of the system under consideration. That is to say, scientists have a choice and they *have to* choose. They have to decide which of the qualitative complexity characteristics of a model and of the corresponding experiment are to be deemed relevant, whereas the selection itself is more or less pragmatic and depends on the specific problem and goal of the modeling issue, i.e., on the objective, matter of interest, situation, context, resource, and the like. The choice of a particular characteristic is the starting point for any empirical test of any unstable model in the field of self-organization theory. Cartwright (1983, 1994) does not speak of characteristics, but in a similar vein of "capacities". Cartwright (1983) elaborates her capacity concept by considering unstable dynamics, such as the motion of a 1000 dollar bill from a tower down to earth. Today, an increasing amount of research is carried out on formulating, exploring, and justifying relevant qualitative capacities or characteristics (Wackerbauer et al. 1994). Not every one of them will fit *all* problems in general. In certain contexts, some complexity characteristics are useless or meaningless. But there will be other characteristics enabling a classification, a detailed description, a partial explanation, or a piecewise prediction of the dynamics of a specific model. The variety of these contextual characteristics is striking.[47]

In sum, the qualitative, which is central to contextual complexity characteristics, is not at all related to subjective impressions or to qualia-like perceptions. On the contrary, it remains mathematical, but from a different perspective. Scientists do not require models to be quantitatively stable, i.e., to show a quantitative robustness of the dynamics or structure, but to possess some complexity characteristics that are invariant under topological transformations, i.e., to be topologically equivalent or conjugate. The qualitative element of the characteristics refers to the appearance, the shape or "Gestalt" of the phenomenon and the geometry of the pattern *after* the process of time evolution—and not solely to the bare equation. It is not grounded on *single* trajectories and *single* measurable events but on *all possible* dynamics of the *whole* dynamical system that can be depicted in the phase space, for instance of a dynamically unstable, chaotic attractor. The global dynamics cannot be described by employing classical quantitative distance measures. Instead, the dynamics is characterized by means of one of the most advanced mathematical theories—differential topology and dynamical systems theory.

[47] Some of these deserve to be mentioned: entropies, information theory parameters, Lyapunov exponent, scaling exponents, lengths of transients, fractal dimensions, renormalization theory parameters, topological characteristics, structure of stable and unstable periodic trajectories, existence of chaos or hyper-chaos, parameters of basin boundaries, types of bifurcations, parameters of chaos control theory, power and Fourier spectra, phenomenological analysis of patterns, etc. Some of these properties are well known from solid state physics, fluid and plasma physics, meteorology, or statistical thermodynamics. What is new is (a) that these properties are necessary to empirically test (complex dynamical) models with just a few degrees of freedom and (b) that the model-theoretical discussion is deepened by the acceptance that nature can be structurally unstable.

Even though instabilities may exist in the case of a particular dynamical system, some of the above listed complexity characteristics will persist under perturbations or small errors—for instance, the instability itself—while others will not and will vary dramatically. Persistence is a necessary requirement for any empirical test and for scientific relevance: the *prevalence of a certain complexity characteristic in the class of models* is a sufficient requirement. Scientists do not require models to be stable but to possess some *prevalent* complexity characteristics. These characteristics constitute the basis for any empirical test of a model. Let us now look in more detail at testability and the central role of calculations in this regard.

Towards a Novel Kind of Calculability and Testability

The fact that instabilities challenge confirmability and testability has been elaborated by Harrell and Glymour (2002). Testing a model or theory involves comparing the theoretically deduced model output with the data measured in the real system. However, from a methodological perspective, *nonlinear data analysis* plays a central role in matching a model's data and the experimental system's data (Takens 1985; Parker and Chua 1989; Sauer et al. 1991; Abarbanel et al. 1993). Nonlinear data analysis is an excellent tool for calculating some of the above listed complexity characteristics from experimental time series of unstable systems. The empirically gained characteristics are confronted with theoretically deduced characteristics generated by the model: If (a) the model characteristic is proven to be prevalent in the model class and if (b) it matches with the characteristics calculated from the experimental data, scientists will accept the model as an adequate description of the empirical phenomenon.

A similar methodological shift from a quantitative to a qualitative focus can be observed in the case of predictability. As explicated above, instabilities restrict the horizon of single trajectory-oriented predictions. In order to deal with this problem, predictability is addressed explicitly and has become a major research topic in self-organization theory and nonlinear dynamics (cp. Aurell et al. 1997; Badii 1991; Jackson 1989). During the last 35 years, scientists have developed qualitative calculation tools, the most important one being the "shadowing lemma" to address issues of predictability when a quantitative approach is impossible (Coven et al. 1988; Peitgen et al. 1992). Besides "nonlinear data analysis", the "shadowing lemma" is the second mathematical tool available to handle general problems of the classical quantitative approach.

The challenge addressed by the shadowing lemma is the following: given a differential equation describing unstable dynamics, e.g., a chaotic attractor, there is no way of calculating and predicting the *true* trajectory. However, the shadowing lemma provides an encouraging way to deal with this situation. In many cases, shadow trajectories that can effectively be calculated do exist; they remain in the neighborhood of the non-predictable, unstable true orbit—although they cannot be derived as any simple approximation to that orbit. So, if we are not interested in an exact (quantitative) prediction, we can describe and calculate the global dynamics more or less in a qualitative way. The shadowing trajectories provide qualitatively similar topological properties that characterize the dynamical systems.[48]

Modeling Turn

A modeling turn in the sciences can be observed: Instabilities entail a model-oriented approach and give rise to modified kinds of explanations. Traditionally, deductive-nomological (and, in particular, reductive) explanations—including condensed descriptions—were and are still associated with universal laws and fundamental theories. As explicated in the preceding subsection, instabilities place limitations on deductive-nomological explanations. That is why scientists, when accounting for unstable objects, hesitate to portray their work as discovering new theories or new laws of nature. They prefer to speak of "models" rather than "theories" or "laws"—regarding their approach as one involving *modeling* systems, *model* construction and *model*-based explanations.[49] In traditional philosophy of science, models play, if any, a minor role, e.g., as mere derivatives of theories in order to match the theoretical account with empirical data. Contrary to the traditional position and in line with the later work of Morgan and Morrison (1999) on "models as mediators", Thom (1975, p. 325), the founder of the catastrophe theory, proposed "a general theory of models". Poser (2001, p. 284) goes even further and identifies "a new general type of physical sciences" based on "model building and simulations". Kellert (1993, p. 85) argues that self-organization theory's "method of understanding" is nothing but "constructing, elaborating, and applying simple dynamical models" and "interpreting" them in the different domains of application.

This again entails a contextualism: in the realm of unstable objects, models can only be tested and justified for a certain context of application, including specific modeling objectives and interests, but not for the "entire world". In the sciences that address phenomena of self-organization and emergence it is very common to discuss issues such as whether a certain "representation or model of the system [... is] a good one", or not (Batterman 2002, p. 57). The criteria for a "good representation" of a model are based on the requirement that the model has to provide some prevalent complexity characteristics.[50] There is no single justification, but various ones.

[48]The assumption required for the existence of such shadow trajectories is that the system is hyperbolic. How serious this restriction to hyperbolicity might be is still a matter of ongoing research.

[49]The shift in terminology is not just rhetoric. Rather, it indicates a modeling turn in the sciences. For a philosophical account of models, see Redhead (1980), Morgan and Morrison (1999), and Hartmann and Frigg (2006). However, these authors already diagnose such a model-orientation in the realm of non-unstable objects.

[50]It is interesting that an epistemological and methodological discussion on the structure of mathematical models is taking place *within* the sciences today, as part of the progress of the sciences themselves. Well-established ideas of philosophy of science are not an add-on—they can be considered as an intrinsic element of sciences. Instability and complexity carry philosophical thought,

One of the interesting aspects in comparing traditional philosophical accounts of explanation with the understanding provided in the realm of instability also results from a mismatch between the item to be explained in the former and the item to be understood in the latter. While, traditionally, philosophers commonly concern themselves with scientific explanations of such things as "facts", or "events", self-organization theory usually studies behaviors, patterns, or bifurcations and it considers processes, time-evolution, the emergence of novelty, and the like (Kellert 1993, p. 81f). A shift in the focus of interest, and hence in what deserves an explanation, can be observed. According to Bergé et al. (1984, p. 102), "the ultimate goal is to understand the origin and characteristics of all kinds of time-evolution encountered, including those which at first seem totally disorganized". The type of explanation required and appreciated in self-organization theory, therefore, differs from traditional science. Self-organization theory, including complex systems theory and nonlinear dynamics, favors a type of explanation "that is holistic, historical, and qualitative, eschewing deductive systems and causal mechanisms and laws" (Kellert 1993, p. 114). This type differs from the deductive-nomological account of explanation.

Decisions ex ante *and Underlying Knowledge Interests*

As mentioned, complex systems and self-organization theory give rise to a shift in the focus of attention—and in knowledge interests. Now, the mere fact that the *context of interest* precedes both the context of discovery and the context of justification is often disregarded by standard positions in philosophy of science.[51] The context of interest shapes any research program, for example, the way in which problems are constituted or identified and objects of scientific inquiry are framed. Decisions are made explicitly or implicitly: which objects and which phenomena are worthy of scientific research? What is of interest? What do we wish to know? What is a relevant and significant problem? In this regard, self-organization theory opens avenues to reflect explicitly upon the context of interest and the purpose of knowledge.

Such reflection turns out to be necessary; decisions are indispensable. According to René Thom (1975, p. 9), a "choice of what is considered scientifically interesting [has to be made]. [...This] is certainly to a large extent arbitrary. [... Traditionally,] [p]hysics uses enormous machines to investigate situations that exist for less than 10^{-23} s [...]. But we can at least ask one question: many phenomena of common experience, in themselves trivial [...]—for example, the cracks in an old wall, the

such as the reflection on the criteria for scientific evidence, truth or objectivity, into the very heart of sciences.

[51]In his seminal book "Knowledge and Human Interest", Habermas (1972) exemplifies this point and encourages a broader perception of the value dimension of research programs in general and among the public. However, in spite of its relevance to our late-modern technoscientific age, in the mainstream of the philosophy of science, Habermas' central point was—and still is—neglected and disregarded.

shape of a cloud, the path of a falling leaf, or the froth on a pint of beer—are very difficult to formalize, but is it not possible that a mathematical theory launched for such homely phenomena might, in the end, be more profitable for science?" Thom addresses these "phenomena of common experience" which have been "generally neglected" throughout the development of modern sciences since the seventeenth century (ibid., 8). The underlying reason for the neglect and disregard is clearly that "these phenomena are highly unstable, difficult to repeat, and hard to fit into a [universal] mathematical theory [...]" (ibid., 9). The objective of Thom's catastrophe theory is to find a remedy. Thom tries to widen the focus and to re-adjust the context of interest in order to draw attention to phenomena of common experience—phenomena that are mostly unstable, dynamical, and complex. These phenomena are at home in the familiar mesocosm of our life-world—not just located in the tiny microcosm or the huge macrocosm; we can observe them with our unaided senses. To gain access to these unstable phenomena in the mesocosm, Thom facilitates a mathematically based morphology—a "general theory of morphogenesis" including the dynamics of forms and the emergence of pattern (ibid., 101/124f).

There are cognate mesocosmic approaches in Benoit Mandelbrot's "fractal geometry" (Mandelbrot 1991), Alfred Gierer's "physics of biological *Gestalt* shaping and structure formation" (Gierer 1981), and Hans Meinhardt's "pattern dynamics of sea shells" (Meinhardt 1995). For instance, Meinhardt (1995, p. vii) shows that "the underlying mechanism that generates the beauty [of sea shells] is eminently dynamic. [...] A certain point on the shell represents a certain moment in its history." The "sea shell patterns are not explicable on the basis of the elementary mechanisms in a straight forward manner" (ibid.). Due to instabilities, no quantitative explanation is feasible, only a qualitative description based on computer simulations and graphical-visual representations. The historical gap between the development of these tools in the 1980s and 1990s and the recognition that there was a need for them by Poincaré and others 80 years before can probably be attributed to the absence of adequate computational power to analyze the unstable systems. The development of computer technology has opened up new opportunities for theorizing and experimentation that enable science to deal with unstable systems. (a) *Theorizing*: The naked model equations say nothing—the phenomena are mostly hidden and cannot be foreseen. They have to be generated by numerical simulations; they are then, often, represented using visualization techniques. Simulations have become an indispensable part of any theory describing unstable systems. (b) *Experimentation*: Computers not only provide new ways of dealing with theoretical problems of unstable systems—most impressive is the emergence of a *new experimental culture*; today, it is common to perform numerical or computer experiments (Lenhard 2007). So, in addition to the tremendous progress on a pragmatic level, numerical experiments and computer simulations are indicative of a shift in the principles of methodology. Poser (2001, p. 282) identifies "new criteria of scientific methodology". According to Küppers (1992, p. 9), numerical experimentation with unstable systems implies a change of "the inner constitution of physical sciences".

It is now undeniable that, although the limitations of stability-oriented methodology cannot be overthrown, self-organization theory has opened up new ways of dealing with instabilities. The classical methodological nucleus of science is being modified—in particular by giving more attention to qualitative instead of quantitative aspects. Most striking is that phenomena, patterns, and structures within the mesocosm—and, more generally, phenomena of emergence–are arousing the interest of physical sciences.

Summary

So, is there anything new under the sun? Indeed, there is. Phenomena of emergence are widely manifest in nature, technology, and society. As we have shown in detail, emergence is today regarded as being inherently linked to the concept of self-organization: hence, self-organization theory has to serve as a foundation of a sound understanding of emergence.

Theories of self-organization—such as complex systems theory, nonlinear dynamics, chaos theory, synergetics, dissipative structures, fractal geometry, catastrophe theory, hyper-cycle theory, autopoiesis, and the like—have rapidly advanced within the natural sciences since the 1960s; from that time on, they have been a central part of the sciences. In fact, we can observe in the physical sciences a structural-paradigmatic *Gestalt* shift that obviously reduces the gap between mathematics and physics on the one hand and between biology and the life-sciences on the other. The strong interdisciplinarity of research in this field is not surprising since the phenomena of self-organization and of emergence are—framed from the perspective of the mathematical structure—not restricted to any one particular discipline; they are interdisciplinary in nature.

In this chapter, we have inquired into ontological, epistemological, and methodological characteristics of the phenomena of emergence. We have asked whether there is an ontological core of emergence. Is there unity in the plurality of different phenomena?—Yes, (0) instabilities are the source (and the underlying condition of the possibility) of self-organization, pattern formation, evolution, in other words: of emergence. Instabilities are central to the three further ontological characteristics of emergence, namely (1) novelty, (2) processuality/temporality, (3) internality ("self"-organization).

At the same time, instabilities challenge the methodological and epistemological nucleus of traditional sciences: (a) When instabilities are present, the usefulness of a theory, a law, or an equation for *predictions* is effectively limited. The principle of superposition does not hold, and—in most cases—the equations cannot be integrated and approximated using analytic tools of mathematics. (b) Insofar as *reproducibility* or *repeatability* is seen as the core of an experiment, the experimental methodology of sciences is challenged. There is no possibility to decompose unstable processes into parts and to isolate these processes from the environment. (c) Furthermore, an issue arises when attempting to match theoretical and experimental phenomena (events, trajectories, data, states). The quantitative *testability* of laws is reduced. (d) In addition, unstable processes cannot be compressed into a simple law. Reductive *explainability* is limited. These limitations also induce, at the same time, the three further methodological and epistemological characteristics of emergence, namely limits regarding (4) reproducibility/repeatability, (5) predictability, and (6) testability and explainability/reducibility.

In sum, the traditional stability-based worldview that made emergent phenomena invisible has been replaced by an approach that puts emphasis on becoming: on novelty, processuality, and internality. Today, self-organization theory is *the* central element of any semantically meaningful understanding of emergence. Instabilities play a crucial role in nature. Essentially, our world is an unstable one.

References

Abarbanel HD, Brown R, Sidorowich JJ, Tsimring LS (1993) The analysis of observed chaotic data in physical systems. Rev Mod Phys 65:1331–1392

Andronov A, Pontryagin L (1937) Systèmes Grossiers. Dokl. Akad. Nauk. (Doklady) SSSR 14, pp 247–251

Andronow A, Witt AA, Chaikin SE (1965/1969) Theorie der Schwingungen, Part I + II. Berlin, Akademie Verlag

Atmanspacher H, Kurths J, Scheingraeber H, Wackerbauer R, Witt A (1992) Complexity and meaning in nonlinear dynamical systems. Open Syst Inform Dyn 1(2):269–289

Aubin D, Dalmedico AD (2002) Writing the history of dynamical systems and Chaos: Longue Durée and revolution. Discipl Cultures Hist Math 29:273–339

Aurell E, Boffetta G, Crisanti A, Paladin G, Vulpiani A (1997) Predictability in the large: an extension of the concept of Lyapunov exponent. J Phys A 30:1–26

Badii R (1991) Quantitative characterization of complexity and predictability. Phys Lett A 160:372–377

Banks J, Brooks J et al (1992) On Devaney's definition of chaos. Am Math Monthly 99:332–334

Batterman R (2002) The devil in the detail: asymptotic reasoning in explanation, reduction and emergence. Oxford University Press, Oxford/New York

Bergé P, Pomeau Y, Vidal C (1984) Order within chaos. Towards a deterministic approach to turbulence. Wiley, New York

Birkhoff GD (1927) Dynamical systems. AMS Colloquium Publications, New York

Böhme G, van den Daele W (1977) Erfahrung als Programm—Über Strukturen vorparadigmatischer Wissenschaft. In: Böhme G, van den Daele W, Krohn W (eds) Experimentelle Philosophie. Frankfurt, Suhrkamp, pp 166–219

Brown R, Chua LO (1996) Clarifying chaos: examples and counterexamples. Int J Bif Chaos 6(2):219–249

Brown R, Chua LO (1998) Clarifying Chaos II: Bernoulli chaos, zero Lyapunov exponents and strange attractors. Int J Bif Chaos 8(1):1–32

Bunge M (1987) Kausalität. Geschichte und Probleme. Mohr, Tübingen

Carnap R (1928) Der logische Aufbau der Welt. Weltkreisverlag, Berlin

Cartwright N (1983) How the laws of physics lie. Oxford University Press, Oxford

Cartwright N (1994) Nature's capacities and their measurement. Oxford University Press, Oxford

Chaitin GJ (1971) Computational complexity and Gödel's incompleteness theorem. ACM SIGACT News 9:11–12

Chaitin GJ (1987) Algorithmic information theory. Cambridge University Press, Cambridge

Chaitin GJ (2001) Exploring randomness. Springer, London

Comte A (2006) Cours de Philosophie positive (1830–1842). Bachelier, Paris

Coven E, Kan I, Yorke JA (1988) Pseudo-orbit shadowing in the family of tent maps. Trans Am Math Soc 308(1):227–241
Crutchfield J, Farmer JD, Packard NH, Shaw RS (1986) Chaos. Sci Am 12:46–57
Darrigol O (2006) Worlds of flow. A history of hydrodynamics from Bernoulli to Prandtl. Oxford University Press, Oxford
Descartes R (1979) Regeln zur Ausrichtung der Erkenntniskraft. Felix Meixner, Hamburg
Devaney RL (1987) An introduction to chaotic dynamical systems. Addison Wesley, Redwood City
Drieschner (2002) Moderne Naturphilosophie. Eine Einführung. Paderborn, Mentis
Duhem P (1991) The aim and structure of physical theory (1906). Princeton Uni Press, Princeton
Ebeling W, Feistel R (1990) Physik der Evolutionsprozesse. Akademie Verlag, Berlin
Einstein A (1917) Kosmologische Betrachtungen zur allgemeinen Relativitätstheorie. Sitzungsberichte der Königlich Preussischen Akademie der Wissenschaften, physikalisch-mathematische Klasse, pp 142–152
Einstein A, Podolsky B, Rosen N (1935) Can quantum-mechanical description of physical reality be considered complete? Phys Rev 47:777–780
Gierer A (1981) Physik der biologischen Gestaltbildung. Naturwiss 68:245–251
Gleick J (1987) Chaos: making a new science. Viking Penguin, New York
Guckenheimer J, Holmes P (1983) Nonlinear oscillations, dynamical systems, and bifurcations of vector fields. Springer, New York
Habermas J (1972) Knowledge and human interest (1968). Heinemann, London
Hacking I (1983) Representing and intervening. Introductory topics in the philosophy of natural sciences. Cambridge University Press, Cambridge
Haken H (ed) (1977) Synergetics. A workshop. Springer, Berlin
Harrell M, Glymour C (2002) Confirmation and chaos. Philos Sci 69:256–265
Hartmann S, Frigg R (2006) Models in science. In: Zalta EN (ed) Stanford encyclopedia of philosophy. Stanford University Press, Stanford
Heidegger M (1986) Nietzsches metaphysische Grundstellung im abendländischen Denken—die ewige Wiederkehr des Gleichen. Vittorio Klostermann, Frankfurt
Hempel CG (1965) Aspects of scientific explanation. Free Press, New York
Hertz H (1963) Die Prinzipien der Mechanik. In neuem Zusammenhang dargestellt (1894). Wiss. Buchgesellschaft, Darmstadt
Hirsch M (1984) The dynamical systems approach to differential equations. Bull Am Math Soc 11:1–64
Holmes P (2005) Ninety plus thirty years of nonlinear dynamics: more is different and less is more. Int J Bif Chaos 15(9):2703–2716
Hooker CA (2004) Asymptotics, reduction and emergence. Brit J Phil Sci 55(3):435–479
Hume D (1990) Enquiries concerning human understanding (1748). Clarendon Press, Oxford
Hund F (1972) Geschichte der physikalischen Begriffe. BI, Mannheim
Jackson EA (1989) Perspectives of nonlinear dynamics. Cambridge University Press, Cambridge
Janich P (1997) Kleine Philosophie der Naturwissenschaften. Beck, München
Kellert S (1993) In the wake of chaos: unpredictable order in dynamical systems. University Chicago Press, Chicago
Krohn W, Küppers G (eds) (1992) Selbstorganisation. Aspekte einer wissenschaftlichen Revolution. Vieweg, Braunschweig/Wiesbaden
Küppers B-O (1992) Naturals Organismus. Schellings frühe Naturphilosophie und ihre Bedeutung für die moderne Biologie. Suhrkamp, Frankfurt
Langer JS (1980) Instabilities and pattern formation. Rev Mod Phys 52:1–28
Lenhard J (2007) Computer simulations: The cooperation between experimenting and modeling. Philos Sci 74:176–194
Lewes GH (1875) Problems of life and mind, first series, vol 2. Trubner, Kegan and others, London
Li T-Y, Yorke JA (1975) Period three implies chaos. Am Math Monthly 82(10):985–992
Lorenz EN (1963) Deterministic nonperiodic flow. J Atmos Sci 20:130–141
Lorenz EN (1989) Computational chaos—a prelude to computational instability. Phys D 35:299–317

Mach E (1988) Die Mechanik in ihrer Entwicklung (1883). Akademie Verlag, Leipzig/Darmstadt
Mainzer K (1996) Thinking in complexity. The complex dynamics of matter, mind, and mankind. Heidelberg/New York: Springer
Mandelbrot B (1991) Die fraktale Geometrie der Natur (1977). Basel: Birkhauser
Maxwell JC (1873) Does the progress of physical science tend to give any advantage to the opinion of necessity (or determinism) over that of the contingency of events and the freedom of the will? In: Campbell L, Garnett W (eds) The life of James Clerk Maxwell. Johnson, New York, pp 434–444
Maxwell JC (1991) Matter and motion (1877). Dover Publications, New York
Meinhardt H (1995) The algorithmic beauty of seashells. Springer, Berlin
Mitchell SD (2002) Integrative Pluralism. Biol Philos 17:55–70
Mittelstraß J (1998) Die Haeuser des Wissens. Suhrkamp, Frankfurt
Morgan MS, Morrison M (eds) (1999) Models as mediators. Perspectives on natural and social sciences. Cambridge University Press, Cambridge
Newton I (1770) Opticks: or, a treatise of the reflections, refractions, inflections and colours of light. Innys, New York, 1730 (1717)
Nicolis G, Prigogine I (1977) Self-organization in nonequilibrium systems. From dissipative structures to order through fluctuations. Wiley, New York/London
Nietzsche F (1930) Die fröhliche Wissenschaft (1887). Leipzig: Kroener
Parker TS, Chua LO (1989) Practical numerical algorithms for chaotic systems. Springer, New York
Pauli W (1961) Aufsaetze und Vortraege über Physik und Erkenntnistheorie. Vieweg, Braunschweig
Peitgen H-O, Jürgens H, Saupe D (1992) Bausteine des Chaos: Fraktale. Springer/Klett-Cotta, Berlin
Pietschmann H (1996) Phaenomenologie der Naturwissenschaft. Springer, Berlin
Poincaré H (1892) Les méthodes nouvelles de la mécanique céleste. Gauthier-Villars et fils, Paris, p 1892
Poincaré H (1914) Wissenschaft und Methode (1908). Teubner, Leipzig
Popper KR (1934) Logik der Forschung. Wien: Julius Springer (English translation: the logic of scientific discovery. Routledge, London)
Poser H (2001) Wissenschaftstheorie. Reclam, Stuttgart
Prigogine I, Glansdorff P (1971) Thermodynamic theory of structure, stability and fluctuation. Wiley, New York/London
Prigogine I (1980) From being to becoming. Time and complexity in the physical sciences. Freeman, New York
Psillos S (1999) Scientific realism. How science tracks truth. Routledge, London/New York
Redhead M (1980) Models in physics. Brit J Phil Sci 31:145–163
Rueger A, Sharp AD (1996) Simple theories of a messy world: truth and explanatory power in nonlinear dynamics. Brit J Phil Sci 47:93–112
Ruelle D (1989) Elements of differentiable dynamics and bifurcation theory. Academic Press, London, p 1989
Ruelle D, Takens F (1971) On the nature of turbulence. Commun Math Phys 20:167–192
Salmon W (1989) Four decades of scientific explanation. In: Kitcher P, Salmon W (eds) Scientific explanation. Minnesota: University Minnesota Press, pp 3–219
Sauer T, Yorke JA, Casdagli M (1991) Embedology. J Stat Phys 77(3/4):579–616
Schmidt JC (2003) Zwischen Berechenbarkeit und Nichtberechenbarkeit. Die Thematisierung der Berechenbarkeit in der aktuellen Physik komplexer Systeme. J Gen Phil Sci 34:99–131
Schmidt JC (2008a) From symmetry to complexity: on instabilities and the unity in diversity in nonlinear science. Int J Bif Chaos 18(4):897–910
Schmidt JC (2008b) Instabilität in Natur und Wissenschaft. Eine Wissenschaftsphilosophie der nachmodernen Physik. De Gruyter, Berlin
Schmidt JC (2015) Das Andere der Natur. Neue Wege zur Naturphilosophie, Stuttgart

Schmidt JC (2017) Science in an unstable world. On Pierre Duhem's challenge to the methodology of science. In: Pietsch W, Wernecke J, Ott M (Hg.) (eds) (2017) Berechenbarkeit der Welt? Philosophie und Wissenschaft im Zeitalter von big data. Springer, Berlin, pp 403–434

Stephan A (2007) Emergenz. Von der Unvorhersagbarkeit zur Selbstorganisation (1999). Mentis, Paderborn

Swinney HL, Gollub J (eds) (1981) Hydrodynamic instabilities and the transition to turbulence. Springer, Berlin

Takens F (1985) Distinguishing deterministic and random systems. In: Barenblatt GI, Ioss G, Joseph D (eds) (1985) Nonlinear dynamics and turbulence. Pitman, Boston, pp 314–333

Thom R (1975) Structural stability and morphogenesis. An outline of a general theory of models. Benjamin, Reading/MA

Vuillemin J (1991) Introduction. In: Duhem P (ed) (1991) The aim and structure of physical theory (1906). Princeton University Press, Princeton, pp xv–xxxiii

Wackerbauer R, Witt A, Atmanspacher H, Kurths J, Scheingraber H (1994) A comparative classification of complexity measures. Chaos, Solit Fract 4(1):133–174

Weizsäcker CFv (1974) Die Einheit der Natur. München: dtv

Wiggins S (1988) Global bifurcations and chaos. Analytical methods. Springer, New York

Woodward J (2000) Explanation and invariance in the special sciences. Brit J Phil Sci 51:197–254

Modular Organization and Emergence in Systems Biology

Marc-Thorsten Hütt

Abstract Understanding, how cellular functions emerge from the interaction of biological components, is the main goal of systems biology. Here, we review the relevance of a modular organization and the emergence of collective dynamical states in systems biology and show that the concept of networks (i.e., the representation of biological systems in terms of nodes and links) allows us to formally define modularity and to quantitatively assess the impact of modularity on the emergent dynamical behaviors.

Introduction

This chapter serves two purposes. On the one hand, we aim at providing a detailed account, how the concepts of modularity and emergence currently impact the research fields of systems biology and systems medicine.

On the other hand, we want to equip the reader with a clear and vivid intuition, what modularity is, how it can be quantified, and how it—as a structural, systemic property—impacts dynamics and, as a consequence, the emergence of functional features.

Even though there is no universally accepted definition of the term "systems biology," some features of this new, emerging discipline have become clear over the last 15 years. *Systems Biology* is the attempt to provide a theoretical framework for biology, as well as the mathematical and computational inventory to analyze and interpret "omics" data (Kitano 2002a, b; Hütt 2014; Badimon et al. 2017).

More recently, the term *Systems Medicine* has become popular for the application of concepts from systems biology to medical data and the understanding of diseases (Auffray et al. 2009; Bauer et al. 2017).

Both, in systems biology and in systems medicine *networks* have proven to be a particularly relevant data structure: Abstracting huge system-level collections of

M.-T. Hütt (✉)
Department of Life Sciences and Chemistry, Jacobs University Bremen, 28759 Bremen, Germany
e-mail: m.huett@jacobs-university.de

L. H. Wegner and U. Lüttge (eds.), *Emergence and Modularity in Life Sciences*,
https://doi.org/10.1007/978-3-030-06128-9_2

biological and medical data as sets of nodes and links has facilitated the organization, analysis, and comparison of diverse biological systems (Barabasi and Oltvai 2004; Barabási et al. 2011). Networks have been successfully employed, among others, for analyzing metabolic systems (Ravasz et al. 2002; Guimera and Amaral 2005; Beber et al. 2012), gene regulatory systems (Milo et al. 2002; Babu et al. 2004; Marr et al. 2010), protein interactions (Jeong et al. 2001; Maslov and Sneppen 2002; Bork et al. 2004), signaling pathways (Li et al. 2006; Helikar et al. 2008), but also for studying the relationships between genes and diseases (Goh and Choi 2012) and among diseases, drugs, and side effects (Hopkins 2008).

In addition to being an efficient data structure for biological information, networks also hold the potential of implementing a systems-level view on biological processes. As pointed out in (Silverman and Loscalzo 2013): "The classical single target-based drug development paradigm [...] tends to neglect the complex perturbations that drugs cause within the cellular molecular network. [...] A recent NIH White Paper on Systems Pharmacology pointed out the importance of viewing drug development within a network context; the cellular molecular network has emergent properties (unique characteristics resulting from the specific combination of network elements) that are not apparent if single molecules are studied in isolation." Understanding, how these "emergent properties" come about in networks and how network structure facilitates dynamical function, is the principal research question in systems biology.

The conceptual transition of the theory of complex systems from the emergence of spatiotemporal patterns and collective behaviors to "network science" is summarized by the following statement: "Reductionism, as a paradigm, is expired, and complexity, as a field, is tired. Data-based mathematical models of complex systems are offering a fresh perspective, rapidly developing into a new discipline: network science" (Barabási 2016).

Modularity

In the diverse attempts to establish relationships between network structure and function in systems biology, *modularity* is a very prominent concept. This structural feature was among the first attempts to relate systemic organization to function in a universal fashion (Hartwell et al. 1999). Broadly speaking, it can be linked to the segmentation of collective tasks into subtasks, i.e., the *modularization* of tasks (Kashtan and Alon 2005; Wagner et al. 2007). Furthermore, in systems biology, the concept of modularity is often augmented by the "bow tie organization" of biological systems as a paradigm denoting a specific form of modular organization, where a high input diversity (one side of the bow tie) enters a narrow processing region with a limited number of (often highly versatile) "data processing" protocols (the knot), leading to a high output diversity (the other side of the bow tie), thus implementing a balance between high flexibility and economical infrastructure (Ma and Zeng 2003; Kitano 2004; Csete and Doyle 2004; Singh et al. 2013).

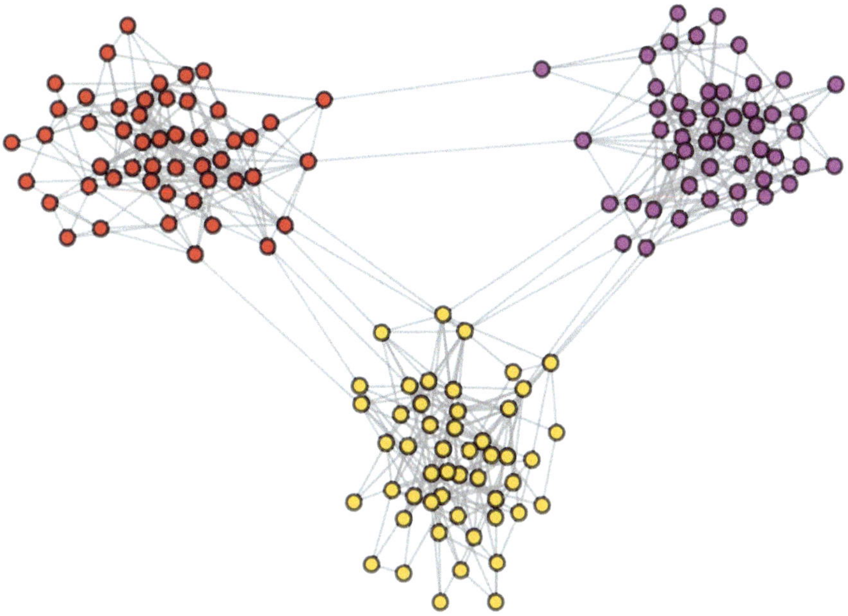

Fig. 1 Example of a modular graph consisting of 150 nodes (vertices) and 470 links (edges), organized into three clearly visible modules. The graph has been generated by sparsely interconnecting three random graphs (created as Erdős-Rényi random graphs, Erdős and Rényi (1959)) of 50 nodes and 150 links with 20 additional, inter-module links. Colors have been assigned via the Girvan–Newman algorithm (Girvan and Newman 2002; Newman and Girvan 2004)

The formal language of networks allows a precise definition of the term "modularity." Qualitatively speaking, modularity is the existence of densely connected regions (groups of nodes) in the graph, which are sparsely interconnected among themselves. The notion of connection density and its local variation also points to the main algorithmic principle underlying the most prominent module detection (or community identification) algorithms (see Fortunato 2010; Fortunato and Hric 2016, for detailed reviews).

Figure 1 shows an example of a modular graph, which will also later allow us to illustrate how this specific form of network architecture systematically influences dynamical processes in such a network.

In Girvan and Newman (2002); Newman and Girvan (2004), a method was introduced for quantifying the modularity of a graph, given a partitioning of nodes into modules. This definition counts, how often two nodes in the same module are connected, compared to the random expectation given the nodes' degrees (where the degree of a node is the number of its neighbors). For a network with N nodes and M links and a given partition into m modules, the modularity Q can be written as (Newman and Girvan 2004; Arenas et al. 2008):

$$Q = \frac{1}{2M} \sum_{i,j} \left(A_{ij} - \frac{k_i k_j}{2M} \right) \delta_{c_i, c_j} , \tag{1}$$

where i and j enumerate the nodes of the graph and k_i is the degree of node i. The quantity c_i denotes the module association of node i, $c_i \in \{1, 2, \ldots, m\}$, and the matrix A_{ij} is the graph's adjacency matrix (where the element A_{ij} is 1, if nodes i and j are linked, and 0 otherwise). Lastly, the function $\delta_{a,b}$ is 1, if $a = b$ and 0 otherwise.

Equation (1) quantifies the modularity of a *given* partition of the nodes of a graph into modules (or communities). We can now turn this view around and ask for the partitioning of nodes into modules that *maximizes* the modularity given in Eq. (1). This approach—module detection as an optimization problem maximizing Q, as described in Girvan and Newman (2002); Newman and Girvan (2004)—is still one of the most prominent module detection algorithms in the literature (Fortunato 2010; Fortunato and Hric 2016). The assignment of colors to the nodes in the graph shown in Fig. 1 has been obtained with this Girvan–Newman algorithm.

It is remarkable how the general intuitive idea of modularity can be (via the language of networks) translated into a precise mathematical definition.

A simple numerical experiment allows us to assess how precisely this quantity Q from Eq. (1) reacts to changes in the modular organization of the graph by applying a sequence of switch-randomization steps, where end points of a pair of randomly selected nodes are removed from the network and re-attached in a cross-wise fashion (swapping the two end points of the links). This operation keeps the degrees of each individual nodes unchanged, but, when iterated, gradually destroys the modular organization of the graph from Fig. 1. The modularity shown in Fig. 2 as a function of the number of consecutive randomization steps has been obtained by running the

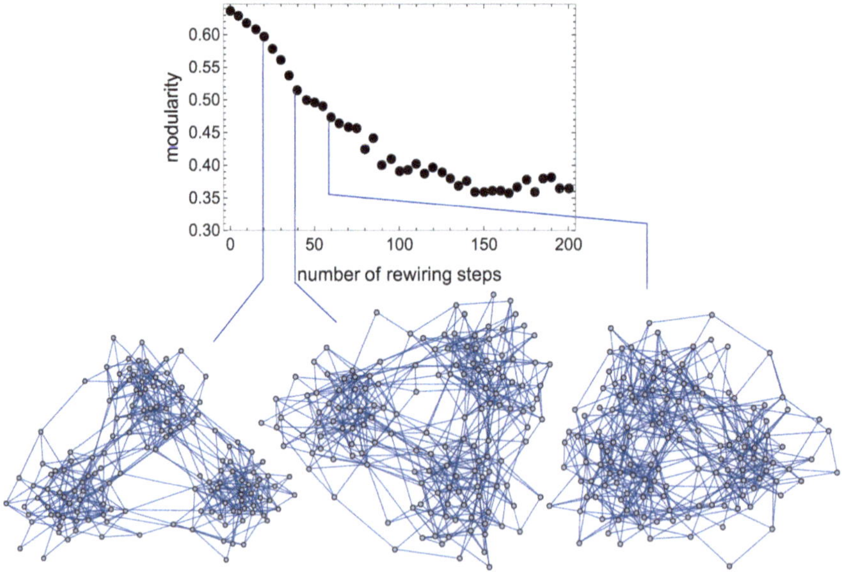

Fig. 2 Modularity Q as a function of the iterative rewiring of the network from Fig. 1 via switch randomization, together with three examples of intermediate networks

Girvan–Newman community detection algorithm and then assessing the modularity for the module structure obtained from that algorithm. We see that already after a few tens of switch randomizations the modularity decreases substantially. Figure 2 also shows a few examples of networks along this randomization path.

Switch randomization has become a prominent tool for investigating statistical properties of complex networks by serving as a null model for real-life networks (Milo et al. 2002) and as a method for studying the interplay of different topological scales (Fretter et al. 2012).

With these tools, the modularity of real-life networks in systems biology and in diverse other disciplines can now be evaluated and discussed (see, e.g., Ravasz et al. 2002; Han et al. 2004; Guimera and Amaral 2005; Voordijk et al. 2006, for a few prominent examples). Furthermore, the dynamical consequences and implications of a modular organization can be studied (see, e.g., Arenas et al. 2006; Parter et al. 2007; Müller-Linow et al. 2008; Beber et al. 2013).

Emergence

Being one of the prominent terms in the theory of complex systems, integrating the notion of emergence into systems biology has been a challenging endeavor. The surprisingly antagonistic nature of systems biology and the theory of complex systems has been pointed out in an early review article by Kitano (2002a): "A popular notion of complex systems is of very large numbers of simple and identical elements interacting to produce "complex" behaviors. [In biology …] large numbers of functionally diverse, and frequently multifunctional, sets of elements interact selectively and nonlinearly to produce coherent rather than complex behaviors."

Like in the case of modularity, already this quote suggests how networks (the capacity to "interact selectively") may allow us to bring together the approaches from the theory of complex systems on the one hand and the conceptual agenda from systems biology on the other.

Again, in order to make use of such a general concept in a quantitative fashion, we can resort to the theory of complex systems for a precise definition.

In the context of self-organization, emergence can be formally defined as the spontaneous onset of a collective ("self-organized") state under variation of an intrinsic (local) parameter. The collective behavior is typically characterized by an order parameter, i.e., a quantity computed from the collective state allowing us to quantitatively discriminate different collective behaviors.

A common view (supported by many well-known examples) of such self-organized systems is that only few distinct collective states (and hence few possible values of the order parameter) are possible for each system.

A paradigmatic example is a system of coupled oscillators, where the amount of synchronization serves as an order parameter and the coupling strength among the oscillators is the intrinsic control parameter. When increasing the coupling in such a system, one sees the spontaneous emergence of a synchronized collective state

(Strogatz 2001, 2000; Rodrigues et al. 2016). In this case, the order parameter can be written down explicitly, quantifying the amount of synchronization in the system. In its simplest form, the system of coupled oscillators is given by phase oscillators

$$\frac{d\phi_j}{dt} = \omega_j + \varepsilon \sum_k A_{jk} \sin(\phi_k - \phi_j), \tag{2}$$

where $\phi_j(t)$ is the phase variable of the jth oscillator at time t, the parameter ω_j is the natural frequency (or "eigenfrequency") of the jth oscillator, ε is the coupling strength and, like before, A_{jk} is the adjacency matrix of the underlying network. The order parameter can now be written as the (time-averaged) length of the vector obtained by considering each phase $\phi_j(t)$ as a vector in the complex plane $\exp(i\phi_j(t))$ and then summing over all oscillators (Kuramoto 1984):

$$R = \left\langle \frac{1}{N} \left| \sum_{j=1}^{N} \exp\left(i\phi_j(t)\right) \right| \right\rangle_t, \tag{3}$$

where the angle brackets ($\langle . \rangle$) indicate the time average. Figure 3 shows the order parameter R as a function of the coupling strength ε for a system of coupled phase oscillators following Eq. (2). Even for this comparatively small network, the sigmoidal change of the order parameter as a function of the control parameter and, hence, the jump-like, spontaneous onset—the *emergence*—of synchronization as a collective network state at a critical value of the coupling strength is clearly visible.

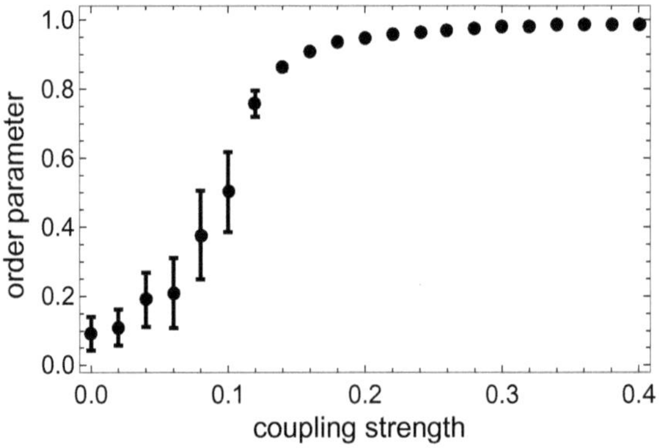

Fig. 3 Order parameter R as a function of the coupling strength ε for a system of 100 phase oscillators coupled according to an Erdős-Rényi random graph with 350 links. Simulations have been performed for 600 time steps, starting from random initial conditions. The eigenfrequencies ω_j (cf. Eq. (2)) have been randomly drawn from a uniform distribution between 0 and 1. The order parameter R has been computed according to Eq. (3) and averaging over the last 500 time steps (thus eliminating an initial transient of 100 time steps)

It is precisely this notion of collective network states spontaneously emerging upon variation of external conditions (or control parameters) that carries a strong appeal for applications in systems biology and systems medicine: "[...] the cellular molecular network has emergent properties" (Silverman and Loscalzo 2013).

Modularity and Emergence

The self-organization of dynamical processes in networks brings together the two main theoretical concepts introduced above: modularity and emergence. It becomes possible to understand how network architecture shapes the dynamical patterns emerging in a network as collective functional states.

Here, we will provide three dynamical processes as examples, illustrating how the modular organization of a network impacts the dynamical patterns arising on such a graph. Each of the three examples—random walks, excitable dynamics, and synchronization of coupled oscillators—is a generic representative of a large class of dynamical processes. In spite of the simplicity of the specific models discussed here, we can thus expect that the tight link between network structure and dynamics established by the network's modular organization is of relevance to diverse application scenarios.

Random walks on graphs are a relevant method for investigating the processing of fluctuating signals by a network. In particular, this model has been used to study scaling relationships between fluctuations and signal strength (De Menezes and Barabási 2004), as well as network effects in such relationships (Kosmidis et al. 2015). Formally, a random walk on a graph is initialized by randomly selecting a starting node $n_0^{(k)}$ for walker k. In all subsequent steps, the next node $n_{i+1}^{(k)}$ is then selected from the neighborhood of node $n_i^{(k)}$. This random walk is simulated for T time steps. When following the "diffusion" of such random walkers through the network over many intervals of size T, the visiting probability f_j of each node j in the network is the relevant dynamical information. This visiting probability is influenced by the internal noise (variance generated by the stochasticity of the random walk itself), the external noise (variance generated by the fluctuations in the number of walkers), and the topological properties of node j (e.g., its degree k_j).

In Fig. 4, we see that a random walk resides in modules for long periods of time, before jumping to another module. This observation that random walks are "trapped" in modules has been made prominent in Rosvall and Bergstrom (2008) with implications for understanding diverse phenomena of information flow and community detection (Rosvall et al. 2014; De Domenico et al. 2015).

For the next exploration of the modular network with a dynamical process, we will use a minimal model of excitation (or activity) patterns. The model is a stochastic cellular automaton with two parameters: the recovery probability p (regulating the refractory phase of the excitable units) and the rate of spontaneous activity f (representing noise in the system). Each node in this network can be susceptible, S,

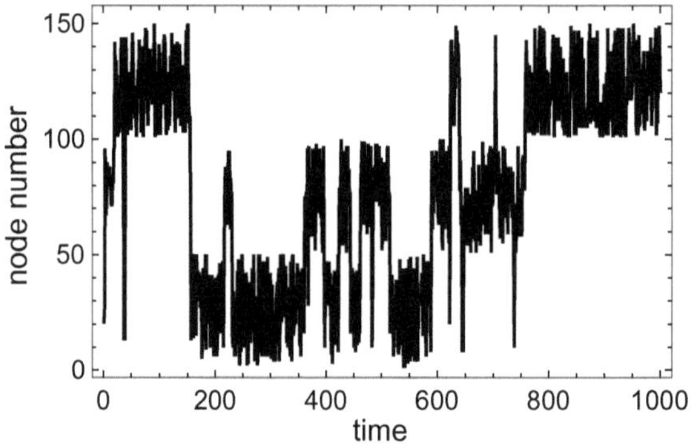

Fig. 4 Trajectory (position as a function of time) of a single random walk on the network from Fig. 1. The position n_i is given by the node occupied by the walk at time i. Note that node numbers for the network from Fig. 1 are organized as follows: module A (red nodes): nodes 1–50, module B (yellow nodes): nodes 51–100, module C (purple nodes): nodes 101–150

excited (or active), E, or refractory, R. These three-state dynamics run on a given graph. Activity at a neighboring node triggers activity at a susceptible node, thus leading to flows of activity through the network. After one time step in the E state, a node enters the R state and recovers from there with probability p in each time step (leading to a geometric distribution of the refractory periods with an average refractory period of $1/p$). Additionally, with probability f each susceptible node is spontaneously activated. This model is particularly suited to study the general principles underlying activity synchronization in networks devoid of any specific details about the system. Such discrete excitable dynamics on graphs have been studied for example in (Müller-Linow et al. 2008; Garcia et al. 2012; Messé et al. 2018).

In order to compare simulated excitation patterns with the network architecture, we compute the co-activation matrix. Let $x_i(t) \in \{S, E, R\}$ be the state of node i at time t. It is convenient to discuss the excitation pattern instead:

$$c_i(t) = \begin{cases} 1, & x_i(t) = E \\ 0, & x_i(t) = S \vee R \end{cases}$$

In this way, we can define the co-activation matrix as

$$C_{ij} = \sum_t c_i(t) c_j(t).$$

In some cases, it is more convenient to consider normalized version of this matrix. Let

$$c_i^* = \sum_t c_i(t)$$

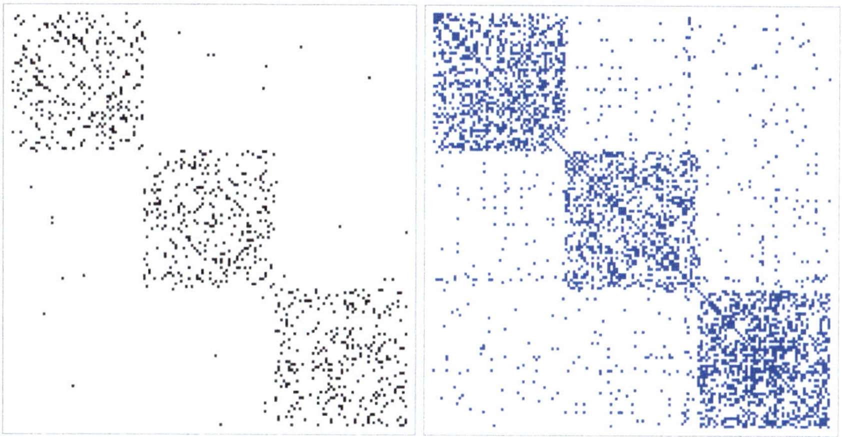

Fig. 5 Adjacency matrix (left) of the modular network from Fig. 1, together with the (binarized) co-activation matrix (right) obtained from running a simulation of 1000 time steps of the model of discrete excitable dynamics and then computing the co-activation matrix C^*_{ij} from the last 500 time steps (thus eliminating an initial transient). In the adjacency matrix, a black dot indicates a 1, while empty (white) matrix elements indicate a 0, thus revealing the modular organization of the network as three blocks enriched in 1s. For the co-activation matrix, a blue dot indicates a (normalized) co-activation value larger than the average value over all node pairs, while a white dot indicates a value smaller than the average

denote the total number of excitations of node i in the time series under consideration. Then, the normalized quantities C^*_{ij} of the co-activation matrix entries are defined as $C^*_{ij} = C_{ij} / \min(c^*_i, c^*_j)$. Quantities in this matrix are normalized such that each entry is between zero and one (see also Hütt et al. 2014; Garcia et al. 2012, for details on these quantities).

In Fig. 5, the normalized co-activation matrix C^*_{ij} obtained from running the discrete model of excitable dynamics on the graph shown in Fig. 1 is compared to the graph's adjacency matrix. Again, like in the case of the random walk, the modular architecture (quadratic blocks of enhanced co-activation) is clearly visible.

The observation that co-activation is enriched in modules, together with its implications for Neuroscience, has been described in Müller-Linow et al. (2008), Garcia et al. (2012), Messé et al. (2015).

For the last exploration of the modular architecture with a dynamical process, we return to the coupled phase oscillators from Eq. (2) and Fig. 3. Figure 6 provides a snapshot of the contributions of each node to the sum in Eq. (3) for the modular network from Fig. 1 at a coupling strength above the critical value. Even though all contributions are nearly identical (indicating the high degree of synchronization in the system), the small remaining differences again reflect the modular organization of the network.

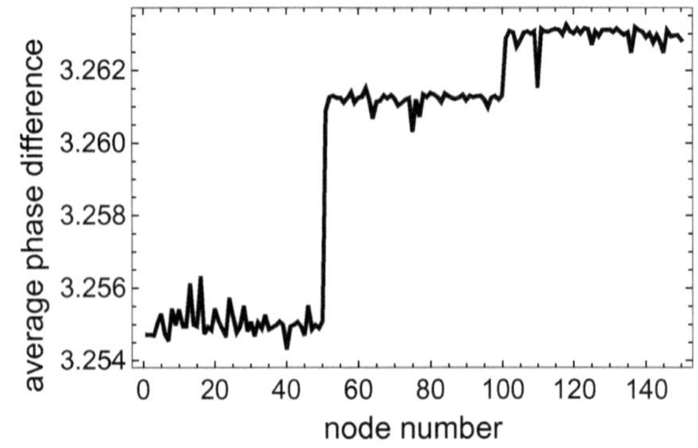

Fig. 6 Snapshot of the contributions $\exp\left(i\phi_j(t)\right)$ to the order parameter R from Eq. (3) for $t = 2000$ and $\varepsilon = 0.3$ as a function of the node number j for the modular network from Fig. 1

The fact that the transient toward full synchronization in coupled oscillators on a graph follows the graph's modular organization has been first described in detail in Arenas et al. (2006) (see also Arenas and Diaz-Guilera 2007; Arenas et al. 2008).

These examples—the entrapment of random walks in modules; the enrichment of co-activation, and therefore high synchronization, of excitable units belonging to the same module; the road toward full synchronization mapping out the modular organization—show how emergent dynamical patterns are facilitated by the modularity of the graph.

Current Research Trends and Outlook

As outlined above, modularity has been analyzed and discussed for in the context of systems biology, with metabolic networks as one prominent example (Ravasz et al. 2002; Guimera and Amaral 2005; Kreimer et al. 2008; Parter et al. 2007). The modular architecture of network has also been studied for brain networks (Meunier et al. 2010), supply chains (Voordijk et al. 2006), scientific communities (Newman 2004), even in networks of scientific hypotheses (Enders et al. 2018).

It was acknowledged that modularity of a system can be evidence of "distributed computation" or an evolutionary consequence of multiple tasks (Kashtan and Alon 2005). Relatedly, modular architectures are often assumed to arise evolutionarily from modular task structures as space and time (Clune et al. 2013). The emergence of modularity in simulated evolution approaches (Kashtan and Alon 2005; Beber et al. 2013) and from selecting on features of the dynamics (Guimera et al. 2004; Damicelli et al. 2017) has therefore become a very prominent research question.

In Costanzo et al. (2016), a fascinating example from systems biology of structure-function relationships derived from modularity is given. There, the modularity of genetic networks in yeast is compared with gene set enrichment, revealing that the modules in genetic coupling systematically map out gene ontology categories.

The few numerical examples discussed above show the impact of a modular organization of a network on the dynamical patterns establishing themselves in the network. This is only one of many examples, how network architecture shapes dynamics and, hence, function. Via the notion of self-organized pattern on graphs (see, e.g., Hütt et al. 2014; Hütt 2014), the terms modularity and emergence pave a way toward using biological networks to interpret ''omics'' data: High-throughput data can be seen as patterns in these networks. First directions toward this data analysis strategy can be seen in Sonnenschein et al. (2011, 2012), Knecht et al. (2016).

The long-term vision of this approach is to use our understanding of patterns in networks to develop a theory of cellular states as "network states." The interplay of network architecture and dynamics, as visualized here using the concepts of modularity and emergence, constitutes a promising starting point on this avenue.

References

Arenas A, Diaz-Guilera A (2007) Synchronization and modularity in complex networks. Eur Phys J Spec Top 143(1):19–25

Arenas A, Diaz-Guilera A, Pérez-Vicente CJ (2006) Synchronization reveals topological scales in complex networks. Phys Rev Lett 96(11):114,102

Arenas A, Díaz-Guilera A, Kurths J, Moreno Y, Zhou C (2008) Synchronization in complex networks. Phys Rep 469(3):93–153

Auffray C, Chen Z, Hood L (2009) Systems medicine: the future of medical genomics and healthcare. Genome Med 1(1):2

Babu MM, Luscombe NM, Aravind L, Gerstein M, Teichmann SA (2004) Structure and evolution of transcriptional regulatory networks. Curr Opin Struct Biol 14(3):283–291

Badimon L, Vilahur G, Padro T (2017) Systems biology approaches to understand the effects of nutrition and promote health. Br J Clin Pharmacol 83(1):38–45

Barabási AL (2016) Network science. Cambridge University Press

Barabasi AL, Oltvai ZN (2004) Network biology: understanding the cell's functional organization. Nat Rev Genet 5(2):101

Barabási AL, Gulbahce N, Loscalzo J (2011) Network medicine: a network-based approach to human disease. Nat. Rev Genet 12(1):56

Bauer CR, Knecht C, Fretter C, Baum B, Jendrossek S, Rühlemann M, Heinsen FA, Umbach N, Grimbacher B, Franke A et al (2017) Interdisciplinary approach towards a systems medicine toolbox using the example of inflammatory diseases. Brief Bioinf 18(3):479–487

Beber ME, Fretter C, Jain S, Sonnenschein N, Müller-Hannemann M, Hütt MT (2012) Artefacts in statistical analyses of network motifs: general framework and application to metabolic networks. J R Soc. Interface p:rsif20120490

Beber ME, Armbruster D, Hütt MT (2013) The prescribed output pattern regulates the modular structure of flow networks. Eur Phys J B 86(11):473

Bork P, Jensen LJ, Von Mering C, Ramani AK, Lee I, Marcotte EM (2004) Protein interaction networks from yeast to human. Curr Opin Struct Biol 14(3):292–299

Clune J, Mouret JB, Lipson H (2013) The evolutionary origins of modularity. Proc R Soc B 280(1755):20122,863

Costanzo M, VanderSluis B, Koch EN, Baryshnikova A, Pons C, Tan G, Wang W, Usaj M, Hanchard J, Lee SD et al (2016) A global genetic interaction network maps a wiring diagram of cellular function. Science 353(6306):aaf1420

Csete M, Doyle J (2004) Bow ties, metabolism and disease. TRENDS in Biotechnology 22(9):446–450
Damicelli F, Hilgetag CC, Hütt MT, Messé A (2017) Modular topology emerges from plasticity in a minimalistic excitable network model. Chaos: An Interdisciplinary Journal of Nonlinear Science 27(4):047,406
De Domenico M, Lancichinetti A, Arenas A, Rosvall M (2015) Identifying modular flows on multilayer networks reveals highly overlapping organization in interconnected systems. Phys Rev X 5(1):011,027
De Menezes MA, Barabási AL (2004) Fluctuations in network dynamics. Phys Rev Lett 92(2):028,701
Enders M, Hütt MT, Jeschke JM (2018) Drawing a map of invasion biology based on a network of hypotheses. Ecosphere 9(3):e02,146
Erdős P, Rényi A (1959) On random graphs, i. Publ Math (Debrecen) 6:290–297
Fortunato S (2010) Community detection in graphs. Phys Rep 486(3–5):75–174
Fortunato S, Hric D (2016) Community detection in networks: a user guide. Phys Rep 659:1–44
Fretter C, Müller-Hannemann M, Hütt MT (2012) Subgraph fluctuations in random graphs. Phys Rev E 85(5):056,119
Garcia GC, Lesne A, Hütt MT, Hilgetag CC (2012) Building blocks of self-sustained activity in a simple deterministic model of excitable neural networks. Front Comput Neurosci 6:50
Girvan M, Newman ME (2002) Community structure in social and biological networks. Proc Natl Acad Sci 99(12):7821–7826
Goh KI, Choi IG (2012) Exploring the human diseasome: the human disease network. Brief Funct Gen 11(6):533–542
Guimera R, Amaral LAN (2005) Functional cartography of complex metabolic networks. Nature 433(7028):895
Guimera R, Sales-Pardo M, Amaral LAN (2004) Modularity from fluctuations in random graphs and complex networks. Phys Rev E 70(2):025,101
Han JDJ, Bertin N, Hao T, Goldberg DS, Berriz GF, Zhang LV, Dupuy D, Walhout AJ, Cusick ME, Roth FP et al (2004) Evidence for dynamically organized modularity in the yeast protein-protein interaction network. Nature 430(6995):88
Hartwell LH, Hopfield JJ, Leibler S, Murray AW (1999) From molecular to modular cell biology. Nature 402(6761supp):C47
Helikar T, Konvalina J, Heidel J, Rogers JA (2008) Emergent decision-making in biological signal transduction networks. Proc Nat Acad Sci 105(6):1913–1918
Hopkins AL (2008) Network pharmacology: the next paradigm in drug discovery. Nat Chem Biol 4(11):682
Hütt MT (2014) Understanding genetic variation-the value of systems biology. Br J Clin Pharmacol 77(4):597–605
Hütt MT, Kaiser M, Hilgetag CC (2014) Perspective: network-guided pattern formation of neural dynamics. Phil Trans R Soc B 369(1653):20130,522
Jeong H, Mason SP, Barabási AL, Oltvai ZN (2001) Lethality and centrality in protein networks. Nature 411(6833):41
Kashtan N, Alon U (2005) Spontaneous evolution of modularity and network motifs. Proc Nat Acad Sci U S A 102(39):13,773–13,778
Kitano H (2002a) Computational systems biology. Nature 420(6912):206
Kitano H (2002b) Systems biology: a brief overview. Science 295(5560):1662–1664
Kitano H (2004) Biological robustness. Nat Rev Genet 5(11):826
Knecht C, Fretter C, Rosenstiel P, Krawczak M, Hütt MT (2016) Distinct metabolic network states manifest in the gene expression profiles of pediatric inflammatory bowel disease patients and controls. Sci Rep 6(32):584
Kosmidis K, Beber M, Hütt MT (2015) Network heterogeneity and node capacity lead to heterogeneous scaling of fluctuations in random walks on graphs. Adv Complex Syst 18(01n02):1550,007

Kreimer A, Borenstein E, Gophna U, Ruppin E (2008) The evolution of modularity in bacterial metabolic networks. Proc Nat Acad Sci 105(19):6976–6981
Kuramoto Y (1984) Chemical oscillations, waves and turbulence
Li S, Assmann SM, Albert R (2006) Predicting essential components of signal transduction networks: a dynamic model of guard cell abscisic acid signaling. PLoS Biol 4(10):e312
Ma HW, Zeng AP (2003) The connectivity structure, giant strong component and centrality of metabolic networks. Bioinformatics 19(11):1423–1430
Marr C, Theis FJ, Liebovitch LS, Hütt MT (2010) Patterns of subnet usage reveal distinct scales of regulation in the transcriptional regulatory network of escherichia coli. PLoS Comput Biol 6(7):e1000,836
Maslov S, Sneppen K (2002) Specificity and stability in topology of protein networks. Science 296(5569):910–913
Messé A, Hütt MT, König P, Hilgetag CC (2015) A closer look at the apparent correlation of structural and functional connectivity in excitable neural networks. Sci Rep 5:7870
Messé A, Hütt MT, Hilgetag CC (2018) Toward a theory of coactivation patterns in excitable neural networks. PLoS Comput Biol14(4):e1006,084
Meunier D, Lambiotte R, Bullmore ET (2010) Modular and hierarchically modular organization of brain networks. Front Neurosci 4:200
Milo R, Shen-Orr S, Itzkovitz S, Kashtan N, Chklovskii D, Alon U (2002) Network motifs: simple building blocks of complex networks. Science 298(5594):824–827
Müller-Linow M, Hilgetag CC, Hütt MT (2008) Organization of excitable dynamics in hierarchical biological networks. PLoS Comput Biol 4(9):e1000,190
Newman ME (2004) Coauthorship networks and patterns of scientific collaboration. Proc Nat Acad Sci 101(suppl 1):5200–5205
Newman ME, Girvan M (2004) Finding and evaluating community structure in networks. Phys Rev E 69(2):026,113
Parter M, Kashtan N, Alon U (2007) Environmental variability and modularity of bacterial metabolic networks. BMC Evol Biol 7(1):169
Ravasz E, Somera AL, Mongru DA, Oltvai ZN, Barabási AL (2002) Hierarchical organization of modularity in metabolic networks. Science 297(5586):1551–1555
Rodrigues FA, Peron TKD, Ji P, Kurths J (2016) The kuramoto model in complex networks. Phys Rep 610:1–98
Rosvall M, Bergstrom C (2008) Maps of random walks on complex networks reveal community structure. Proc Nat Acad Sci 105:1118–1123
Rosvall M, Esquivel AV, Lancichinetti A, West JD, Lambiotte R (2014) Memory in network flows and its effects on spreading dynamics and community detection. Nat Commun 5:4630
Silverman EK, Loscalzo J (2013) Developing new drug treatments in the era of network medicine. Clin Pharmacol Ther 93(1):26–28
Singh S, Samal A, Giri V, Krishna S, Raghuram N, Jain S (2013) Flux-based classification of reactions reveals a functional bow-tie organization of complex metabolic networks. Phys Rev E 87(5):052,708
Sonnenschein N, Geertz M, Muskhelishvili G, Hütt MT (2011) Analog regulation of metabolic demand. BMC Syst Biol 5(1):40
Sonnenschein N, Dzib JFG, Lesne A, Eilebrecht S, Boulkroun S, Zennaro MC, Benecke A, Hütt MT (2012) A network perspective on metabolic inconsistency. BMC Syst Biol 6(1):41
Strogatz S (2001) Exploring complex networks. Nat 410(6825):268–76
Strogatz SH (2000) From kuramoto to crawford: exploring the onset of synchronization in populations of coupled oscillators. Phys D: Nonlinear Phenom 143(1):1–20
Voordijk H, Meijboom B, de Haan J (2006) Modularity in supply chains: a multiple case study in the construction industry. Int J Oper Prod Manag 26(6):600–618
Wagner GP, Pavlicev M, Cheverud JM (2007) The road to modularity. Nat Rev Genet 8(12):921

The Emergence of Sustainability

Fabio Rubio Scarano

Abstract I use Mario Bunge's approach to the interplay between emergence, submergence and convergence to address the emergence of sustainability. In line with this approach, my argument is that sustainability emerges as a novelty out of the convergence between independent lines of inquiry, such as sociology, economics and ecology in the science realm, and out of the tension between environmentalism and development as social movements or ideologies. It is also related to the emergence of a new ethics based on intergenerational justice. The convergence of sustainability as science, as policy and as moral value, in turn, implies submergence or even full disappearance of the old. For instance, out of the modules from which sustainability emerged, namely biological conservation (in science) and environmentalism (its social movement counterpart) are undergoing a crisis that might precede their submergence if not full disappearance. The notion of socio-economic development, one of the modules that contributed to the emergence of sustainability, is also in crisis as alternative views collectively framed as 'postdevelopment' appear. Sustainability, as an emerging new whole, still requires 'glue concepts' (sensu Bunge) to bind together its social, economic and environmental dimensions—this applies equally to sustainability as policy, as moral value and as science. Sustainability emerges as a normative goal (i.e. a desire based on beliefs and values) and as a principle of collective good and, as such, it is a new utopia. Finally, I discuss how the emergence of sustainability furthers the vitality of Lovelock's Gaia hypothesis. Failure or success of sustainability might determine the fate of humankind on the planet.

Introduction

Sustainability emerged as a concept and as a moral value in the Anthropocene (see Crutzen and Stoermer 2000), and particularly in response to the Great Acceleration triggered in the 1950s (Steffen et al. 2015a). Social movements and science embraced this concept, and it became the object of political agreements and decision-making at

F. R. Scarano (✉)
Universidade Federal do Rio de Janeiro, Rio de Janeiro, Brazil
e-mail: fscarano@biologia.ufrj.br

L. H. Wegner and U. Lüttge (eds.), *Emergence and Modularity in Life Sciences*,
https://doi.org/10.1007/978-3-030-06128-9_3

global, national and local levels. The Brundtland Report on Sustainable Development (WCED 1987) was a key milestone in putting sustainability on the global agenda, giving birth to global policy agreements—such as those related to biodiversity, climate change, combat against desertification and, ultimately, the Sustainable Development Goals (SDG)—and triggering the advent of sustainability science (Bettencourt and Kaur 2011).

I use the theoretical framework of Argentinian philosopher Bunge (2003) to address the emergence of sustainability. Although there are many connotations to emergence (Banzhaf 2014), Bunge's approach to the interplay between emergence and submergence, convergence and divergence, seems particularly relevant to address sustainability. It has also been used to address issues as diverse as technology, social sciences, ecology and evolutionary biology (Trosper 2005; De Haan 2006; Kaidesoja 2009). Bunge (2003) extensively focuses on emergence as novelty: it "occurs every time a qualitative new whole appears". This novelty, in turn, often implies submergence or even full disappearance of the old. Emergence calls for convergence between "initially separate approaches and fields". Convergence in this sense is synonym to unification, merger, integration and therefore requires what Bunge calls 'glue' concepts or hypotheses to bind together different components or modules. Conversely, emergence can also take place as a result of divergence or splitting of a whole entity.

In line with Bunge's framework, my argument is that sustainability emerges as a novelty out of independent lines of inquiry, such as sociology, economics and ecology in the science realm, and out of the tension between environmentalism and development as social movements or ideologies. In other words, sustainability emerges from cross-disciplinary collaboration in the science realm and multisectoral dialogue in the policy arena. Meanwhile, out of the modules from which sustainability emerged, namely biological conservation (in science) and environmentalism (its social movement counterpart) are undergoing a crisis that might precede their submergence if not full disappearance. While not the focus of this book, I will also briefly discuss how the development branch that partly promoted the emergence of sustainability is also in crisis. Moreover, I argue that sustainability, as an emerging new whole, still requires a 'glue' to bind together its social, economic and environmental dimensions—this applies equally to sustainability as policy, as moral value and as science. Thus, for now, sustainability emerges as a normative goal (i.e. a desire based on beliefs and values) and as a principle of collective good (Geels 2010). I argue that it is a new utopia, which needs further convergence of its various social, economic and environmental modules to be actually realized. Finally, I will discuss how the emergence of sustainability furthers the vitality of Lovelock's Gaia hypothesis (Lovelock 1979). As society approaches the sustainability utopia, one can even think of a 'deep' sustainability state. Figure 1 shows a schematic representation for the rationale of this chapter.

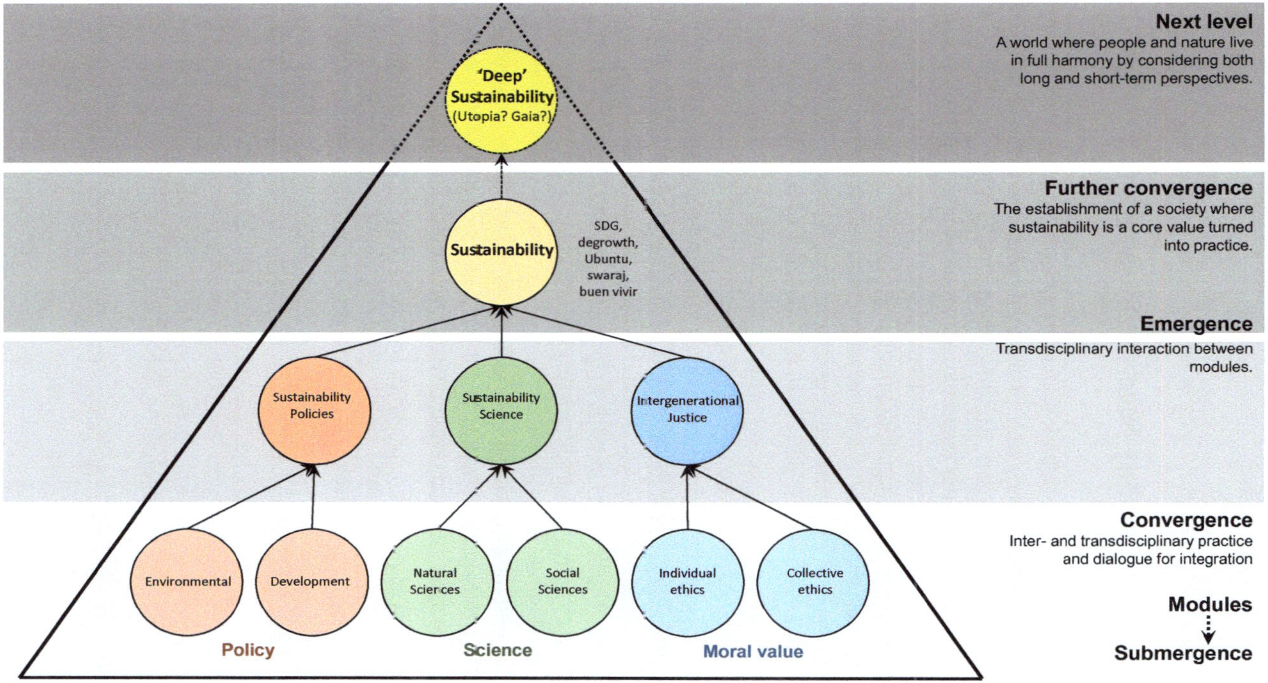

Fig. 1 Schematic representation of the emergence of sustainability. The integration and convergence of distinct modules related to science, policy and moral values, resulted in the emergence of sustainability. These original modules are to some extent experiencing submergence. Sustainability, however, requires further convergence of modules—including principles, values and practices belonging to different worldviews (e.g. African *Ubuntu*, Andean *Buen Vivir* and Asian *Swaraj*, alongside with SDG—Sustainable Development Goals, and *degrowth*)—for becoming a new 'normal'. This new state of things is a new utopia, open in format and definition. The main principle of Lovelock's Gaia hypothesis, of the biosphere in the planet interacting as an indivisible whole, is an example of the need of individually and collectively developing a new mindset to achieve sustainability goals

This chapter is then divided in three main sections: emergence, submergence and convergence. The fourth component of Bunge's framework, divergence, is not dealt with here since clearly, in the case of sustainability, emergence results from convergence and not divergence.

Emergence: Three Sustainability Facets in Response to the Planetary Crisis

While concepts such as Anthropocene and planetary boundaries are still being debated (e.g. Malm and Hornborg 2014; Montoya et al. 2017), there is growing scientific consensus that recognizes the magnitude of human impact on Earth and its climate (IPCC 2014). "Anthropocene" is the term originally assigned by Crutzen and Stoermer (2000) to the present human-dominated geological epoch, when the impact of human activities on the Earth System supplemented the Holocene (see also Crutzen 2002). Despite some criticism (e.g. Malm and Hornborg 2014), the term is now broadly used. For some, it starts with the invention of the steam engine by James Watt in 1784, which also coincides with the beginning of growing global carbon dioxide and methane found by analyses of air trapped in polar ice (Crutzen 2002). For others (Steffen et al. 2015a), the beginning of the Anthropocene coincides with the onset of the post-1950 "Great Acceleration", when Earth System indicators show shifts beyond the range of variability of the Holocene, driven by human activities. Arguably, the Great Acceleration has driven humanity beyond a safe operating space, by transgressing planetary boundaries especially as regards climate change and biosphere integrity (Steffen et al. 2015b).

Although concerns with the environment go back to early civilizations, environmentalism as a social movement consolidated in the 1960s (Dauvergne 2009), partly as a resistance to the changes provoked by the Great Acceleration. By the 1970s, environmentalism and conservationism were nearly synonyms, while conservation science was still taking form (Wiens and Hobbs 2015). By then, a lot of the environmentalism was 'ecocentric', nature-focused, and thus defied the notion of the human–nature divide that grants humans greater intrinsic value than all other non-human species (Shoreman-Ouimet and Kopnina 2015). This vision was met with strong opposition from social scientists for various reasons, including the impact of conservation on local peoples and cultures, and connections with the capitalist notion of progress and development (e.g. West and Brockington 2012). Out of this tension, sustainability began to emerge (see timeline on Table 1).

Table 1 Timeline related to the emergence of sustainability as policy, moral value and science

Year	Events
1784	Anthropocene begins
1950s	Great acceleration begins
1960s	Environmentalism consolidates as a social movement
1970s	Early days of conservation science
1972	Stockholm: UN conference on human environment
1979	Lovelock's Gaia hypothesis is published
1987	Brundtland report: sustainable development emerges as a concept
1987	Early days of sustainability science
1988	IPCC is launched
1992	Rio de Janeiro: UN conference on environment and development
1992	Rio de Janeiro: the three UN conventions are created: climate, biodiversity, desertification
1994	World Bank report: sustainability consolidates as a concept
1997	Kyoto protocol is signed
2000	Sustainability science framework begins to consolidate
2000s	Sustainability spreads in governments' manifestos and large corporations' mission statements
2006	Lovelock's "The Revenge of Gaia" published, on more pessimistic tone
2007	IPCC wins Nobel Peace Prize
2012	Rio de Janeiro: UN conference on sustainable development
2015	SDG announced by the united nations
2015	Paris agreement of the UN framework convention on climate change

Acronyms: *UN* United Nations, *IPCC* Intergovernmental Panel on Climate Change, *SDG* sustainable development goals

Sustainability as Policy

By contrasting the topics of three milestone UN Conferences—on Human Environment: 1972; on Environment and Development: 1992; on Sustainable Development: 2012—it becomes clear that sustainable development and sustainability as policy emerged from the convergence between the environmental policy agenda and the development policy agenda. In 1972, global awareness about environmental risks led to the organization of the United Nations Conference on Human Environment in Stockholm. The conference concluded that environmental issues and development issues should be dealt with jointly. Fifteen years later, in 1987, the Brundtland report was published, coining the term 'sustainable development' defined as "a development that meets the needs of the present without compromising the ability of future generations to meet their own needs" (WCED 1987). Later, Asheim (1994) in a World Bank report defines sustainability as "a requirement of our generation to manage the resource base such that the average quality of life that we ensure ourselves can potentially be shared by all future generations". These documents are evidence of the emergence of sustainability as a concept and as a pathway for policy. It is interesting, however, that despite the political aspiration to converge environmentalism and development expressed in these documents, the diplomatic decision at the (aptly named) UN Conference on Environment and Development (Rio de Janeiro, 1992) was to create three sectoral conventions, namely the UN Framework Convention on Climate Change (UNFCCC), the UN Convention on Biological Diversity (CBD) and the UN Convention to Combat Desertification (UNCCD) (Tolefson and Gilbert 2012). It took twenty years for these conventions to converge and meet again, now under the name UN Conference on Sustainable Development, also in Rio de Janeiro, 2012. On this occasion, heads of states signed a document called "The Future We Want" (see Griggs et al. 2013)—partly a sequel to the Brundtland report also known as "Our Common Future" (WCED 1987).

This process translated into the Sustainable Development Goals (SDG) announced by the UN in 2015. The SDG comprise 17 goals to be achieved globally by 2030. They combine biosphere conservation goals (life on land, life on water, climate action, clean water), with social goals (no poverty, no hunger, good health and well-being, gender equity, clean energy, quality education, sustainable cities, peace and justice), with economic goals (economic growth and decent jobs, sustainable production and consumption, sustainable infrastructure, reduced inequalities), and partnerships between all actors to achieve all goals. Although no clear path has been defined on how to achieve such progress in 15 years, the goals comprise 169 targets and indicators that are expected to guide implementation (Costanza et al. 2016a, b).

I argue that the SDG need a 'glue'—using Bunge's terminology from the start of this chapter—to transform all goals and indicators into the 'indivisible whole' it aspires to be (see also Nilsson et al. 2017). A number of studies reflect on what is needed to deliver these ambitious goals by 2030. Aspects of global and local governance (Biermann et al. 2017; Bowen et al. 2017) and the need to eliminate trade-offs and to promote synergies between goals (Campagnolo and Davide 2017; Nilsson et al. 2017; Pradhan et al. 2017) are among the top concerns. I will come back to that later, when I discuss convergence.

Sustainability as Moral Value

From both Brundtland's (WCED 1987) and Asheim's (1994) definitions above, it can be concluded that the concept of sustainability is inextricably linked with intergenerational justice (Clayton et al. 2016), which by itself is a moral value. Perhaps for this reason, sustainability has become one of the most popular terms used in public and scientific discourse (WCED 1987): it appears in manifestos of governments and of large corporations, in a vast quantity of scientific publications and in various initiatives across civil society (Bettencourt and Kaur 2011). Of course, there is considerable misuse of the term, which has been associated to 'green washing' rethoric (Saha and Darnton 2005; Robinson 2012) and to scepticism of consumers regarding green or sustainable labels (Leonidou and Skarmeas 2017). However, even when the term sustainability is used for illegitimate purposes, it confirms the notion that it became a value: it is a term used to try to convince people about quality or value associated to a given product or action. Thus, this profusion of the concept as value exemplifies how the relationship between societies' present and future needs and rights has gained increased attention (Clayton et al. 2016). Indeed, as early as 1972 (and in parallel to the Stockholm Conference), German philosopher Hans Jonas argued for the need of a new global ethics directed towards future generations—an ethics of the species, since traditional ethics are locally based and present time—an ethics of the individual (Jonas 1979, 2017). The new ethics of responsibility that Jonas proposes is directed towards future generations, people who are not living yet, and who, in present, do not have a voice or a 'lobby', in his words. Thus, I propose that sustainability emerges as a new moral value that combines humanitarian and environmental values related to intergenerational justice.

Sustainability as a moral value increasingly implies a moral conduct towards Earth itself. Many argue that the ongoing environmental degradation is a result of the fact that Western philosophy places nature outside the moral community, viewing nature as a collection of objects to be used according to the benefit of property owners (Koons 2008). This has provoked the emergence of ecocentric principles such as environmental ethics (e.g. Barrett and Grizzle 1999) and the so-called Earth Jurisprudence (see Koons 2008). There are now institutions and centres for Earth Jurisprudence in the UK, South Africa, Australia, New Zealand, the USA, among others (Kauffman and Martin 2018). Related trends can also be observed in Latin America. In 2008, Ecuador was the first country to recognize the rights of nature in its constitution (Beling et al. 2018; Kauffman and Martin 2017). In Bolivia, the legislation related to The Rights of Mother Earth appeared shortly after that and has components related to (1) right to life and the diversity of life; (2) right to stabilize concentrations of greenhouse gases in the atmosphere at a level that would prevent dangerous anthropogenic interference with the climate system, and in sufficient time to allow the components of Mother Earth to adapt naturally to climate change; (3) non-commodification of the environmental functions of Mother Earth; (4) right to support the restoration and regeneration capabilities of all its components that enables the continuity of life cycles; and (5) right to clean air and to live without contamination

(Pacheco 2014). This logic begins to spread. In Brazil, for the first time in history, a river (Rio Doce, in the State of Minas Gerais), represented by an NGO (Associação Pachamama), has entered a lawsuit asking for the recognition of its rights to life and demanding a plan for disaster risk reduction for the local population in the watershed (Scarano et al. 2018). This took place in 2017, two years after the river was impacted by the worst environmental disaster in Brazil's history, with the collapse of a dam and the spill of 40–62 million m^3 of mining tailings in the river (Garcia et al. 2017; Pires et al. 2017).

Nevertheless, the debate around 'weak' (social component predominant) versus 'strong' sustainability (environmental component predominant) is an evidence that sustainability as a moral value remains in dispute, largely because its three dimensions—social, economic and environmental—still did not overcome trade-offs and conflicts (Beckerman 1995; Clayton et al. 2016; Jia et al. 2017). Beckerman (1995) admits that the so-called 'strong' sustainability can be seen as a moral value due to "some intrinsic value in non-sentient forms of natural capital". But he also warns that "democratic societies should be very wary of those who claim, without full explanation, that the activities that they happen to prefer should be elevated to some over-riding moral value to which individuals should willingly sacrifice themselves". As we have seen for sustainability as policy, given all the controversies and competing interests, sustainability as a moral value has emerged, but clearly needs maturing.

Sustainability as Science

Sustainability science (see definitions on Table 2) emerged 30 years ago, following the Brundtland report, and it took some twenty years for a new conceptual and practical whole to consolidate out of the common methodologies designed to connect knowledge and methods from a variety of traditional disciplines (Bettencourt and Kaur 2011). Its emergence as a new and robust research framework is evidenced by the existence of a number of important scientific journals and conferences entirely or partly dedicated to the theme (Clark 2007; Bettencourt and Kaur 2011; Spangenberg 2011). It has been argued that sustainability science is developed in "constructive tension between a descriptive-analytical and a transformational mode" (Wiek et al. 2012), or between 'critical' and 'problem-solving' approaches (Jerneck et al. 2011). The descriptive-analytical mode (critical) seeks understanding of sustainability challenges in coupled human–environment systems, whereas the transformational mode (problem-solving) searches for practical solutions to those problems. To achieve this, sustainability science is highly integrative in nature, comprising integration between theoretical and applied approaches and disciplines (contained in the term interdisciplinarity), integration between research and its application to policy (contained in the term transdisciplinarity), as well as to examine and act in both developed and developing societies (Ascher 2007; Aronson 2011; Bettencourt and Kaur 2011; Brink et al. 2017). The aim to be transformational demands a routine of generating actionable knowledge, and intense dialogue and action outside academia, both incorporating

Table 2 Some definitions of and statements about sustainability science

Definitions	References
"seeks to understand the fundamental character of interactions between nature and society"	Kates et al. (2001)
"transcends the concerns of its foundational disciplines and focuses instead on understanding the complex dynamics that arise from interactions between human and environmental systems"	Clark (2007)
has "its own specific body of knowledge and framework with which to address sustainability issues, even while retaining relationships with other disciplines (…) currently a work in progress, and therefore one may argue that it is still too early to discuss what sustainability science is"	Kajikawa (2008)
"a dynamic and evolving transdisciplinary effort addressing symbiosis between human activity and the environment, providing visions and scenarios indicating transition pathways towards global sustainability while elucidating relevant decisions and agents (…) research providing the necessary insights to make the normative concept of sustainability operational, and the means to plan and implement adequate steps towards this end"	Spangenberg (2011)
"a solution-oriented endeavour" that "must address two additional streams of research questions: first, the normative question of how coupled human–environment systems would function and look like in compliance with a variety of value-laden goals and objectives (…); and, second, the strategic and operational questions that explore which transition pathways are viable for coupled human–environment systems and strategies that find what solutions to sustainability problems could be"	Wiek et al. (2012)
"seeks to address the major challenges facing society while ensuring that human well-being is undiminished and the basic Earth systems continue to operate"	Redman (2014)
"probes interactions between global, social, and human systems, the complex degradation mechanisms of these systems, and the concomitant risks to human well-being"	Saito et al. (2017)

non-scientific knowledge and dealing with different values and political interests (Jahn et al. 2012; Wiek et al. 2012; Popa et al. 2015). In this context, sustainability as policy and as value function as attractors to science, creating a transdisciplinary arena where conventional disciplines meet and feed back into policy and into moral and ethics.

In that sense, transdisciplinarity is more than a research approach well suited to cope with complex issues such as sustainability: it is the very interface between science and society, as proposed by Jahn et al. (2012). In a thorough analysis of the literature on the topic, these authors have shown that transdisciplinarity is still emerging since it remains far from academically established, despite increased popularity (see also Lang et al. 2012). These similarities suggest that sustainability and transdisciplinarity are likely to coevolve. For instance, there is a parallel between

Jahn et al.'s (2012) three types of knowledge framework for transdisciplinarity and our three facets of sustainability: (1) systems knowledge, or the knowledge involved in the understanding of an issue is addressed by sustainability science; (2) orientation knowledge, or the knowledge required for decision-making is addressed by sustainability policy; and (3) transformation knowledge, or the ways and means of practically realizing such decision, is the gap to be filled, but is driven by the imperative of the moral value that sustainability represents. Jahn et al.'s (2012) framework is already being applied to sustainability issues such as ecosystem-based adaptation to climate change and municipal ecosystem services and adaptation planning (Brink et al. 2016, 2017).

However, despite the growing abundance of theoretical and structural studies (e.g. Kumazawa et al. 2009; Jerneck et al. 2011; Miller et al. 2014b) and of empirical studies (e.g. Gruen et al. 2008; Ostrom 2009; Pohl et al. 2017), impact of sustainability research on societal transformations also seems smaller than desired (Van der Leeuw et al. 2012; Wiek et al. 2012; Lang et al. 2017). It still falls short of expectations about integration of its three parts: economic, social and environmental: for instance, by 2012, the economic pillar had the fewest papers published but was the most integrative, while the environmental pillar, on the contrary, had the most articles but drew the least from outside disciplines (Schoolman et al. 2012). Moreover, there is only a handful of higher education programmes fully dedicated to the topic (Wiek et al. 2011).

Submergence: System Dismantling

Bunge (2003) says that the glue that gives rise and holds together a new system often explains the dismantling of another system. In other words, he continues, "emergence explains submergence". The emergence of sustainability provokes (or derives from) a crisis in some of its pillars. In this section, I will discuss submergence and system dismantling by briefly introducing the ongoing crises in conservation science, environmentalism, and of the capitalist notion of progress and development as the driver of the economic–political system.

Crisis of Conservation Science

Colloff et al. (2017) declare that 'normal' conservation science no longer 'works' because the world itself has changed so much, and therefore conservation science should embrace 'postnormal' times—our present times, characterized by chaos, contradiction and complexity (see Funtowicz and Ravetz 1993; Sardar 2010). This crisis of conservation science is also apparent from the debate of 'traditional' versus 'new' conservation science (see Soule 2013; Doak et al. 2014; Kareiva 2014; Miller et al. 2014a). For new conservation scientists, traditional conservation (1) can give rise to socio-economic issues by emphasizing biodiversity protection without considering human well-being; (2) is based on the myth of pristine nature, which in fact no longer exists; (3) assumes, often wrongly, that nature is inherently fragile and incapable to recover from human interventions; and (4) failed to protect biodiversity, since 13% of the planet's cover is within protected areas and extinction rates continue to be alarmingly high. On the other hand, traditional conservation scientists claim that new conservation science (1) has a week scientific background; (2) has ethical rather than scientific priorities; (3) has economic motivations; and (4) shows no evidence that it can correct eventual flaws of traditional conservation (Doak et al. 2014).

This 'friendly fire' between conservation scientists, in short, seems to be based on accusations that traditional conservation scientists are 'nature-centric', while new conservation scientists are 'anthropocentric' and 'utilitarian'. Colloff et al. (2017) argue that conservation science should move away from the nature-centric versus anthropocentric debate, and instead build on the links and causal relations between nature and human well-being. In their view, redirection of conservation science would involve: (1) building new framings of the links between ecosystems and society; (2) developing new relationships and roles for conservation science; (3) developing new models of how conservation links to society and social change and (4) a search for new approaches to promote conservation outcomes (e.g. addressing conflict resolution, power and intergenerational equity in decision). I argue that if conservation moves in the direction proposed by Colloff et al. (2017), or even in the direction of the somewhat similar proposition of the Intergovernmental Panel on Biodiversity and Ecosystem Services (IPBES; see Díaz et al. 2015; Pascual et al. 2017), it will move closer to sustainability science. Altogether, these new trends and expectations seem to indicate a possible gradual future submergence of conservation science as we know it.

Crisis of Environmentalism and of Development

There are contrasting arguments in the literature that environmentalism is dead (Shellenberg and Nordhaus 2004; Blühdorn 2011), or, conversely, that environmentalism persists (Dauvergne 2009) and that sustainability is a new regime of environmentalism (Bothello and Djelic 2015). All recognize, however, that environmentalism moved from marginal, in its early days, to mainstream, influencing policy, civil society and even science, with the consolidation of conservation science. On the other hand, although mainstreamed in various societal fronts, environmentalism always faces three major obstacles: (1) the disconnect between popular perception and the concept of environment; (2) its incapacity to overcome the broader development culture of consumerism; (3) and the denial discourse (Anderson 2010). These difficulties turned environmentalism into a 'zombie' movement, to use Anderson's (2010) analogy. His argument is that, to thrive, environmentalism needs reframing and alignment with other discourses and identities, "while retaining the utopian stimulus of conventional environmentalism": a more productive framing of environmental practice, or a 'coyote' approach, in his language. Similarly to what I have discussed about conservation science, if environmentalism moves in the direction suggested by Anderson (2010), it will be a lot closer to the sustainability discourse.

Interestingly, although Anderson (2010) argues that consumerism and denial are hurdles that the development culture imposes upon environmentalism, it is also clear that environmentalism has produced some cracks on the wall of the capitalist logic of development. The emergence of sustainability is evidence of that; especially the alternative visions of sustainability that begin to compete/dialogue. Sustainable development is only one of these visions, but other options emerge. Since 'development' led to profound impact on how Asia, Africa, and Latin America came to be seen and treated as underdeveloped since post-World War II, a movement called 'postdevelopment' emerged to question the ability of capitalism to fully and naturally occupy the economy (Escobar 2015). Re-emerging trends in the Global South, such as *buen vivir*, *Ubuntu* and ecological *Swaraj* (or Radical Ecological Democracy), can be seen as postdevelopmental phenomena. More typically in the Global North, another movement emerges, known as '*degrowth*'. These various alternatives are interesting to highlight because they arise from different social and cultural contexts, they either represent a revival of ancient worldviews of indigenous peoples or have emerged from recent social and environmental movements, but essentially they share similar principles (Kothari et al. 2014).

Although no universal definition of *buen vivir* (which means, 'living well') has been attained yet, this vision typical of Andean regions in Bolivia and Ecuador, includes as key elements: (1) harmony with nature; (2) vindication of the principles of marginalized peoples; (3) the State as guarantor of basic needs, social justice and equality; and (4) democracy (Beling et al. 2018). As we have discussed before, this approach has been incorporated to the legislation of Bolivia and Ecuador. Three basic tenets are love of self, of community and of nature. In this respect, it bears great similarity to the African *Ubuntu* (which means 'humanness') metaphysics that connects self, society and nature (Shumba 2011; Ontong and Le Grange 2014). Ubuntu's philosophy has been to some extent incorporated into business and corporate practices in Africa (Khomba et al. 2013). In India, the ecological *Swaraj* (which loosely means self-rule or self-reliance) "is a framework that respects the limits of the Earth and the rights of other species, while pursuing the core values of social justice and equity" (Kothari et al. 2014). These authors explain that this worldview that emerges from practices in the Indian sub-continent places communities and collectives at the centre of decision-making process and governance, which is also similar to *buen vivir* (Beling et al. 2018) and to *Ubuntu* (Shumba 2011).

Degrowth started around 2008 as an activist movement that rejects economic growth as the only alternative. Turned into an academic discipline, it starts from the premise that economic growth cannot be sustained ad infinitum on a resource constraint planet (Kothari et al. 2014). Scientific output on degrowth is largely European, but US and Canada also have a relevant contribution to this vision and it has presence, even as a movement, in countries such as Colombia and Cuba (Weiss and Cattaneo 2017). Escobar (2015) argues that a closer dialogue between postdevelopment and degrowth movements and scholars are much needed, while clearly demonstrating scepticism with capitalism and sustainable development. Beling et al. (2018), on the other hand, seem more inclusive and suggest that synergies between sustainable development, *buen vivir* and degrowth can compensate for the caveats of each discourse and eventually open pathways towards what they call 'a global new Great Transformation' towards a new state of things. This aimed new state of things—sustainability—can be seen as a new utopia that unlocks new possibilities and alternatives for the future.

Convergence: Consolidating Sustainability and Moving to the Next Level

As we have discussed earlier, referring to Bunge (2003), for convergence to result in emergence 'glue concepts' are required. Sustainability has emerged, but given the challenges described to consolidation of its three facets (as policy, as moral value and as science) further convergence is needed. Mathematical formula or rationale is a desirable glue in Bunge's approach, but we still seem far from that as regards sustainability. Interestingly, two of the concepts often described as providing such 'glue' are ecosystem-based. Ecosystem services are increasingly used to couple human and natural systems, and conservation science with social sciences (Díaz et al. 2015; Crouzat et al. 2018; Pascual et al. 2017). Ecosystem-based adaptation to climate change and disaster risk are also used as glue concepts and practices to couple ecosystem health and poverty reduction, as an integrated tool to reduce societal vulnerability to climate change and natural disasters (Brink et al. 2016; Kasecker et al. 2018; Scarano 2017). As glue concepts are applied for convergence towards sustainability, one can imagine what a desirable future state would be like (a new utopia), and one can even wonder about how the 'next level' might look like.

Sustainability as Utopia

Funtowicz and Ravetz (1994) pointed out that the main contradiction society lives in the current postnormal times is "(…) the incompatibility between the individual drive for material comfort, convenience and safety, and the ecological consequences of this being achieved even for a significant minority of humanity". It is perhaps no coincidence that the main utopic and dystopic concerns of the twenty-first century have an ecological background, given the mounting evidence of the risk of catastrophes related to climate and land use change (Claeys 2013). Many of the utopias of the twentieth century (e.g. communism, the hippie movement) submerged, while possibly the main concrete current pattern—progress and consumerism—is in steep decline, since it led to the existing ecological crisis. Indeed, the ecological crisis replaced totalitarianism as the main dystopia of our times (Claeys 2013).

The decline of the utopias of the twentieth century has led some to announce the 'death' of utopia (e.g. Gray 2007), or the 'end of the future', as Zygmunt Bauman declared in a conference in Rio de Janeiro, 2014, in reference also to projected futures that were not accomplished (http://epoca.globo.com/ideias/noticia/2014/02/bzygmunt-baumanb-vivemos-o-fim-do-futuro.html). Claeys (2013) says that if utopias are not designed as religion or as a psychological state, but as a 'discourse of voluntary sociability', they can avoid the mistakes that normally define dystopias and, as a consequence, perhaps guide a common future. Similarly, Bregman (2017) says that there are two types of utopia. The one that 'predicts' the future with immutable rules (that he calls 'blueprint' utopia), and the one that unlocks the future (that I call 'open' utopia). The latter, he argues, is the one that inspires new common futures. He also claims that this approach is what inspired Thomas More's 1516 book, 'Utopia' (Logan and Adams 2009), which coined the term.

There are two points discussed previously in this chapter that I would like to address as examples. First, the SDG: with their 17 objectives and 169 targets and indicators, they can be seen as a 'blueprint' utopia. A kind of official 'recipe', which despite tremendous adherence by governments, corporations and science, still receives much criticism for some of its assumptions, especially that of continued economic growth (see Holden et al. 2016; Evans and Musvipwa 2017; Halvorsen 2017; Reid et al. 2017). On the other hand, precisely because of the complexity that their many objectives and targets entail, the SDG can also be perceived as an 'open utopia', applied more as a 'toolbox' than as a 'manual' or a 'roadmap' (see Stafford-Smith et al. 2017; Halvorsen 2017). In fact, the SDG may gradually shift the needle from open utopia to somewhere between open and blueprint utopia, as science furthers the understanding of the interconnections between the goals, and society advances a narrative to describe shifts and policy reforms necessary to achieve the goals (Costanza et al. 2016a, b).

Secondly, alternative approaches to sustainability—such as sustainable development (that the SDG encompass), degrowth, *buen vivir*, *Ubuntu*, ecological *Swaraj*—differ in various aspects, including their visions of the future, but I agree with Escobar (2015) and Beling et al. (2018) that dialogue and synergies between them can compensate for individual caveats and open new pathways towards sustainability. The IPBES is to some extent an effort in this direction (Díaz et al. 2015; Pascual et al. 2017; Scarano et al. 2018). This integrative model, where supposed differences are openly discussed towards finding potential synergies for a common future, I would argue that better fits the 'open utopia' model, which enters into focus as glue concepts appear.

Gaia: The Next Level?

Sustainability is a normative goal (i.e. a desire based on beliefs and values), thus convergence towards sustainability will demand debates about the relative importance of its various social, economic and environmental modules. In parallel, it is also an issue related to collective, common goods, and as such individual actors have no immediate incentive to address sustainability problems (Geels 2010). These facts create a conundrum, where on the one hand public authorities and civil society are key drivers for a transition to sustainability, while on the other hand individual actors—which largely define action and attitude of public authorities and civil society—may not feel incentivized for changes, for instance, in production and consumption. Clearly then, as suggested by Hans Jonas' ethics of responsibility, and as we have hinted throughout this chapter, transition towards sustainability demands changes in individual mindset towards more collective concerns that include present and future generations of human and non-human beings.

In that sense, reference to Lovelock's Gaia hypothesis (Lovelock 1979) can be particularly relevant. It postulates that Earth is a living system with humanity as part of its large web of life. However, with or without humankind on the planet, the evolution of Gaia will continue (Lovelock 2006; see also Peroff 2008). Nearly forty years of advancement in Gaia's thought and research led to the understanding that organisms (including humankind) and their environment are a coupled system and that they do not evolve separately. Lovelock (2010) even speculates that things might have been different had Darwin considered Gaia (the whole organism rather than only its biological modules) as part of his Theory of Evolution by Natural Selection.

Coupled human–natural systems are central to the agenda of sustainability science and to UN panels such as the IPCC (Intergovernmental Panel on Climate Change) and the IPBES (Díaz et al. 2015; Liu et al. 2015). Research on socio-ecological systems has 'resilience' as an important glue concept (Sterk et al. 2017). Resilience—the capacity of systems (natural and human) to cope with hazardous events responding or reorganizing so that structure, function and identity are maintained, while capacity for adaptation, transformation and learning is also maintained (Field et al. 2014)—is in many ways close to the "Gaian approach of stimulating the Earth to cure itself", highlighted by Lovelock (2010). It is also evident in a number of recent conceptual (e.g. the concept of 'stem species' as 'regenerants of Gaia' in Scarano and Garbin 2013; Lüttge et al. 2013) and applied studies (e.g. natural and spontaneous regeneration of tropical forests after degradation in Rezende et al. 2015; Crouzeilles et al. 2017). Further dialogue between these different approaches and languages (Gaia, resilience, natural regeneration, etc.) to address potentially similar phenomena can possibly promote fertile ground for convergence towards sustainability.

Perhaps, if the sustainability utopia becomes realized one day, society may reach another level, where individuals perceive themselves as important modules of a large web of life—as the Gaia hypothesis proposes and indeed many traditional beliefs suggest.

Concluding Remarks: The Ticking Clock

Postnormal times are a transition period between one old 'normal' and a new 'normal' (Sardar and Sweeney 2016). In this paper, I argue that sustainability is the new 'normal' society is largely aiming for. However, perhaps the greatest challenge of our current postnormal times is that there are clear deadlines for society to effectively bring transition to an end and definitely move into a sustainability normal. The SDG and the Paris Agreement of the Climate Convention envisage a sustainable future by 2030 (although they are not precise about how to get there). In parallel, under a "business-as-usual" scenario, climate change models indicate a global mean temperature increase of >2 °C in 2050 compared with pre-industrial times, and consequent dramatic changes in life as we know it (IPCC 2013; Rogelj et al. 2016). Thus, 2030 and 2050 are crucial deadlines and potential tipping points. In other words, society has some twelve years to perform its transition to sustainability, so as to deliver on the moral value of intergenerational justice and avoid a warmer future with dramatic consequences to future generations. Given the shortage of time for such profound changes, one can be inclined to imagine that ruptural transformation (i.e. a sharp break with existing institutions, structures, mindsets) will be required, whereas symbiotic transformation (i.e. through broader social participation in decision-making processes) and interstitial transformation (i.e. that takes places in specific niches, often marginal to mainstream) would just not meet the supposed deadlines. However, Wright's (2010) recommendation for combining all three approaches to turn utopias real is likely also very applicable to the case of sustainability. There is as much need for rupture with sectoral policies and thinking, and also with individualistic mindset, as there is for a greater symbiosis between different actors to share decision-making responsibilities, and for new sustainability models and examples at local scale to inspire change at larger scales.

Souza and Lüttge (2015) remind us that "menace and hope are currently deeply attached to sustainability", and therefore failure or success of sustainability might determine the fate of humankind on the planet. Thus, the clock is ticking: from emergence, to convergence, to new normal, humanity will need to move fast with transformations to make sure it continues to be part of Gaia.

Acknowledgements I dearly thank Ebba Brink for critically reading the manuscript, for sharing her reflections about emergence and convergence, about transdisciplinarity, and for our constructive discussions about sustainability science—all which helped give final shape to this paper. I also thank Aliny Pires for artwork in Fig. 1, for her inspiring capacity to transform ideas in images and for critically reading the manuscript. My warmest thanks also to Prof. Ulrich Lüttge for two decades of partnership, for his insightful coaching all these years and, more recently, for critically reading this manuscript and introducing me to the philosophy of Hans Jonas. My studies on sustainability are supported by the Brazilian Platform on Biodiversity and Ecosystem Services (BPBES), funded by CNPq Grant Number 405593/2015-5.

References

Anderson J (2010) From 'zombies' to 'coyotes': environmentalism where we are. Environ Politics 19:973–991

Aronson J (2011) Sustainability science demands that we define our terms across diverse disciplines. Landscape Ecol 26:457–460

Ascher W (2007) Policy sciences contributions to analysis to promote sustainability. Sustain Sci 2:141–149

Asheim GB (1994) Sustainability: ethical foundations and economic properties. World Bank Policy Research Working Paper 1302
Banzhaf W (2014) Genetic programming and emergence. Genet Program Evolvable Mach 15:63–73
Barrett CB, Grizzle R (1999) A holistic approach to sustainability based on pluralism stewardship. Environ Ethics 21(1):23–42
Beckerman W (1995) How would you like you 'sustainability', Sir? Weak or strong? A reply to my critics. Environ Values 4(2):169–179
Beling AE, Vanhulst J, Demaria F, Rabi V, Carballo AE, Pelenc J (2018) Discursive synergies for a 'great transformation' towards sustainability: pragmatic contributions to a necessary dialogue between human development, degrowth, and buen vivir. Ecol Econ 144:304–313
Bettencourt LMA, Kaur J (2011) Evolution and structure of sustainability science. Proc Natl Acad Sci 108:19540–19545
Biermann F, Kanie N, Kim RE (2017) Global governance by goal-setting: the novel approach of the UN sustainable development goals. Curr Opin Environ Sustain 26–27:28–31
Blühdorn I (2011) The politics of unsustainability: COP15, post-ecologism, and the ecological paradox. Organ Environ 24:34–53
Bothello J, Djelic M-L (2015) Transnational path generation: the institutionalization patterns of modern environmentalism. In: Academy of management proceedings. https://doi.org/10.5465/ambpp.2015.13168abstract
Bowen KJ, Cradock-Henry NA, Koch F, Patterson J, Häyhä T, Vogt J, Barbi F (2017) Implementing the "Sustainable Development Goals": towards addressing three key governance challenges—-collective action, trade-offs, and accountability. Curr Opin Environ Sustain 26–27:90–96
Bregman R (2017) Utopia for realists: how can we build the ideal world, 1st ebook edition. Little, Brown and Company, New York
Brink E, Aalders T, Ádám D, Feller R, Henselek Y, Hoffmann A, Ibe K, Matthey-Doret A, Meyer M, Negrut NL, Rau A-L, Riewerts B, von Schuckmann L, Törnros S, von Wehrden H, Abson DJ, Wamsler C (2016) Cascades of green: a review of ecosystem based adaptation in urban areas. Glob Environ Change 36:111–123
Brink E, Wamsler C, Adolfsson M, Axelsson M, Beery T, Björn H, Bramryd T, Ekelund N, Jephson T, Narvelo W, Ness B, Jönsson KI, Palo T, Sjeldrup M, Stålhammar S, Thiere G (2017) On the road to 'research municipalities': analysing transdisciplinarity in municipal ecosystem services and adaptation planning. Sustain Sci 13:765–784
Bunge M (2003) Emergence and convergence: qualitative novelty and the unity of knowledge. University of Toronto Press, Toronto
Campagnolo L, Davide M (2017) Can the Paris deal boost SDGs achievement? An assessment of climate mitigation co-benefits or side-effects on poverty and inequality. Working Paper 048.2017. Fondazione Enrico Mattei, Venice
Claeys G (2013) Utopia: A História de uma Ideia. Translation by Pedro Barros. Edições SESC, São Paulo
Clark (2007) Sustainability science: a room of its own. Proc Natl Acad Sci 104:1737–1738
Clayton S, Kals E, Feygina I (2016) Justice and environmental sustainability. In: Sabbagh C, Schmitt M (eds) Handbook of social justice theory and research. Springer, New York, pp 369–386
Colloff MJ, Lavorel S, Van Kerkhoff LE et al (2017) Transforming conservation science and practice for a post-normal world. Conserv Biol 31:1008–1017
Costanza R, Daly L, Fioramonti L, Giovannini E, Kubiszewski I, Mortensen LF, Pickett KE, Ragnarsdottir KV, De Vogli R, Wilkinson R (2016a) Modelling and measuring sustainable wellbeing in connection with the UN sustainable development goals. Ecol Econ 130:350–355
Costanza R, Fioramonti L, Kubiszewski I (2016b) The UN sustainable development goals and the dynamics of well-being. Front Ecol Environ 14:59
Crouzat E, Arpin I, Brunet L, Colloff MJ, Turkelboom F, Lavorel S (2018) Researchers must be aware of their roles at the interface of ecosystem services science and policy. Ambio 47:97–105

Crouzeilles R, Ferreira MS, Chazdon RL, Lindenmayer DB, Sansevero JBB, Monteiro L, Iribarrem A, Latawiec AE, Strassburg BBN (2017) Ecological restoration success is higher for natural regeneration than for active restoration in tropical forests. Sci Adv 3:e1701345
Crutzen PJ (2002) Geology of mankind. Nature 415:23
Crutzen PJ, Stoermer EF (2000) The "Anthropocene". IGBP Glob Chang Newsl 41:17–18
Dauvergne P (2009) Historical dictionary of environmentalism. Scarecrow Press, Plymouth, UK
De Haan J (2006) How emergence arises. Ecol Complex 3:293–301
Díaz S, Demissew S, Carabias J, Joly C, Lonsdale M, Ash N, Larigauderie A, Adhikari JA, Arico A, Báldi A et al (2015) The IPBES conceptual framework: connecting nature and people. Curr Opin Environ Sustain 14:1–16
Doak DF, Bakker VJ, Goldstein BE, Hale B (2014) What is the future of conservation? Trends Ecol Evol 29:77–81
Escobar A (2015) Degrowth, post development, and transitions: a preliminary conversation. Sustain Sci 10:451–462
Evans H-C, Musvipwa R (2017) The sustainable development goals, the paris agreement and the addis agenda: neo-liberalism, unequal development and the rise of a new imperialism. In: Halvorsen T, Ibsen H, Evans H-C, Penderis S (eds) Knowledge for justice: critical perspectives from southern African-nordic research partnerships. African Minds and Sanord, Cape Town, pp 37–56
Field CB, Barros VR, Mach KJ et al (2014) Technical summary. In: Field CB et al (eds) Climate change 2014: impacts, adaptation, and vulnerability. Part A: Global and sectoral aspects. Contribution of working group II to the fifth assessment report of the intergovernmental panel on climate change. Cambridge University Press, Cambridge, pp 35–94
Funtowicz SO, Ravetz JR (1993) Science for the post-normal age. Futures 25:739–755
Funtowicz SO, Ravetz JR (1994) Emergent complex systems. Futures 26:568–582
Garcia LC, Ribeiro DB, de Oliveira Roque F, Ochoa-Quintero JM, Laurance WF (2017) Brazil's worst mining disaster: corporations must be compelled to pay the actual environmental costs. Ecol Appl 27:5–9
Geels FW (2010) Ontologies, socio-technical transitions (to sustainability), and the multi-level perspective. Res Policy 39:495–510
Gray J (2007) Black mass: apocalyptic religion and the death of Utopia. Penguin Books, London
Griggs D, Stafford-Smith M, Gaffney O et al (2013) Sustainable development goals for people and planet. Nature 495:305–307
Gruen RL, Elliott JH, Nolan ML, Lawton PD, Parkhill A, McLaren CJ, Lavis JN (2008) Sustainability science: an integrated approach for health-programme planning. Lancet 372:1579–1589
Holden E, Linnerud K, Banister D (2016) The imperatives of sustainable development. Sustain Dev. https://doi.org/10.1002/sd.1647
Holvorsen T (2017) The sustainable development goals, knowledge production and the global struggle over values. In: Halvorsen T, Ibsen H, Evans H-C, Penderis S (eds) Knowledge for justice: critical perspectives from Southern African-nordic research partnerships. African Minds and Sanord, Cape Town, pp 13–36
IPCC (2013) Climate change 2013: the physical science basis. Contribution of working group I to the fifth assessment report of the intergovernmental panel on climate change. Cambridge University Press, Cambridge
IPCC (2014) Climate change 2014: impacts, adaptation, and vulnerability. Part A: Global and sectoral aspects. Contribution of working group II to the fifth assessment report of the intergovernmental panel on climate change. Cambridge University Press, Cambridge
Jahn T, Bergmann M, Keil F (2012) Transdisciplinarity: between mainstreaming and marginalization. Ecol Econ 79:1–10
Jerneck A, Olsson L, Ness B, Anderberg S, Baier M, Clark E, Hickler T, Hornborg A, Kronsell A, Lövbrand E, Persson J (2011) Structuring sustainability science. Sustain Sci 6:69–82
Jia F, Soucie K, Alisat S, Curtin D, Pratt M (2017) Are environmental issues moral issues? Moral identity in relation to protecting the natural world. J Environ Psychol 52:104–113

Jonas H (1979) Das Prinzip Verantwortung. Versuch einer Ethik für die technologische Zivilisation, Insel Verlag, Frankfurt
Jonas H (2017) Tecnologia e responsabilidade: reflexões sobre as novas tarefas da ética. In: Jonas H (ed) Ensaios Filosóficos: Da Crença Antiga ao Homem Tecnológico (Portuguese translation by Lopes WES). Editora Paulus, São Paulo, Brazil, pp 23–49
Kaidesoja T (2009) Bhaskar and Bunge on social emergence. J Theory Soc Behav 39(2):300–322
Kajikawa Y (2008) Research core and framework of sustainability science. Sustain Sci 3:215–239
Kareiva P (2014) New conservation: setting the record straight and finding common ground. Conserv Biol 28:634–636
Kasecker TP, Ramos-Neto MB, Silva JMC, Scarano FR (2018) Ecosystem-based adaptation to climate change: defining hotspot municipalities for policy design and implementation in Brazil. Mitig Adapt Strateg Glob Change 23:981–993
Kates RW, Clark WC, Corell R, Hall JM, Jaeger CC, Lowe I, McCarthy JJ, Schellnhuber HJ, Bolin B, Dickson NM, Faucheux S, Gallopin GC, Grübler A, Huntley B, Jäger J, Jodha NS, Kasperson RE, Mabogunje A, Matson P, Mooney H, Moore B III, O'Riordan T, Svedin U (2001) Sustainability science. Science 292:641–642
Kauffman CM, Martin PL (2017) Can rights of nature make development more sustainable? Why some Ecuadorian lawsuits succeed and others fail. World Dev 92:130–142
Khomba JK, Bakuwa RC, Kangaude-Ulaya EC (2013) Shaping business ethics and corporate governance: an inclusive African Ubuntu philosophy. Glob J Manag Bus Res A 13(5). https://globaljournals.org/GJMBR_Volume13/4-Shaping-Business-Ethics-and.pdf
Koons JE (2008) Earth jurisprudence: the moral value of nature. Pace Environ Law Rev 25(2):263–339
Kothari A, DeMaria F, Acosta A (2014) Buen vivir, degrowth and ecological swaraj: alternatives to sustainable development and the green economy. Development 57(3–4):362–375
Kumazawa T, Saito O, Kozaki K, Matsui T, Mizoguchi R (2009) Toward knowledge structuring of sustainability science based on ontology engineering. Sustain Sci 4:99–116
Lang DJ, Wiek A, Bergmann M, Stauffacher M, Martens P, Moll P, Swilling M, Thomas CJ (2012) Transdisciplinary research in sustainability science: practice, principles, and challenges. Sustain Sci 7(Supplement 1):25–43
Lang DJ, Wiek A, von Wehrden H (2017) Bridging divides in sustainability science. Sustain Sci 12:875–879
Leonidou CN, Skarmeas D (2017) Gray shades of green: causes and consequences of green skepticism. J Bus Ethics 144(2):401–415
Liu J, Mooney H, Hull V, Davis SJ, Gaskell J, Hertel T, Lubchenco J, Seto KC, Gleick P, Kremen C, Li S (2015) Systems integration for global sustainability. Science 347. https://doi.org/10.1126/science.1258832
Logan GM, Adams RM (2009) Utopia: Thomas More (Portuguese translation by Camargo JL, Cipolla MB), 3rd edn. WMF Martins Fontes, São Paulo, Brazil
Lovelock J (1979) Gaia. A new look at life on earth. Oxford University Press, Oxford
Lovelock J (2006) The revenge of Gaia. Penguin Books, London
Lovelock J (2010) Our sustainable retreat. In: Crist E, Rinker HB (eds) Gaia in turmoil: climate change, biodepletion, and earth ethics in an age of crisis. MIT Press, Cambridge, pp 21–24
Lüttge U, Garbin ML, Scarano FR (2013) Evo–Devo–Eco and ecological stem species: potential repair systems in the planetary biosphere crisis. Prog Bot 74:191–212
Malm A, Hornborg A (2014) The geology of mankind? A critique of the anthropocene narrative. Anthropocene Rev 1:62–69
Miller B, Soulé ME, Terborgh J (2014a) 'New conservation' or surrender to development? Anim Conserv 17:509–515
Miller TR, Wiek A, Sarewitz D, Robinson J, Olsson L, Kriebel D, Loorbach D (2014b) The future of sustainability science: a solutions-oriented research agenda. Sustain Sci 9:239–246
Montoya JM, Donohue I, Pimm SL (2017) Planetary boundaries for biodiversity: implausible science, pernicious policies. Trends Ecol Evol. https://doi.org/10.1016/j.tree.2017.10.004

Nilsson M, Griggs D, Visbeck M (2017) Map the interactions between sustainable development goals. Nature 534:320–322
Ontong K, Le Grange L (2014) The role of place-based education in developing sustainability as a frame of mind. South Afr J Environ Educ 30:27–38
Ostrom E (2009) A general framework for analyzing sustainability of social-ecological systems. Science 325:419–422
Pacheco D (2014) Living-well in harmony and balance with Mother Earth: a proposal for establishing a new global relationship between human beings and Mother Earth. Universidad de la Cordillera, La Paz. Available at http://www.ucordillera.edu.bo/descarga/livingwell.pdf
Pascual U, Balvanera P, Díaz S, Pataki G, Roth E et al (2017) Valuing nature's contributions to people: the IPBES approach. Curr Opin Environ Sustain 26:7–16
Peroff NC (2008) Complexity theory, Lovelock's Gaia and tribal peoples. Soc Evol Hist 7(2):26–39
Pires APF, Rezende CL, Assad ED, Loyola R, Scarano FR (2017) Forest restoration can increase the Rio Doce watershed resilience. Perspect Ecol Conserv 15:187–193
Pohl C, Krütli P, Stauffacher M (2017) Ten reflective steps for rendering research societally relevant. Gaia 26:43–51
Popa F, Guillermin M, Dedeurwaerdere T (2015) A pragmatist approach to transdisciplinarity in sustainability research: from complex systems theory to reflexive science. Futures 65:45–56
Pradhan P, Costa L, Rybski D, Lucht W, Kropp JP (2017) A systematic study of sustainable development goal (SDG) interactions. Earth's Future 5. https://doi.org/10.1002/2017ef000632
Redman CL (2014) Should sustainability and resilience be combined or remain distinct pursuits? Ecol Soc 19(2):37. https://doi.org/10.5751/es-06390-190237
Reid AJ, Brooks JL, Dolgov L, Laurich B, Sullivan BG, Szekeres P, Wood SLR, Bennett JR, Cooke SJ (2017) Post-2015 sustainable development goals still neglecting their environmental roots in the anthropocene. Environ Sci Policy 77:179–184
Rezende CL, Uezu A, Scarano FR, Araujo DSD (2015) Atlantic forest spontaneous regeneration at landscape scale. Biodivers Conserv 24:2255–2272
Robinson JG (2012) Common and conflicting interests in the engagements between conservation organizations and corporations. Conserv Biol 26(6):967–977
Rogelj J, den Elzen M, Höhne N, Fransen T, Fekete H, Winkler H, Schaeffer R, Sha F, Riahi K, Meinshausen M (2016) Paris agreement climate proposals need a boost to keep warming well below 2 °C. Nature 534:631–639
Saha M, Darnton G (2005) Green companies or green con-panies: are companies really green, or are they pretending to be? Bus Soc Rev 110(2):117–157
Saito O, Managi S, Kanie N, Kauffman J, Takeuchi K (2017) Sustainability science and implementing the sustainable development goals. Sustain Sci 12:907–910
Sardar Z (2010) Welcome to postnormal times. Futures 42:435–444
Sardar Z, Sweeney JA (2016) The three tomorrows of postnormal times. Futures 75:1–13
Scarano FR (2017) Ecosystem-based adaptation to climate change: concept, scalability and a role for conservation science. Perspect Ecol Conserv 15:65–73
Scarano FR, Garbin ML (2013) Stem species: plant species that function as regenerating cells of Gaia. Nova Acta Leopoldina NF 114:317–324
Scarano FR, Garcia K, Diaz-de-Leon A, Queiroz HL, Rodríguez Osuna V, Silvestri LC, Díaz CF, Pérez-Maqueo O, Rosales M, Salabarria DM, Zanetti EA, Farinaci JS (2018) Options for governance and decision-making across scales and sectors. In: Rice J, Seixas CS, Zaccagnini ME, Bedoya-Gaitán M, Valderrama N (eds), IPBES 2018: the IPBES regional assessment report on biodiversity and ecosystem services for the Americas. Secretariat of the intergovernmental science-policy platform on biodiversity and ecosystem services, Bonn, Germany, pp 644–744
Schoolman ED, Guest JS, Bush KF, Bell AR (2012) How interdisciplinary is sustainability research? Analyzing the structure of an emerging scientific field. Sustain Sci 7:67–80
Shellenberger M, Nordhaus T (2004) The death of environmentalism: global warming politics in a post-environmental world. http://www.grist.org/article/doe-reprint/

Shoreman-Ouimet E, Kopnina H (2015) Reconciling ecological and social justice to promote biodiversity conservation. Biol Conserv 184:320–326
Shumba O (2011) Commons thinking, ecological intelligence and the ethical and moral framework of Ubuntu: an imperative for sustainable development. J Media Commun Stud 3(3):84–96
Soulé ME (2013) The new conservation. Conserv Biol 27:895–897
Souza GM, Lüttge U (2015) Stability as a phenomenon emergent from plasticity–complexity–diversity in eco-physiology. In: Lüttge U, Beyschlag W (eds) Progress in botany, vol 76. Springer, Switzerland, pp 211–238
Spangenberg JH (2011) Sustainability science: a review, an analysis and some empirical lessons. Environ Conserv 38(3):275–287
Stafford-Smith M, Griggs D, Gaffney O, Ullah F, Reyers B, Kanie N, Stigson B, Shrivastava P, Leach M, O'Connell D (2017) Integration: the key to implementing the sustainable development goals. Sustain Sci 12:911–919
Steffen W, Broadgate W, Deutsch L, Gaffney O, Ludwig C (2015a) The trajectory of the anthropocene: the great acceleration. Anthropocene Rev 2:81–98
Steffen W, Richardson K, Röckstrom J et al (2015b) Planetary boundaries: guiding human development on a changing planet. Science 347. https://doi.org/10.1126/science.1259855
Sterk M, van de Leemput IA, Peeters ETHM (2017) How to conceptualize and operationalize resilience in socio-ecological systems? Curr Opin Environ Sustain 28:108–113
Tolefson J, Gilbert N (2012) Rio report card. Nature 486:20–23
Trosper RL (2005) Emergence unites ecology and society. Ecol Soc 10(1):14. http://www.ecologyandsociety.org/vol10/iss1/art14/
Van der Leeuw S, Wiek A, Harlow J, Buizer J (2012) How much time do we have? Urgency and rethoric in sustainability science. Sustain Sci 7(Supplement 1):115–120
WCED—World Commission on Environment and Development (1987) Our common future. Oxford University Press, Oxford
Weiss M, Cattaneo C (2017) Degrowth—taking stock and reviewing an emerging academic paradigm. Ecol Econ 137:220–230
West P, Brockington D (2012) Introduction: capitalism and the environment. Environ Soc Adv Res 3(1):1–3
Wiek A, Withycombe L, Redman CL (2011) Key competencies in sustainability: a reference framework for academic program development. Sustain Sci 6:203–218
Wiek A, Ness B, Schweizer-Ries P, Brand FS, Farioli F (2012) From complex systems analysis to transformational change: a comparative appraisal of sustainability science projects. Sustain Sci 7(Supplement 1):5–24
Wiens JA, Hobbs RJ (2015) Integrating conservation and restoration in a changing world. Bioscience 65:302–312
Wright EO (2010) Envisioning real Utopias. Verso, London

Part II
From Modules to Emergent Holistic Properties in Living Organisms

Modularity *Versus* Emergence: How to Cope with Complexity in Whole-Plant Physiology?

Lars H. Wegner

Abstract Complex, multigenic traits like growth, yield, or photosynthetic carbon assimilation are of central importance for our understanding of plant biology and are highly relevant for plant breeding, agriculture, and other applied plant sciences. Therefore, scientific description of complexity and the elucidation of its molecular basis have recently received increasing interest. Surprisingly though, less attention is given to the classical concepts of 'modularity' and 'emergence'. Modular features of a biological system result from more or less autonomously acting (molecular) components, whereas novel, 'emergent' features emanate from their interaction. In the very limited work published on this issue, modularity and emergence were considered as strictly opposing concepts. Here, a new, semi-quantitative, heuristic approach is introduced describing the contribution of individual proteins to a complex trait ('protein-trait relationships'). On a phenomenological basis, criteria are defined that allow to decide whether a protein is affecting the trait rather in a modular or emergent way, treating them as gradual, rather than mutually exclusive, features. The approach is exemplified by case studies related to photosynthetic carbon assimilation, making use of empirical data that are available from the literature.

Keywords CO_2 · Carbon assimilation · Rubisco · Control theory · Growth · Plant systems biology · Photosynthesis: carbon reactions

Introduction

Since the wide spectrum of molecular biological techniques became generally available some 30 years ago, research on plant physiology has seen a revolution, and a dramatic shift of focus from studying traits of higher functional units (ranging from the cell to the whole organism) to the protein level. Numerous proteins, which

L. H. Wegner (✉)
Karlsruhe Institute of Technology, Institute for Pulsed Power and Microwave Technology (IHM), Campus North, Building 630, Hermann v. Helmholtz Platz 1, 76344 Eggenstein-Leopoldshafen, Germany
e-mail: lars.wegner@gmx.net

L. H. Wegner and U. Lüttge (eds.), *Emergence and Modularity in Life Sciences*,
https://doi.org/10.1007/978-3-030-06128-9_4

form the 'basic units' of physiological function, have been identified, sequenced, and physiologically characterised and their spatio-temporal expression patterns have been unravelled. However, with the complete sequencing of the genomes of various (plant) species, the era of 'molecular discoveries' will sooner or later come to an end, and the interest of the scientific community will gradually shift to the functional integration of proteins at higher organisational levels in an evolutionary and ecological context (Souza et al. 2016). Systems biology has emerged as a new biological discipline, which is concerned with the integration of genes and proteins with respect to physiological function. Starting point of systems biology approaches is the database that molecular techniques provide; a complete inventory of a system is performed with respect to a class of components (be it proteins, metabolites, etc.). By mathematical, mostly statistical operations, interactions between these components are quantified, thus attempting to understand their systemic integration (compare this book, Chap. Modular Organization and Emergence in Systems Biology, by M. Th. Hütt). Usually, systems biology follows a typical bottom-up approach: a system is disintegrated into its parts and then re-assembled to understand how it works (see, e.g., Lucas et al. 2011 who also discuss some top-down approaches, though).

Systems biology has become very popular among plant scientists (Cui et al. 2008; Yuan et al. 2008; Fukushima et al. 2009; Usadel et al. 2009; Lucas et al. 2011), but the approach has also recently received some fundamental criticism from an epistemological viewpoint. In a series of publications, Lüttge (2012, 2013) argued that systems biology with its mostly bottom-up approach treats the plant as an assembly (or, rather, by-product) of molecular 'modules' and their interaction. As pointed out by Lüttge, these concepts tend to undervalue and neglect synergistic emergence as a key feature of living organisms. At a higher organisational level, biological systems exhibit completely new features that cannot be traced back to the properties of the elements they consist of. In fact, this is characteristic of life itself—it transcends its physico-chemical basis in a way notoriously confronting us with our conceptual deficits when studying biological systems (but this is certainly not exclusive to biological systems—scaling up is *always* accompanied by the introduction of emergent features, see Laughlin 2005 for examples from physics). Emergence in complex systems and its ongoing resistance against all attempts of mathematical or any other formal representation (Goldstein 2002; Kauffman and Clayton 2006) require us to establish and improve 'top-down' approaches emanating from the complex feature of the intact organism (photosynthesis, growth, etc.), which will be addressed here as 'complex traits'. Initially, this term has been introduced by geneticists to denote multigenic traits 'in which a one-to-one relationship between genotype and phenotype does not exist' (Darvasi 1998), in contrast to traits that follow classical Mendelian heritage scheme. Meanwhile, the term is also commonly used by plant breeders, e.g., when 'quantitative trait loci' are determined to analyse the genetic architecture of a particular phenotype (Bernardo 2008; Cooper et al. 2009). Here, the term 'complex trait' is used in a way that is compatible with its definition in the (applied) genetics context, although this study focuses on protein-trait relationships and genetic aspects remain unconsidered. It has to be kept in mind that just these complex traits are of particular interest for agriculture, silviculture, and other applied aspects of plant biology and that molecular research has still relatively little impact on practical agriculture, biogenic energy production, or wood technology in view of the huge scientific and financial effort that has been invested (Sinclair and Purcell 2005). There is no doubt that recurrence to the molecular level is required and that it will extend our knowledge of organismic features and can improve their predictability. However, at our current status of knowledge, we need criteria to identify those cases where molecular data will directly make a contribution to our understanding of a complex physiological trait. In other cases, strong emergence still provides a principal barrier to extrapolation of systemic features from the molecular units (i.e., in most cases, proteins; do Amaral and Souza 2017).

Here, heuristic principles to develop criteria are worked out that allow us to make the distinction between these cases. It is the goal of this study to suggest semi-quantitative guidelines for organismic plant physiology; they will serve as useful operational tools until comprehensive quantitative models may become available that allow us to predict emergent properties from molecular data (for which only very few pertinent examples exist so far, e.g. Gajdanowicz et al. 2011; Blatt et al. 2014; Albert et al. 2017). It is conceivable, though, that the complexity of many physiological processes will preclude this chance for ever, e.g., because it exceeds limits of computability (Noble 2012).

Modularity and Emergence: General Remarks

Some more basic considerations on modularity and emergence as contrasting concepts for understanding complex systems will be required before we can envisage their reconciliation within a practical concept for research on organismic plant biology. Modularity is a reductionist concept in that the features of a system are supposed to result from its structural and functional parts that act more or less semi-autonomously (De Kroon et al. 2005; Lüttge 2012). Therefore, a system is claimed to be fully characterised by studying its individual components, or 'modules', it is composed of. The opposite concept of dealing with multicomponent, complex systems is systemic and takes account of 'emergence'. Following Goldstein (2002), emergence represents 'the arising of novel and coherent structures, patterns and properties during the process of self-organization in complex systems'. Hence, one tends to miss crucial features of a system by just studying its components. While it is intuitively clear that life in general is a strongly emergent phenomenon, the concept of emergence is still undervalued in biology. One reason is that the concept of emergence is highly elusive and so far defies any (mathematical) formalisation. Various approaches to 'capture' the very nature of emergent phenomena were summarised by Goldstein (2002), including those of information theory, thermodynamics, and formal logics. But all these approaches apparently fell short of fully characterising the phenomenon and did not match our intuitive understanding of emergence. Although a comprehensive treatise could not be achieved, some important features of emergent systems became apparent in Goldstein's review: *firstly* (and not surprisingly), the parts of a system with emergent properties show strong 'non-trivial' interactions that shape the properties of the system more than the individual features of its parts. This implies that a mathematical and/or methodological framework adequately describing the underlying functional level must fail at least partly when dealing with the emergent ensemble, as exemplified by Prokopenko et al. (2009) for the relationship between statistical mechanics and (classical) thermodynamics. This statement may readily be transferred to the relationship of the biochemistry of proteins and the biological units they form part of. *Secondly*, emergent systems tend to be rather stable with respect to fluctuations (or 'noise') occurring at the hierarchically underlying organisational level (Souza and Lüttge 2015; this does of course not mean that they are invariant to those fluctuations).

Modularity and Emergence in Plant Biology

Although controversial discussions about modularity and emergence have persisted in various scientific disciplines, including physics (Anderson 1972; Laughlin 2005), it is obvious that this issue is of particular relevance for the biological sciences, and notably for the plant sciences. Unlike animals, plants do not possess a separate organ co-ordinating all parts of the organism suggesting that the parts are endowed with a greater autonomy (unless one follows the 'root brain hypothesis' of Charles and Frances Darwin; see Baluška et al. 2009). Therefore, Haukioja et al. (1991) contended that plants essentially are modular organisms. An individual is composed of 'repetitive multicellular meristems and adjacent differentiated tissue act mostly independently as an integrated functional unit (IFU)'. De Kroon et al. (2005) extended Haukioja's concept to environmental plasticity in general. According to these authors, 'whole-plant plasticity is the sum of all environmentally induced modular responses, plus all interaction effects that are due to communication and behavioural integration of modules'. Various examples for localised modular responses to environmental changes are discussed in their publication, including acclimation of root growth to local nutrient availability, and shaded versus sun-exposed leaves.

The concept of predominantly modular organisation of higher plants was strongly opposed by Lüttge (2012, 2013). He pointed out that emergence is the central key feature of life in general. The existence of a self-organising metabolism as a concerted, highly co-ordinated interaction of multiple enzymes, e.g., is a criterion separating biotic from abiotic nature. Furthermore, the phenotype of a plant is always an emergent phenomenon with respect to the genetic information that is encoding it, including epigenetic control of gene expression (Lüttge 2012). He also presented numerous examples for the integration of physiological and biochemical processes at the level of the individual plant (Lüttge 2013), most prominently the co-ordination of carbon source and sink and the vascular system as an 'integral unit' of higher plants. A prerequisite for the concerted action of modules is an efficient and rapid transfer of information. Therefore, this aspect is given much attention and space in Lüttge's review. A high degree of co-ordination also becomes apparent when studying water and nutrient transport in plants (Wegner 2015a).

This juxtaposition of modularity *versus* emergence evokes the impression that these concepts are strictly mutually exclusive. However, this is not the case! De Kroon et al. (2005), e.g., do consider integration of local responses to environmental stimuli on the level of the whole plant (without explicitly referring to the concept of emergence). Conversely, Lüttge accepts the concept of a modular structure of life when modules are considered as knots (or nodes) in a network (Lüttge and Hütt 2009; Lüttge 2012)—stressing the intensities of the ties with neighbouring knots to be highly variable. Therefore, modularity and emergence may be seen as gradually interrelating, rather than absolute, mutually exclusive features of higher plants, which can vary across modules, and in time and space. For practical purposes, it would be required to somehow quantify modularity and emergence in order to decide on the extent to which information obtained at the molecular scale will directly contribute to our understanding of complex traits such as photosynthetic carbon gain, growth, and yield. Such an approach will be suggested for critical examination in the following chapter.

The approach chosen here may seem somewhat uncommon to plant scientists who tend to handle fundamental aspects of their subject in a rather intuitive way (as most experimental scientists prefer to do). Therefore, issues that are discussed in a systematic way in the following section have been partly anticipated in many experimental studies, without providing a conceptual framework, however, of science theory. The latter will be striven for and discussed now in a systematic way, as it is the conviction of the author that whole-plant physiology, as a genuine subject within the plant sciences, needs an explicated epistemological basis (Souza et al. 2016; see also Volkov (2014)). This study is intended to make a significant step towards that goal.

Synthesis

In order to arrive at a synthesis of the concepts of modularity and emergence, we will focus on complex plant traits that are readily scalable, like photosynthetic carbon assimilation, growth, and yield. These can be quantified for the whole plant, and in case of carbon assimilation and yield also for plant organs (being morphological 'modules' sensu Haukioja et al. (1991)). These three traits depend on numerous molecular components with proteins playing a central role as enzymes, membrane transporters, and receptors. In case a protein Pr_i (that may include several isoforms) affects the trait t in a strictly modular way, we would expect that t scales with the abundance of Pr_i, at least within a certain range. Consider, e.g., as the most simple setting, a reaction catalysed by a single enzyme under in vitro conditions, without any limitation by the availability of substrates, nor accumulation of products. Under these (quite hypothetical) conditions, the reaction rate, ν, is linearly dependent on the concentration of an enzyme (or, more generally, of a protein i, $[Pr_i]$), and we can write:

$$v = v_0 * [Pr_i] \quad (1)$$

with v_0 being the reaction rate at a standard concentration. Under more realistic conditions, metabolite concentrations will soon become rate limiting, and v will tend to saturate with increasing enzyme concentration. Moreover, other metabolic steps will exert control and prevent linear increase ad infinitum when the catalytic reaction is part of a metabolic network (which is always the case in a cellular environment). These considerations can be extended from enzymology to scalable physiological traits in general (see also further below). Hence, dependence of a trait t on protein content will be satisfactorily described by a saturating function. Among these, an exponential equation of the following form will be adequate under most conditions:

$$t = t_{max}\left(1 - \exp\left(-\frac{[Pr_i]\ln 2}{[Pr_i]_{0.5}}\right)\right) \quad (2)$$

with $[Pr_i]_{0.5}$ being the protein content at which the trait attains half-maximum rates (Fig. 1a). Occasionally, at low concentrations, the protein-trait relationship may be more linear than predicted by Eq. 2, but this will just lead to minor deviations. Most importantly, this relationship is purely phenomenological and will hold irrespective of any mechanistic aspects of interaction, and irrespective of the structure of the underlying 'modular network' (Lüttge and Hütt 2009), unless the system is organised in a redundant way (a case to be discussed further below, section 'Redundancy'). For example, for a certain enzyme, the throughput rate will depend on its concentration as described by Eq. 2, both when it is part of a 'simple' metabolic pathway (with no feedback or other control) and when just one catalytic step is involved.

Whereas in the case of modularity the contribution of one part to the whole process is readily quantified, this is not so straightforward for an emergent contribution. As stated above, emergent systems may be quite invariant (or 'stable') with respect to fluctuations of its parts. Hence, we may expect t to be constant with respect to $[Pr_i]$ at least over a certain range. However, Pr_i is also supposed to be constitutive with respect to t. Hence, t will approach zero when Pr_i is completely eliminated (at the same time making the difference to the trivial case in which t is simply independent of Pr_i). Note that eventually a quantitative relationship may result as described by Eq. 2 (Fig. 1b). So what would essentially be the difference between a modular and an emergent process, and how can it be formalised? In the case of modularity, we would expect that $[Pr_i]$ in the wild type, $[Pr_i]_{WT}$, is in the range where changes in $[Pr_i]$ would affect t, or $[Pr_i]$ would at least be close to this range. This can be expressed mathematically in a simple way by relating the constant $[Pr_i]_{0.5}$ to $[Pr_i]_{WT}$ as follows:

$$[Pr_i]_{0.5} = [Pr_i]_{WT} * x + \Delta[Pr_i] \quad (3)$$

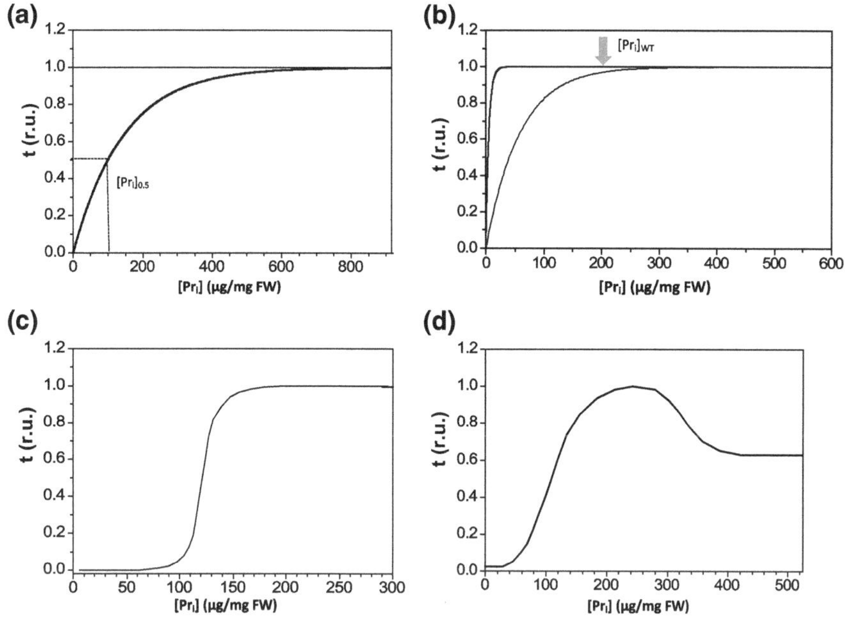

Fig. 1 Graphs showing the hypothetical dependence of a trait t on $[Pr_i]$, the content of protein i normalised to fresh weight. For $[Pr_i]$, arbitrary values were chosen supposed to mimic 'real activities', but experimental data may deviate considerably, or other units may be used. **a** represents the modular case described in Eq. 2. The trait attains half-maximum values at a protein content of 100 µg/g FW ($[Pr_i]_{0.5}$). The curve converges towards t = 1 (thin line). **b–d** shows emergent protein-trait relationships according to the definition that is given here, with t being stable with varying $[Pr_i]$ (**b**, upper line), increasing strongly above a threshold value of $[Pr_i]$ (**c**, ~ 100 µg/g FW), or passing through a maximum (**d**; at about 250 µg/g FW). The lower line in b represents the fit of the data by Eq. 2, with $[Pr_i]_{0.5}$ specified according to Eq. 3 ($[Pr_i]_{WT} = 200$ µg/mg FW). The 'best fit' with $\Delta[Pr_i] = 0$ does not adequately represent the data, confirming that (**b**) is not in accordance with a modular protein-trait relationship

with x being a factor that can vary between 0 and 1. For t scaling with $[Pr_i]$, at least potentially (which serves as a criterion for modularity, see above), $[Pr_i]_{WT}$ should not exceed $[Pr_i]_{0.5}$ by a factor of about 5. Hence, x is initially set to 0.2, and for a modular contribution to a complex trait, we would expect $\Delta[Pr_i] \geq 0$ under these conditions. For negative values of $\Delta[Pr_i]$, fluctuations of $[Pr_i]_{WT}$ would not have any direct effect on the trait. Hence, by combining Eqs. 2 and 3, we obtain an expression that can potentially serve as a mathematical criterion for modularity. In the case of Eq. 2 describing the relationship between t and $[Pr_i]$ adequately (with $[Pr_i]_{0.5}$ specified according to Eq. 3), we would say that t is highly modular with respect to $[Pr_i]$.

As explained above, emergence of a trait with respect to a certain protein (or set of proteins) could imply that this trait is rather stable with respect to gradual changes in $[Pr_i]$ (Fig. 1b), but it could also be otherwise. In fact, non-linear relationships of various complexities are conceivable, including a sudden increase in t within a small increment of $[Pr_i]$ (Fig. 1c; in fact, t may just come into existence by this transition, even by a discontinuity in the function). Alternatively, t may pass through a maximum when plotted as a function of $[Pr_i]$ (Fig. 1d). There may also be cases that cannot be described adequately by the suggested mathematical form. All these possibilities have in common that they are not adequately described by Eqs. 1 or 2, and hence, *emergence* is best characterised negatively as '*non-modularity*' which is associated with a poor representation by these equations in the way specified above. However, it is not really satisfactory to characterise something 'by what it is not', as stated already by Goldstein (2002).

Apart from this problem of defining emergence in a negative way, it may also be argued that the quantitative criterion for modularity as defined here does not reflect the very essence of what is meant by modularity (and, in a negative sense, by emergence) as expressed, e.g., by Goldstein's definition.

Uneasiness about these points may be overcome by introducing another criterion that appears to be more straightforward and in accordance with the original meaning of 'modularity' and 'emergence', at least to a first approximation (see also Fig. 2). This approach elaborates on control theory that was originally designed to quantify the extent to what an enzyme is rate limiting with respect to metabolic flux in a certain pathway (Kacser and Burns 1973; Fell 1992), but was later shown to be suitable for quantifying the contribution of an enzyme to a complex, quantitative trait, including photosynthetic carbon assimilation in plants (Stitt and Schulze 1994) and bacterial growth (Jensen et al. 1993) as well as regulatory interactions (Kahn and Westerhoff 1991). Relative changes of a trait t with the concentration of a protein Pr_i are described by the control coefficient, c:

$$c \equiv \frac{[\mathrm{Pr}_i]}{t}\left[\frac{\partial t}{\partial[\mathrm{Pr}_i]}\right] \tag{4}$$

Note that this is also an alternative approach to the one expressed in Eq. 2 for quantifying the dependence of a (complex) trait on the abundance of a certain protein. Now, if a complex trait is, in a strict sense, modular with respect to a molecular

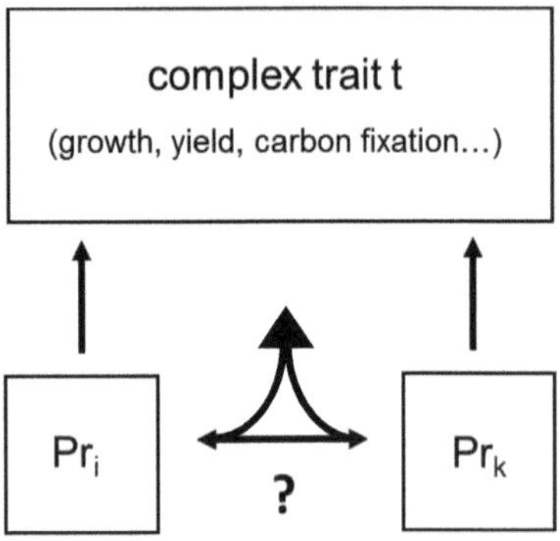

Fig. 2 Simple scheme showing the dependence of a complex trait on two proteins (Pr_i and Pr_k). These proteins may affect the trait independently (modular case), or they may contribute to t in an interactive way (emergent case)

component (i.e. a protein Pr_i), it is acting autonomously. Hence, the impact of a change in the abundance of this component will be strictly additive to those of other proteins ($Pr_{n \neq i}$), and there will be no interaction among them. Hence, we can write:

$$\frac{\partial t}{\partial [\mathrm{Pr}_i]} \neq f\left(\left[\mathrm{Pr}_{n \neq i}\right]\right) \tag{5}$$

Or, in a different notation:

$$\frac{\partial}{\partial \left[\mathrm{Pr}_{n \neq i}\right]} \left(\frac{\partial t}{\partial [\mathrm{Pr}_i]} \right) = 0 \tag{6}$$

This case is illustrated in Fig. 3a.

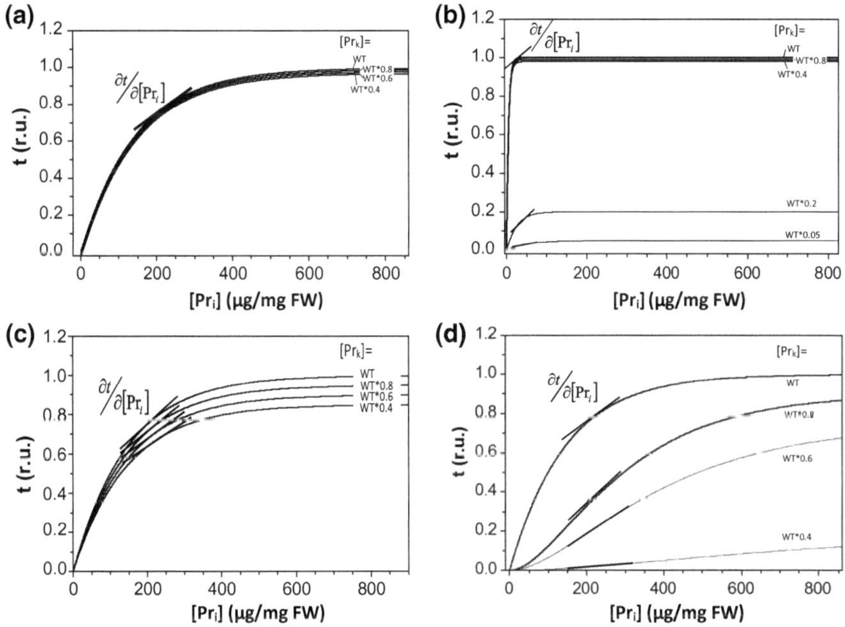

Fig. 3 Graphs showing the hypothetical dependence of a trait t on $[Pr_i]$, as affected by the abundance of a second protein Pr_k which is supposed to be down-regulated to different degrees (WT to WT * 0.4). In **a**, the first derivative, $\partial t/\partial[Pr_i]$, remains unaffected by varying Pr_k corresponding to the modular case (Eq. 6; note that the curves were slightly shifted with respect to each other to make them optically discernible). The short tangent represents the slope of the curve at a particular $[Pr_i]$. **b** Shows that Eq. 6 (supposed to characterise modular behaviour) also holds in case of synergistic emergence supporting the homeostasis of t over a wide range of $[Pr_k]$. Only at very low values for both proteins (curves for WT_k * 0.2 and WT_k * 0.05), the slope could be affected. **c** Represents a case of 'weak emergence', since Pr_k affects the slope of the curve, but just modifies the protein-trait relationship, without changing its basic features, whereas **d** is rather an example of 'strong emergence'. For more explanations, see the text

Any deviation would be taken as evidence for emergent behaviour. This is expressed here for 'pairs' of proteins I and k (concentrations of other proteins remaining constant):

$$\left(\frac{\partial t}{\partial[\mathrm{Pr}_i]}\right)_{[\mathrm{Pr}_{n\neq i,k}]} = f([\mathrm{Pr}_k]) \tag{7}$$

In order to end up with dimensionless parameters, ∂t and $\partial[\mathrm{Pr_i}]$ are divided by absolute values as in Eq. 4:

$$\frac{[\mathrm{Pr}_i]}{t}\left(\frac{\partial t}{\partial[\mathrm{Pr}_i]}\right)_{[\mathrm{Pr}_{n\neq i,k}]} = f([\mathrm{Pr}_k]) \tag{8}$$

And, correspondingly:

$$\frac{[\mathrm{Pr}_i]*[\mathrm{Pr}_k]}{t}\frac{\partial}{\partial[\mathrm{Pr}_k]}\left(\frac{\partial t}{\partial[\mathrm{Pr}_i]}\right)_{[\mathrm{Pr}_{n\neq i,k}]} = \kappa \quad \text{with}\,\kappa \neq 0 \tag{9}$$

For practical reasons, it could be useful to normalise the differential changes to values measured under wild-type conditions:

$$\frac{[\mathrm{Pr}_i]_{WT}*[\mathrm{Pr}_k]_{WT}}{t_{WT}}\frac{\partial}{\partial[\mathrm{Pr}_k]}\left(\frac{\partial t}{\partial[\mathrm{Pr}_i]}\right)_{[\mathrm{Pr}_{n\neq i,k}]} = \kappa \tag{10}$$

It has to be noted, though, that the criterion for modularity defined above is not unambiguous. A trait will also be unaffected by changes in the content of interacting proteins (with a possible exception at very low abundance of both proteins) in case emergent behaviour favours homeostasis, which is in accordance with Eqs. 5 and 6 (Fig. 3b, compare also Fig. 1b). Hence, for an unequivocal definition of modularity, an additional criterion has to be introduced in order to exclude this case:

$$\frac{\partial t}{\partial[\mathrm{Pr}_i]} \neq 0 \tag{11}$$

in the relevant range of $[\mathrm{Pr_i}]$. Note that this corresponds to $c > 0$ (Eq. 4).

Turning to emergence now, it should be noted that its characterisation by Eq. 8 implies that regulation of protein function by another type of protein is *always* interpreted as emergent behaviour. We end up with a somewhat far-reaching interpretation of emergence, since protein activity is always subject to regulation of some kind, indicating that 'true' modularity as expressed by Eq. 6 will rarely be found in nature. It should be noted, though, that the quantitative criteria for defining 'modularity' versus 'emergence' expressed by (in)Eqs. 5–10 formally imply that both concepts are strictly mutually exclusive. If we accept that there is a gradual transition from modularity to emergence, the (in)equations provide a guideline, rather than a definition. Most importantly, taking this viewpoint as a quantitative approach, defining a 'degree of emergence' κ can be established based on Eqs. 9 and 10. Until more empirical data are available (see the discussion further below), we have to abide with very broad, qualitative categories. Minor interactions among proteins will be taken as evidence for modularity sensu *latiore*, rather than emergence. Generally, interactions that evoke mere quantitative changes will be considered a 'weak' criterion for emergence (Fig. 3c). 'Weak emergence' in this sense is compatible (though not identical) with the definition of Bedau (1997) who first introduced this term. If the interaction of molecular components entails a functional, qualitative change instead of merely leading to quantitative modifications, this would be taken as a 'strong' indication for emergence. In terms of a graphical representation according to Eq. 9, a sudden 'jump' in the slope or a discontinuity would be incited at a critical threshold, e.g. for $[Pr]_i$. Upon an increase of $[Pr_k]$ beyond that threshold value, the complex trait (and its dependence on $[Pr_i]$) is boosted. Other patterns may occur, but 'strong' emergence always implies a highly non-linear relationship (Fig. 3d).

The foregoing considerations have been quite theoretical, and it is time now to come up with some palpable examples. In search of a model case for a protein affecting a complex trait in a *modular* way, we come across Rubisco and photosynthetic carbon assimilation as a paradigmatic example. Because of its primary role in CO_2 fixation, Rubisco has a key function in plant life, and this protein and its regulation have been investigated in much detail. Therefore, it is selected here for illustration of the above analytical approach (despite its heteromeric structure with large and small subunits encoded in the plastid and the nucleus, respectively, that may be seen as a complication). Dependence of photosynthetic carbon fixation on the level of Rubisco expression has extensively been studied and analysed, e.g., by Stitt and Schulze (1994). The expression of the enzyme was varied by antisense technology (which has been the method of choice for manipulating protein abundance for a long time; more recently, other techniques like the RNAi and CRISPR-Cas9 technology, riboswitches, and inducible promoters became available. A more detailed report on technical aspects is beyond the scope of this communication). Carbon assimilation was shown to depend on Rubisco activity over a wide range (Fig. 4) until it tended to saturate when the protein content approached the native level (Quick et al. 1991a, b). This behaviour persisted when experimental conditions were varied with respect to irradiation, nutrient supply, and other factors. Stitt and Schulze interpreted their data in the context of control theory (for a review, see Fell (1992)). Usually, the formalism of control theory is most suitable for a detailed characterisation of a metabolic pathway or network of a known structure (Raines 2003), a state of scientific advancement that is frequently still out of reach in the field of plant physiology. This is probably the reason why control theory has rarely been applied in the plant sciences so far.

Given its paramount importance in photosynthesis, and whole-plant physiology in general, it is not surprising that Rubisco is a tightly regulated enzyme. Control of its activity by various environmental factors, including light intensity, temperature, and CO_2 partial pressure, has been studied in much detail (Salvucci 1989; Quick et al. 1991a, b; Stitt and Schulze 1994; von Caemmerer and Quick 2000; Spreitzer and Salvucci 2002). In the presence of CO_2, Rubisco is converted from an inactive to an active form by a carbamylation reaction. Carbamylation can occur spontaneously

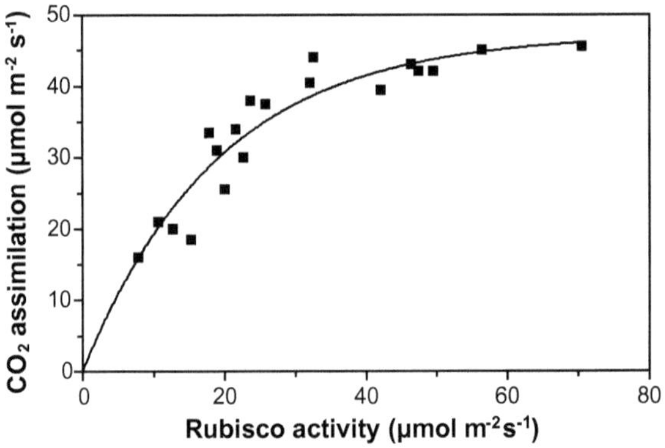

Fig. 4 Dependence of photosynthetic CO_2 assimilation on the Rubisco activity in *Nicotiana tabacum* L leaves (data adopted from Stitt and Schulze 1994). Under the conditions employed here, Rubisco activity is directly proportional to the protein content of the leaf. In the wild type, Rubisco activity ranged between 40 and 70 μmol m^{-2} s^{-1} corresponding to a protein content of 5–7 g protein m^{-2}. Data were fitted with Eq. 2; best fit (continuous line) was obtained with a maximum CO_2 assimilation (corresponding to t_{max}) and $[Pr_i]_{0.5}$ of 47 μmol m^{-2} s^{-1} and 13 μmol m^{-2} s^{-1}, respectively

and is favoured by an increase in light intensity. Moreover, carbamylation is strongly enhanced in the presence of a regulatory enzyme, the Rubisco activase (RCA), that is also light dependent (Portis 1995, 2003). In order to study regulation of Rubisco (and, in turn, carbon assimilation) by RCA, molecular tools have been applied since the early nineties of the last century when molecular biology first became available in plant physiology (Andrews et al. 1995). Therefore, an impressive body of experimental evidence is now available. Again, RCA expression was down-regulated by antisense technology to a varying extent (Andrews et al. 1995; Eckardt et al. 1997). Assessing the latter effect on Rubisco activity and gas exchange parameters, it was found that a severe suppression of RCA content (below 10% of wild-type level) was required to see an impact on these parameters (Fig. 5a). In *Arabidopsis* as well as in young tobacco seedlings, the abundance of other leaf proteins (including Rubisco) was hardly affected by RCA suppression and so was total leaf protein content, with RCA just contributing ~ 4% in wild type. Only in older RCA antisense tobacco plants, an up-regulation of Rubisco expression was observed (without any increase in Rubisco activity, though); this special case is not discussed in more detail here. Even though excellent experimental studies on Rubisco, RCA, and CO_2 assimilation are available, the data do not allow an exact analysis according to Eqs. 7–10 and the hypothetical cases depicted in Fig. 3 (that would require breeding of double mutants). Nonetheless, with respect to the criteria defined above, we can clearly identify a case of *weak emergence* here, since features of Rubisco are not fundamentally changed by the expression level of RCA. Merely its activity level is modified when RCA is severely down-regulated. Consistently, the response of the enzyme to abiotic factors (light, CO_2 partial pressure) remains basically unaffected in the RCA antisense

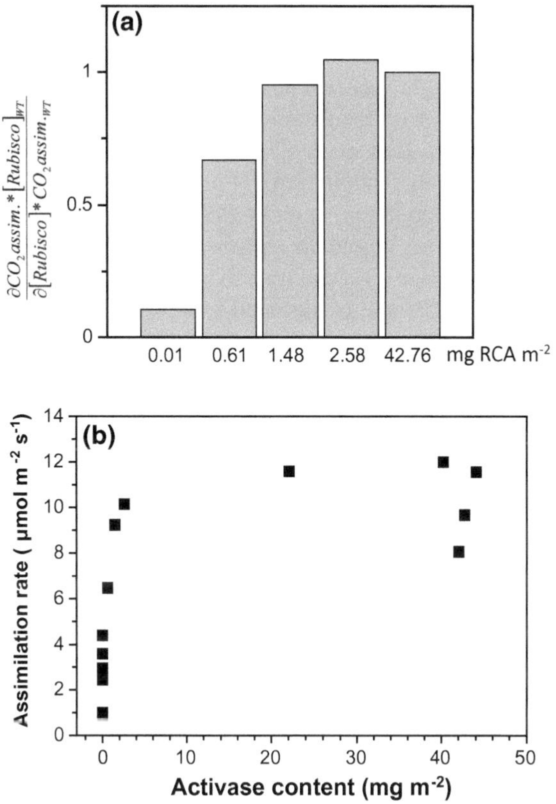

Fig. 5 Regulation of CO_2 assimilation rate by Rubisco activase. **a** Change of CO_2 assimilation (normalised to the CO_2 assimilation in the wild type; here 9.7 μmol m^{-2} s^{-1}) with Rubisco content at various activities of Rubisco activase (normalised to leaf area). $\partial CO_2assim/\partial[Rubisco]$ was linearly interpolated (Eq. 1). **b** Plot of CO_2 assimilation against Rubisco activase content. Data reproduced from Andrews et al. (1995), with permission

plants. Finally, properties of Rubisco in intact leaves agree fairly well with those obtained from in vitro tests of the enzyme. Other mechanisms of Rubisco regulation have been identified, but they are also in agreement with, at most, weakly emergent interactions. Hence, it can be stated that Rubisco exerts its central function within carbon assimilation in a weakly emergent way (as far as interaction with other proteins is considered) and in a modular way with respect to its contribution to the trait itself. Strikingly, though, when the dependence of *carbon assimilation* on the *RCA expression level* is considered (with Rubisco being intermediate), the relationship is clearly emergent, according to the above definition (Fig. 5b, compare Fig. 1b). Carbon assimilation is largely invariant with RCA content over a wide range and only collapses at some 10% of wild-type level. So the regulatory enzyme maintains homeostasis of the complex biological function, in this case with CO_2 assimilation.

Another interesting type of emergent pattern stabilising complex biological function has been described for CO_2 fixation in C4 plants. In mesophyll cells of these plants, malate is formed by carboxylation of phosphoenolpyruvate (PEP) which is transported into bundle sheath to increase efficiency of CO_2 fixation in these cells (for a review, see Hatch 1987). Synthesis of PEP, in turn, is brought about by pyruvate orthophosphate dikinase (PPDK), an enzyme that exerts strong control on the metabolic C4 cycle. Experimental data suggest that malate synthesis is tightly regulated by light, but remains insensitive to the expression level of PPDK (and the concentration of its substrate pyruvate and its product PEP) over a wide range (Ohta et al. 2006). Hart et al. (2011) suggested a refined model to explain stability of PPDK activity (and, in turn, C4 metabolic pathway). The enzyme is activated by an autophosphorylation step from PPDK0 to PPDK1; in this state, either Pi can be transferred to pyruvate and the enzyme returns to the PPDK0 state, or the enzyme can be phosphorylated a second time to produce PPDK2 by an enzyme termed 'regulatory enzyme' (RP). RP is supposed to be bimodal, regulating both phosphorylation of PPDK1 and subsequent dephosphorylation of PPDK2 (thus regenerating the active form PPDK1). RP tends to form ternary complexes with 4 PPDK molecules simultaneously binding to RP. PPDK1 formation both by autophosphorylation of PPDK0 and by dephosphorylation of PPDK2 are linked to each other by producing PPi and consuming Pi. Hart et al. could show that under these conditions, the formation of PEP is largely independent of the concentrations of PPDK and RP (Fig. 6). However, an up-regulation of malate formation is observed when the PPDK-to-RP ratio exceeds a threshold level. These features have been identified as characteristic of emergent processes. Arguably, the purely phenomenological description as suggested in this study adds little to the understanding of a metabolic process when a detailed quantitative model is available (as in this particular case; see also further below). But for other, less well-understood physiological phenomena, developing a typology of emergent processes, according to the above criteria, may be very helpful to identify underlying regulatory schemes, similar to the one described by Hart et al. (2011; that scheme proved successful for describing further cases of metabolic homeostasis, too). Hence, it will be very instructive to study other 'real' cases of emergent protein-trait relations in plants and to discuss them in more detail in this context. Unfortunately, however, only few relevant experimental data have been published that allow to link complex traits to protein abundance (or, alternatively, protein expression levels). Just one case was found in the literature that corresponded to the scenario depicted in Fig. 1c: fitness of yeast cells was shown to depend in this way on LCB2, a factor involved in sphingolipid synthesis (Fig. 2c in Rest et al. 2013). No empirical data could be found equalling Fig. 1d. After more than 20 years of extensive research in the field of molecular plant biology, this is a puzzling gap of knowledge that seems hard to explain at first sight. The most likely explanation is that (i) it was not considered a task of pre-eminent importance to systematically study the dependence of complex traits such as growth, yield, or photosynthetic carbon assimilation on the expression levels of key proteins, and (ii) emergent protein-trait relations tend to produce 'irregular' curves that are hard to interpret and were considered to convey little relevant information, so they remained unpublished. In fact, it is an important goal of this publication to change that view and to highlight the importance of pertinent data.

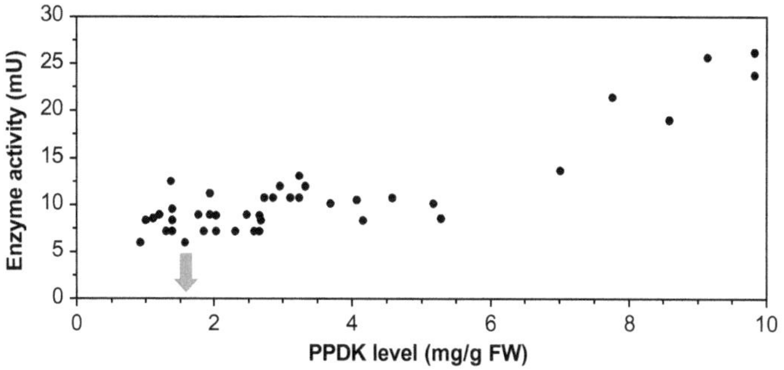

Fig. 6 PPDK enzyme activity (being proportional to the rate of phosphoenolpyruvate formation) as a function of the enzyme content in transgenic maize. The arrow indicates the wild-type level. Data adopted from Hart et al. (2011), with permission

Redundancy

The above criteria to differentiate between modularity and emergence when dealing with physiological processes appear to be quite straightforward and, due to the phenomenological approach, are universally applicable to quantifiable traits. However, redundancy of metabolic pathways or physiological processes may lead to a misinterpretation of experimental data. A rather modular organisation of one process will be masked in case that (a) parallel process(es) exist(s), taking over when the one to be investigated is down-regulated in transgenic plants. In this case, results would suggest emergent behaviour ensuring homeostasis, or even evoke the false impression that a molecular component is irrelevant for a particular trait (even though it is not, in the wild type). This is an important restriction, since redundancy frequently occurs in physiological processes stabilising a physiological trait (Wagner 1999; Souza and Lüttge 2015). Processes redundant to the one in focus would have to be blocked by a selective inhibitor, or by a knockout mutation to perform the above type of analysis (provided that redundant processes have been identified before, of course).

In addition to this specific mechanism of redundancy, any knockout bears the risks of misinterpretation, because any 'missing' process will have unpredictable impact on the complex, highly interrelated feedback and feedforward regulation network at the whole-system level and its performance.

Implications for Future Basic and Applied Research in the Plant Sciences

After these fundamental thoughts on 'modularity' and 'emergence' in complex plant traits (and evaluating some examples), it remains to be discussed how basic and applied research in the plant sciences could directly benefit from these considerations. Obviously, whole-plant physiology could directly profit from molecular studies, when a protein contributes in a (at least predominantly) modular way to a complex trait. In this case, intrinsic, molecular features of the protein will have a direct impact on that trait. Hence, molecular data would directly contribute to the understanding of the complex trait, with the important corollary that modular protein-trait relationships offer a promising starting point for genetic engineering. When features of a protein acting in a modular (or weakly emergent) way are modified, a direct and predictable effect on the complex trait is anticipated. Hence, promising candidates for successful genetic engineering could systematically be identified using the criteria for modular or weakly emergent organisation outlined in this study.

It is not by accident that Rubisco was an early object of interest in molecular plant physiology, followed by suggestions on how photosynthesis could be made more efficient by 'improving' this key enzyme using the genetic engineering toolbox (Spreitzer and Salvucci 2002). However, Rubisco is also an instructive example demonstrating that superior physiological performance observed in the laboratory is not necessarily preserved in field-grown plants (Sinclair and Purcell 2005). Obviously, difficulty increases with emergent protein-trait relationships. Overexpression of a certain protein, e.g., would not enhance the process of interest (compare Fig. 1b) or could even be counter-productive (depending on the type of emergent behaviour; Fig. 1d). This is not to say that emergent features would necessarily *preclude* establishing any linkage between a complex physiological trait and its molecular basis. The only way to be successful in this case is by rigorous quantitative modelling, taking into account the interplay of the molecular components that are involved in the process. Based on a well-founded mathematical model, it can putatively be decided whether—and how—a physiological process can be streamlined by modulating the expression of a protein (or, more likely, a group of proteins), or by introducing new ones. This implies an advanced knowledge of the system that evolves a complex trait. Most importantly, the physiological process and the 'molecular machinery' contributing to it need to be separated from other 'neighbouring' processes, an operation that always remains somewhat arbitrary when dealing with (by definition) open systems. In some cases, defining the molecular entities contributing to a complex trait in a comprehensive way may seem impossible, or far out of reach from the viewpoint of our current state of knowledge. Even the definition of an organism, as a functional unit, is not as straightforward as it may seem at first glance. This is reflected by the recent discussion about the holobiont and multispecies holobiont-like systems (zu Castell et al. 2015; see also this book, Chap. Gaia—A Holobiont-like System Emerging From Interaction, by zu Castell et al.).

Still, there are impressive examples for the successful reconstruction of emergent properties of a complex system from molecular features of the involved proteins. An instructive case is the variable size of the stomatal pore. Being proportional to the stomatal conductance in a species-dependent way (Buckley et al. 2003; Franks and Farquhar 2007), stomata are of central importance for leaf gas exchange and water use efficiency. Stomatal aperture is regulated by osmotic swelling or shrinking of the two guard cells that enclose the stomatal pore. Guard cells proved to be a well-suited model system for studying mechanisms of uptake and release of osmotica by plant cells (mainly KCl), e.g., in response to environmental stimuli (Roelfsema and Hedrich 2005). Detailed information on the involved transport proteins (particularly ion channels) located in the guard cell plasma membrane and tonoplast, which has been compiled throughout three decades of extensive research on this system, forms the basis for a computer program termed 'OnGuard' that also includes information on metabolic processes in guard cells (Hills et al. 2012). The underlying model is sufficiently detailed to predict emergent properties of ion transport and stomatal control in the wild type, as well as in knockout mutants that are deficient of certain proteins (Chen et al. 2012; Blatt et al. 2014). Modelling of these processes is favoured by the advanced state of knowledge and the fact that we have a well-defined functional unit with two highly specialised cell types, i.e. guard cells and subsidiary cells. Moreover, the quantitative description of osmotic processes and the underlying metabolic as well as membrane transport processes is well established for decades now (Zimmermann and Steudle 1978; Maggio et al. 2006; but see also Wegner 2014, 2015a, b, 2017). For other physiological processes, the situation is by far less favourable and well defined, and more complicated, branched regulatory networks can be involved. In these cases, modelling with Boolean mathematics may provide an adequate approach (Helikar et al. 2008). The identified molecular components of a network '(usually genes or proteins, respectively)' function as 'nodes', and their interactions are denoted as 'edges' (Lucas et al. 2011). Recently, Albert et al. (2017) used this approach to model the molecular mechanism of stomatal closure by the phytohormone ABA. Boolean networks allow to identify, e.g., feedback loops of regulation that may otherwise have been undetected and can explain counter-intuitive effects reported in the literature. It is a disadvantage of this approach, though, that interactions can only be represented in a qualitative way; Bayesian networks, which also allow to consider the strength of the interactions, may be superior, but are more difficult to apply (Lucas et al. 2011). Alternatively, principal component analysis, a statistical tool, can be used for top-down analysis of complex traits (Helikar et al. 2008). Further approaches, including those derived from computational neuroscience, are discussed by Pezzulo and Levine (2016). Common to all these approaches is that they require an advanced state of knowledge of the system to be analysed, a prerequisite that is frequently not fulfilled in plant biology. In those cases, phenomenological approaches as the one outlined in this publication may be a heuristic tool paving the way to understand emergent properties of complex physiological traits. Such a heuristic tool should be employed to develop systematics of emergent phenomena (following the advice of Goldstein 2002) in the organisation and control of complex traits. It is an obvious problem that emergent phenomena always represent 'individual solutions' to a physiological problem. The quantitative description of stomatal function cannot readily be transferred to other complex traits (or, at best, in a limited and fragmentary way). However, we may identify patterns of emergent organisation of molecular units that lead to certain traits. Only very few attempts towards that goal have been undertaken so far. One example is the identification of bifunctional enzymes as the one described by Hart et al. (2011) that tend to stabilise a metabolic process. Other—yet unidentified—patterns are likely to await elucidation. It is conceivable, e.g., that enzymatic regulation by phosphorylation/dephosphorylation cycles and cascades shapes and stabilises such patterns of emergent protein interaction. Consistently, Nijhout et al. (2003) modelling MAP kinase cascades obtained protein-trait relationships closely resembling that in Fig. 1b. In the near future, the theoretical approach that is presented here (Eqs. 5–10) to formalise single protein-trait relationships has to be extended to sets of n proteins shaping a complex trait. Currently, this approach is still lacking practicability, since the experimental database is unavailable. However, technical advance in this field is fast, and simultaneous control of the expression of several proteins and screening of the resulting phenotype is within reach.

In conclusion, it is the main goal of this communication to stimulate discussion on how the contribution of single proteins to complex traits can be formalised with respect to the categories of modularity and emergence. A comprehensive treatise of this issue, which is beyond the scope of this paper, would also require consideration of the genetic aspect. Recently, the genetic basis ('genetic architecture') of complex traits has received much interest. The link to the approach discussed here appears straightforward, since each protein is encoded by a gene, but only at first sight. Complications arise from the fact there is only moderate correlation between protein and mRNA levels (Wu et al. 2013), e.g. due to post-transcriptional regulatory mechanisms, and it has to be kept in mind that elucidation of the genetic architecture of a complex trait does not convey any information on functional interactions of proteins. Hence, it is of high interest for breeders, but less relevant for plant physiologists. By contrast, this communication is concerned with the question of how quantitative interaction between proteins results in complex physiological function. Finding answers to this question will be of paramount importance for future successful handling of complexity in (plant) physiology.

Acknowledgements I would like to thank Prof. Dr. Rainer Matyssek, Weihenstephan, Germany; Prof. Dr. Sergey Shabala, Hobart, Australia; and Dr. Vadim Volkov, Davis, USA, for discussion and for critical reading of a draft version of the manuscript.

References

Albert R, Acharya BR, Jeon BW, Zañudo JGT, Zhu M, Osman K et al (2017) A new discrete dynamic model of ABA-induced stomatal closure predicts key feedback loops. PLoS Biol 15(9):e2003451

do Amaral MN, Souza GM (2017) The challenge to translate OMICS data to whole plant physiology: the context matters. Front Plant Sci 8:2146

Anderson PW (1972) More is different. Science 177:393–396

Andrews TJ, Hudson GS, Mate CJ, von Caemmerer S, Evans JR, Arvidsson YB (1995) Rubisco: the consequences of altering its expression and activation in transgenic plants. J Exp Bot 46:1293–1300

Baluška F, Mancuso S, Volkmann D, Barlow P (2009) The 'root-brain' hypothesis of Charles and Francis Darwin: revival after more than 125 years. Plant Signal Behav 4:1121–1127

Bedau MA (1997) Weak emergence. In: Tomberlin J (ed) Philosophical perspectives. Mind, causation and world, vol 11. Blackwell Malden, MA, US, pp 375–399

Bernardo R (2008) Molecular markers and selection for complex traits in plants: learning from the last 20 years. Crop Sci 48:1649–1664

Blatt MR, Wang Y, Leonhardt N, Hills A (2014) Exploring emergent properties in cellular homeostasis using OnGuard to model K+ and other ion transport in guard cells. J Plant Physiol 171:770–778
Buckley TN, Mott KA, Farquhar GD (2003) A hydromechanical and biochemical model of stomatal conductance. Plant Cell Environ 26:1767–1785
von Caemmerer S, Quick WP (2000) Rubisco: physiology in vivo. In: Leegood RC, Sharkey TD, von Caemmerer S (eds) Photosynthesis. Springer, Heidelberg, pp 85–113
zu Castell W, Fleischmann F, Heger T, Matyssek R (2015) Shaping theoretic foundations of holobiont-like systems. In: Canovas FM, Lüttge U, Matyssek R (eds) Progress in botany. Springer, Cham, Germany, pp 219–244
Chen Z-H, Hills A, Bätz U, Amtmann A, Lew VL, Blatt MR (2012) Systems dynamic modeling of the stomatal guard cell predicts emergent behaviors in transport, signaling, and volume control. Plant Physiol 159:1235–1251
Cooper M, van Eeuwijk FA, Hammer GL, Podlich DW, Messina C (2009) Modeling QTL for complex traits: detection and context for plant breeding. Curr Opin Plant Biol 12:231–240
Cui J, Li P, Li G, Xu F, Zhao C, Li Y, Shi T et al (2008) AtPID: Arabidopsis thaliana protein interactome database—an integrative platform for plant systems biology. Nucleic Acids Res 36:D999–D1008
Darvasi A (1998) Experimental strategies for the genetic dissection of complex traits in animal models. Nat Genet 18:19–24
De Kroon H, Huber H, Stuefer JF, Van Groenendael JM (2005) A modular concept of phenotypic plasticity in plants. New Phytol 166:73–82
Eckardt NA, Snyder GW, Portis AR Jr, Ogren WL (1997) Growth and photosynthesis under high and low irradiance of Arabidopsis thaliana antisense mutants with reduced ribulose-1, 5-bisphosphate carboxylase/oxygenase activase content. Plant Physiol 113:575–586
Fell DA (1992) Metabolic control analysis: a survey of its theoretical and experimental development. Biochem J 286:313–330
Franks PJ, Farquhar GD (2007) The mechanical diversity of stomata and its significance in gas-exchange control. Plant Physiol 143:78–87
Fukushima A, Kusano M, Redestig H, Arita M, Saito K (2009) Integrated omics approaches in plant systems biology. Curr Opin Chem Biol 13:532–538
Gajdanowicz P, Michard E, Sandmann M, Rocha M, Corrêa LGG, Ramírez-Aguilar SJ, Dreyer I, et al (2011) Potassium (K+) gradients serve as a mobile energy source in plant vascular tissues. Proc Natl Acad Sci USA 108:864–869
Goldstein J (2002) The singular nature of emergent levels: suggestions for a theory of emergence. Nonlinear Dyn Psychol Life Sci 6:293–309
Hart Y, Mayo AE, Milo R, Alon U (2011) Robust control of PEP formation rate in the carbon fixation pathway of C4 plants by a bi-functional enzyme. BMC Syst Biol 5:171
Hatch MD (1987) C4 photosynthesis: a unique blend of modified biochemistry, anatomy and ultrastructure. Biochimica Biophys Acta (BBA)—Rev Bioenerg 895:81–106
Haukioja E, Grubb PJ, Brown V, Bond WJ (1991) The influence of grazing on the evolution, morphology and physiology of plants as modular organisms [and Discussion]. Philos Trans R S Lond B Biol Sci 333:241–247
Helikar T, Konvalina J, Heidel J, Rogers JA (2008) Emergent decision-making in biological signal transduction networks. Proc Natl Acad Sci USA 105(6):1913–1918
Hills A, Chen Z-H, Amtmann A, Blatt MR, Lew VL (2012) OnGuard, a computational platform for quantitative kinetic modeling of guard cell physiology. Plant Physiol 159:1026–1042
Jensen PR, Michelsen O, Westerhoff HV (1993) Control analysis of the dependence of Escherichia coli physiology on the H(+)-ATPase. Proc Natl Acad Sci USA 90:8068–8072
Kacser H, Burns J (1973) The control of flux. Symp Soc Exp Biol 27:65–104
Kahn D, Westerhoff HV (1991) Control theory of regulatory cascades. J Theor Biol 153:255–285
Kauffman S, Clayton P (2006) On emergence, agency, and organization. Biol Philos 21:501–521
Laughlin RB (2005) A different universe: reinventing physics from the bottom down. Basic Books, New York

Lucas M, Laplaze L, Bennett MJ (2011) Plant systems biology: network matters. Plant Cell Environ 34:535–553
Lüttge U (2012) Modularity and emergence: biology's challenge in understanding life. Plant Biol 14:865–871
Lüttge U (2013) Whole-plant physiology: synergistic emergence rather than modularity. In: Progress in botany. Springer, pp 165–190
Lüttge U, Hütt M-T (2009) Talking patterns: communication of organisms at different levels of organization—an alternative view on systems biology. Nova Acta Leopoldina NF 96:161–174
Maggio A, Zhu J-K, Hasegawa PM, Bressan RA (2006) Osmogenetics: Aristotle to Arabidopsis. Plant Cell 18:1542–1557
Nijhout HF, Berg AM, Gibson WT (2003) A mechanistic study of evolvability using the mitogen-activated protein kinase cascade. Evol Dev 5:281–294
Noble D (2012) A theory of biological relativity: no privileged level of causation. Interface Focus 2:55–64
Ohta S, Ishida Y, Usami S (2006) High-level expression of cold-tolerant pyruvate, orthophosphate dikinase from a genomic clone with site-directed mutations in transgenic maize. Mol Breed 18:29–38
Pezzulo G, Levine M (2016) Top-down models in biology: explanation and control of complex living systems. J R Soc Interface 13:20160555
Portis AR (1995) The regulation of Rubisco by Rubisco activase. J Exp Bot 46:1285–1291
Portis AR (2003) Rubisco activase—Rubisco's catalytic chaperone. Photosynth Res 75:11–27
Prokopenko M, Boschetti F, Ryan AJ (2009) An information-theoretic primer on complexity, self-organization, and emergence. Complexity 15:11–28
Quick WP, Schurr U, Fichtner K, Schulze ED, Rodermel SR, Bogorad L, Stitt M (1991a) The impact of decreased Rubisco on photosynthesis, growth, allocation and storage in tobacco plants which have been transformed with antisense rbcS. Plant J 1:51–58
Quick WP, Schurr U, Scheibe R, Schulze E-D, Rodermel SR, Bogorad L, Stitt M (1991b) Decreased ribulose-1, 5-bisphosphate carboxylase-oxygenase in transgenic tobacco transformed with 'anti-sense' rbcS. Planta 183:542–554
Raines CA (2003) The Calvin cycle revisited. Photosynth Res 75:1–10
Rest JS, Morales CM, Waldron JB, Opulente DA, Fisher J, Moon S, Bullaughey K, Dedousis D et al (2013) Nonlinear fitness consequences of variation in expression level of a eukaryotic gene. Mol Biol Evol 30:448–456
Roelfsema MRG, Hedrich R (2005) In the light of stomatal opening: new insights into 'the Water-gate'. New Phytol 167:665–691
Salvucci ME (1989) Regulation of Rubisco activity in vivo. Physiol Plant 77:164–171
Sinclair TR, Purcell LC (2005) Is a physiological perspective relevant in a 'genocentric' age? J Exp Bot 56:2777–2782
Souza GM, Lüttge U (2015) Stability as a phenomenon emergent from plasticity–complexity–diversity in eco-physiology. In: Lüttge U, Beyschlag W (eds) Progress in botany, vol 76. Springer, Cham, Germany, pp 211–239
Souza GM, Prado CHBA, Ribeiro RV, Barbosa JPRAD, Goncalves AN, Habermann G (2016) Toward a systemic plant physiology. Theor Exp Plant Physiol 28(4):341–346
Spreitzer RJ, Salvucci ME (2002) Rubisco: structure, regulatory interactions, and possibilities for a better enzyme. Ann Rev Plant Biol 53:449–475
Stitt M, Schulze D (1994) Does Rubisco control the rate of photosynthesis and plant growth? An exercise in molecular ecophysiology. Plant Cell Environ 17:465–487
Usadel B, Poree F, Nagel A, Lohse M, Czedik-Eysenberg A, Stitt M (2009) A guide to using MapMan to visualize and compare Omics data in plants: a case study in the crop species, maize. Plant Cell Environ 32:1211–1229
Volkov V (2014) How to integrate biological research into society and exclude errors in biomedical publications? Progress in theoretical and systems biology releases pressure on experimental research. Commun Integr Biol 7:e27966

Wagner A (1999) Redundant gene functions and natural selection. J Evol Biol 12:1–16
Wegner LH (2014) Root pressure and beyond: energetically uphill water transport into xylem vessels? J Exp Bot 65:381–393
Wegner LH (2015a) Interplay of water and nutrient transport: a whole-plant perspective. In: Lüttge U, Beyschlag W (eds) Progress in botany, vol 76. Springer, Cham, Germany, pp 109–141
Wegner LH (2015b) A thermodynamic analysis of the feasibility of water secretion into xylem vessels against a water potential gradient. Funct Plant Biol 42:828–835
Wegner LH (2017) Cotransport of water and solutes in plant membranes: the molecular basis, and physiological functions. AIMS Biophys 4(2):192–209
Wu L, Candille SI, Choi Y, Xie D, Jiang L, Li-Pook-Than J, Tang H, Snyder M (2013) Variation and genetic control of protein abundance in humans. Nature 499:79–82
Yuan JS, Galbraith DW, Dai SY, Griffin P, Stewart CN (2008) Plant systems biology comes of age. Trends Pant Sci 13:165–171
Zimmermann U, Steudle E (1978) Physical aspects of water relations of plant cells. Adv Botanical Res 6:45–117

Emergence in Biomimetic Materials Systems

Thomas Speck and Olga Speck

Abstract The importance of emergent characteristics in the knowledge transfer from biological role models to biomimetic materials systems is discussed. After a brief definition of biomimetics and a description of the two basic approaches to develop biomimetic materials systems and products, we attempt to discern different types of emergence important in biomimetics. Emergent characteristics occurring in the materials systems may concern the same physical property (type 1) or a different physical property (type 2) compared to the properties of the single elements. Additional emergent characteristics important in biomimetics are related to societal values. These emergent values include aesthetic and symbolic values of the biomimetic products (type 3) and the potential contribution to moral values such as sustainability (type 4).

Introduction

Emergence is understood as the appearance of novel characteristics of a system as a result of the interaction of its single elements whereby the new emergent characteristics of the system cannot (obviously) be traced back to the characteristics of the isolated elements building up the system. The focus of this book chapter is on the emergent characteristics of biological and biomimetic materials systems such as physical properties, structures, functions or societal values. If we look at our bio-

T. Speck (✉) · O. Speck
Plant Biomechanics Group, Botanic Garden, University of Freiburg, Schänzlestraße 1, 79104 Freiburg, Germany
e-mail: thomas.speck@biologie.uni-freiburg.de

T. Speck · O. Speck
Freiburg Center for Interactive Materials and Bioinspired Technologies (FIT), Georges-Köhler-Allee 105, 79110 Freiburg, Germany

L. H. Wegner and U. Lüttge (eds.), *Emergence and Modularity in Life Sciences*,
https://doi.org/10.1007/978-3-030-06128-9_5

logical and technical environment, striking examples of emergent characteristics can be found in many biological and technical materials and structures. However, we often do not realise that the excellent mechanical and other physical properties of well known and widely used materials and structures are based on the emergent behaviour of their individual components causing more than a mere additive effect when combined. It should be mentioned that the terms "material" and "structure" blur in biology and often also in biomimetics. All biological materials, e.g. wood, bamboo, seed coats, bone, muscle, are hierarchically structured. Typically, five to nine hierarchical structural levels can be discerned that may cover a size range of up to twelve orders of magnitude. Therefore, all biological and many biomimetic materials are intrinsically structured and the terms "material" and "structure" lose their distinctive character (Fratzl and Weinkamer 2007; Speck and Burgert 2011; Gibson 2012; Wegst et al. 2015). For this reason and with a focus on the topic of the book chapter, we speak here of materials systems.

Striking examples for emerging mechanical properties are fibre-reinforced composite materials that are widespread in biological and technical materials systems. Their excellent mechanical properties are brought about by the combination of (purely) tension-stiff fibres embedded in a (purely) compression-stiff matrix (see section "Biomimetic Composite Materials"). Also in many other bio-inspired and biomimetic materials systems, emergent characteristics occur which may be "inherited" from the biological role model and often widen the potential field of application considerably. In this context, it is worthwhile to analyse whether the development of novel bio-inspired materials systems based on the inspiration of biological role models also transfers or maybe lose the emergent characteristics of the biological role model. Or otherwise, whether this process might even include the emergence of new characteristics that did not occur in biological models or were of no ecological or evolutionary importance.

If emergent characteristics found in bio-inspired or biomimetic products already exist in the biological role models, the emergent characteristics have been conserved during the transfer process. In these cases, it is interesting to understand if these emergent characteristics are of known selective evolutionary advantage for these organisms or if they occur—so to say—inevitably caused by structural requirements originated during the evolution of other selectively important structures or functions (see section "Biomimetic Micro-laminates").

Other emerging characteristics of bio-inspired or biomimetic products belong to a completely different realm than the physical–mechanical ones and seem to have no functional or selective meaning in the biological models. They are only of interest if human emotions or values are involved in the assessment of these products and only arise when bio-inspired or biomimetic products are used in interaction with humans and society. These characteristics have to be discussed with reference to their emotional (aesthetics and symbolic character) and normative (contribution to sustainability) aspects (see sections "Biomimetic Façade Shading Systems" and "Bio-inspired Sustainability (Assessment)") (Speck et al. 2017).

Biomimetic Materials Systems

"Biomimetics combine the disciplines of biology and technology with the goal of solving technical problems through the abstraction, transfer, and application of knowledge gained from biological models".

The above sentence is the definition of biomimetics as given in 2012 in the VDI Guideline "Biomimetics: Conception and strategy—Differences between biomimetic and conventional methods/products" (VDI 6220 2012). Less technically, biomimetics can be considered as a transfer of scientific findings from biology to technical applications. It represents a creative transfer from biology to technology, i.e. a "new-invention" inspired by living nature and is never a pure technical blueprint of the biological role model. Biomimetic developments typically include several modification and abstraction levels.

Before introducing the two basic types of biomimetic development processes, it should be briefly defined how the often confused terms "bio-inspired", "biomorphic", "biomimetic" and "biomimetically optimised" are used in this article. Following the classification suggested by Speck et al. (2017) "bio-inspired developments" are technical solutions that trace back to the transfer of an inspiration originating from a biological idea provider (VDI 6226 2015; Whitesides 2015). If the morphology of biological models was the basis for an additional knowledge transfer based on the original inspiring idea, then one has a "biomorphic development". If functional principles were the basis for an additional knowledge transfer "biomimetic developments" originate, and the application of biomimetic algorithms results in a "biomimetically optimised product". It should also be kept in mind that combinations of various types of knowledge transfer are possible and that inspiration might be gained from more than one biological role model.

In biomimetic development processes, two basic procedures can be distinguished, which can be described as the "bottom-up approach" (biology push) and "top-down approach" (technology pull). There are many transitions between these two basic biomimetic approaches, which are used as needed in each individual case. Biomimetic product developments typically are carried out in highly interdisciplinary teams in which biologists and other natural scientists (physicists, chemists, etc.) work together with engineers, materials researchers, designers, architects, computer scientists and mathematicians (Speck and Speck 2008; Masselter et al. 2012; VDI 6226 2015).

In the "bottom-up approach", the starting point is a new result from basic biological research promising for biomimetic implementation. The first step in the process of, for example, developing a biomimetic materials system is the quantitative analysis of the physical, mechanical and/or chemical properties of interest and of the functional morphology of the biological system. In the next step, these results lead to a comprehensive understanding of the form–structure–function relationship of the biological role model. This is followed by abstraction, i.e. the deciphering of the found functional principles and the detachment from the biological model. The technical implementation usually takes place first on a laboratory scale and then in the next stage on a pilot plant scale, i.e. with methods and production techniques that are already established in industry in this way or in a similar manner. Within the framework of industrial development, biomimetic products are improved in terms of production processes and production costs and finally brought to market readiness with the cooperation of and often lead by industrial partners.

In the "top-down approach", biomimetic innovations are sought for already existing technical products that often are already successfully established on the market. The cooperation focuses on the improvement or further development of an existing product, i.e. the treatment of a concrete technical problem. For a promising biomimetic development, the technical challenge and its boundary conditions must first be precisely defined. In the next step, the biologist is searching for natural models that have developed solutions to challenges in the course of evolution that can be regarded as analogies to the technical question. From these the most promising role models are then selected, experimentally investigated and quantitatively characterised. Similar to the "bottom-up approach", the next step in the "top-down approach" is abstraction. After the successful transfer of knowledge in terms of functional principles, the technical implementation possibilities are checked and suitable production methods are determined. The first biomimetic prototypes are then manufactured and—if successful—their development is continued right up to industrial production. As in the "bottom-up approach", also in the "top-down approach" this is followed by industrial development and market introduction by the industrial partner.

The simulations typically applied in the biomimetic transfer process, the usually necessary scaling processes and the use of other than biological materials often lead to new insights concerning the form–structure–function relationship (Knippers et al. 2016). In the so-called process of "reverse biomimetics", these contribute to a more detailed examination and better understanding of the biological model based on the extended knowledge base. This allows a deeper comprehension of biological model organisms and structures to be achieved and provides an improved basis for further transfer into biomimetic products to be developed. Ideally, this creates a heuristic spiral of knowledge for all scientific disciplines involved (Speck and Speck 2008).

The following section discusses whether emergent characteristics can be found in biomimetic products and whether they are important for the success of these products. It is also discussed whether these characteristics are already present in the biological concept generators and were conserved in a biomimetic product during the transfer process or whether they occur in the biomimetic product "de novo" and are of interest only for its specific field of application in the human environment.

Emergence in Biomimetic Materials Systems

If considering emergent characteristics of biomimetic materials systems, various "types" of emergence can be discerned (Table 1). Firstly, emergence may affect the same kind of physical properties, e.g. when combining tension-stiff fibres with a

Table 1 Overview of various types of emergence in biomimetics

Types of emergence in biomimetics	Short description	Examples
First type of emergence	Combining individual elements with specific physical properties (e.g. stiffness in tension with stiffness in compression) causes emergence of an additional physical property of the same class (e.g. stiffness in bending)	Biomimetic composite materials (e.g. fibre-reinforced polymers), see section "Biomimetic Composite Materials"
Second type of emergence	Combining individual elements with specific physical properties (e.g. stiff but brittle with tough but elastic) causes emergence of an additional physical property of another class (e.g. optical property)	Biomimetic micro-laminates (e.g. materials systems inspired by nacre or insect cuticles), see section "Biomimetic Micro-laminates"
Third type of emergence	Societal values emerge and are attributed to the biomimetic product, which also can be attributed to the biological role models (e.g. aesthetic value, symbolic value)	Biomimetic façade shading systems (e.g. flectofin®, flectofold), see section "Biomimetic Façade Shading Systems"
Fourth type of emergence	Societal values emerge and are attributed to the biomimetic product, which cannot be attributed to biological role models (e.g. moral values)	Bio-inspired sustainability and bio-inspired sustainability assessment, see section "Bio-inspired Sustainability (Assessment)"

compression-stiff matrix bending and torsion stability emerge which represent additional mechanical properties (see section "Biomimetic Composite Materials"). Secondly, the emergent characteristic may belong to another kind of physical property than the one for which the bio-inspired materials system was originally developed. This is for example the case in biomimetic micro-laminates inspired by the ultrastructure of nacre as found in many mollusc shells. Here the combination of high toughness and stiffness is brought about by "gluing together" sub-micrometre thick stiff but brittle mineral layers with even thinner layers of a tough but flexible organic matrix. In addition to the superior mechanical properties of the micro-laminate also an interesting optical property occurs, the famous nacreous lustre. The latter represents an additional non-mechanical emergent property caused by interference (see section "Biomimetic Micro-laminates"). Thirdly, also qualities may emerge or be conserved belonging to an entirely different realm than the physical–mechanical ones for which the biomimetic materials system was originally developed and for which originally inspiration from biological role models was taken. An example is the "beauty" of many biological role models, which are used as concept generators in biomimetics. Beauty is an abstract concept, which is in many aspects involved in human life and always has a valuing connotation. Therefore, it might be better to use the term "aesthetical" if a biomimetic structure is described that has borrowed its "beauty" from the biological role model (Speck 2015). Examples include the aesthetics of motion principles, which has been conserved during the development of biomimetic façade shading systems as the flectofin® (inspired by the perch of the bird-of-paradise flower) or the flectofold (inspired by the snap-trapping leaves of the waterwheel plant) (Poppinga et al. 2016) (see section "Biomimetic Façade Shading Systems"). In these examples, the motion patterns of the plant organs which humans find aesthetical have been conserved during the biomimetic transfer and reinvention process. Aesthetics represents a different type of emergent values that is only valid in the human perspective (Budd 1998; Sepp and Embree 2010).

Fourthly, especially in the context of biomimetic materials systems, sustainability must be discussed, because nowadays this is a very important feature and one of the most discussed topics in politics and society. Based on a misinterpretation of the so-called biomimetic promise, it is often thought and said that biomimetic products, because of the inspiratory flow from living nature, are per se sustainable. Consequently, this would mean that sustainability is an emergent characteristic of biomimetic products brought about alone by the inspiration from living nature on which the development of such products is based and could be considered as a fourth type of emergence possible in biomimetics products. As discussed in detail in section "Bio-inspired Sustainability (Assessment)", this statement cannot be upheld in a generalised manner, since living nature cannot be regarded as sustainable in the sense of the mission statement created by man. On the contrary, the contribution to sustainability has to be tested individually for each biomimetic product, best compared to a suitable conventional product. Therefore, sustainability which represents a purely man-made concept cannot simply be considered as an emergent characteristic of biomimetic materials systems in general. However, several aspects important during evolution, as e.g. material and energy efficiency of biological materials systems, can lead to a higher contribution to sustainability of biomimetic products (von Gleich 2010; Antony et al. 2014; Horn et al. 2016; Speck et al. 2017).

Selected Examples for Emergence in Biomimetics

Biomimetic Composite Materials

Fibre composite materials are widespread in biological and technical materials systems (Fratzl and Weinkamer 2007; Wegst et al. 2015). They owe their outstanding mechanical properties to a combination of (purely) tensile-stiff fibres embedded in a (purely) compression-stiff matrix (Fig. 1). The mechanical behaviour of fibre-reinforced composite materials is superior to their components not only in tension and compression but they are additionally stiff in bending and torsion, which represent novel emergent mechanical properties existing not or only to a very limited extent in the individual components (Milwich 2006; Masselter and Speck 2011).

Fig. 1 Arrangement of highly lignified vascular bundles in the branching region of a dragon tree (*Dracaena marginata*) serving as role model for branched biomimetic fibre-reinforced composite materials with emergent material properties (left). Branched bio-inspired pillar with a braided outer jacket of fibre-reinforced polymers shown in an exhibition of the State Museum of Natural History Stuttgart (SMNS), Germany (right)

These novel mechanical properties emerge only when combining the two components and are often accompanied by a benign fracture behaviour, which can be seen as an additional emergent mechanical property and represents examples for the first type of emergence found in fibre-reinforced composite materials.

Many if not most structural materials in biology are fibre-reinforced composites as, e.g. wood, bone or insect cuticle (Fratzl and Weinkamer 2007; Wegst et al. 2015). Fibre-reinforced composites can be found in biology on different hierarchical length scales spanning several orders of magnitudes. They exist from the molecular/ultra-structural level (e.g. cellulose micro-fibril/lignin composites in plant cell walls) over the tissues/organ level (e.g. vascular bundles in a sclereid matrix in seed coats and fruit walls) to the organismic level of the entire plant (e.g. highly lignified vascular bundles in palm stems) (Speck and Burgert 2011; Gibson 2012). In living beings, the fibres often form complex "interwoven" systems embedded in a matrix, which may look like a two- or three-dimensional woven fabric due to fusing of single fibres or fibre bundles. However, in living nature never real woven, braided or knitted fibrous structures occur in which the fibres are interconnected by processes equivalent to those in technical waving, braiding or knitting. All natural fibre arrangements have to be considered—from a technical point of view—as non-woven independent of the complexity of their areal or spatial arrangement.

Here, it has to be pointed out that not all technical fibre composite materials are biomimetic in the direct sense and developed in a way as described in section "Biomimetic Materials Systems". However, iron/steel-reinforced concrete the first widely commercially used composite material taking advantage of the emergent superior mechanical properties occurring by embedding tension-stiff (rod-shaped) materials in a compression-stiff matrix was bio-inspired. This macroscopic composite material was invented and commercialised by the French gardener Joseph Monier in the second half of the nineteenth century (patent 1867). He (most probably) took the inspiration from the wood segments in the flattened shots of *Opuntia* cacti, which are spatially arranged in a kind of 3D-net and embedded in parenchymatous tissue. Since Monier never realised the functional importance of the spatial arrangements of the iron tension bracings in full detail, his invention should not be called biomimetic (Frey et al. 2011; Speck et al. 2017). Whether modern fibre-reinforced composite materials, in which various types of fibres or fibre bundles typically with a diameter of a few to several tens of micrometres are embedded in a polymer matrix, are also bio-inspired, is an ongoing debate.

In addition to mechanical emergence (first type of emergence), biological fibre-reinforced composites may also show other emergent physical properties that are caused by the pronounced anatomical inhomogeneity and mechanical anisotropy of their composition. Examples for this second type of emergence include structural colours in insect cuticles (see section "Biomimetic Micro-laminates") and hygroscopically actuated passive motion as found for example in pine cone scales or grass awns (Burgert and Fratzl 2009; Poppinga et al. 2013, 2017). In the latter two examples, a hygroscopically sensitive tissue matrix shows an undirected 3D-swelling or shrinking, which is transferred into complex directional motions by the arrangement and orientation of fibres (one- or two-phase bending motion in cone scales, helicoidal motions in wheat awns). Also these emergent characteristics have been used in biomimetics. Examples are hygroscopically actuated autonomous flaps inspired by scales of pine cone that can be used for sun shading and/or air conditioning in roofs and building skins (Menges and Reichert 2015; Reichert et al. 2015; Rüggeberg and Burgert 2015; Poppinga et al. 2018).

Biomimetic Micro-laminates

Micro-laminates in which micro- or sub-micrometre thick layers with different mechanical properties are stacked together exist in many biological materials systems including seed coats, fruit shells, insect and crustacean cuticles and mollusc shells. The layered arrangement of alternating stiff but (relatively) brittle material layers with tough but (relatively) flexible material layers causes materials systems with excellent mechanical properties. The combination of high stiffness with high toughness can only be reached by the micro-laminate arrangement and is way superior to the properties of each of the individual material components. Additionally, these micro-laminates are highly fracture-insensitive which can be considered as a first type emerging mechanical property that does not exist in the individual materials but only occurs in the micro-laminate arrangement.

It is not surprising that these biological micro-laminates have attracted interest from biomimetics. Some of the first biomimetic micro-laminates that are still subject of pronounced ongoing research were inspired by the ultra-structure of nacre that is found as an inner layer in the shells of many mollusc groups (Fig. 2). Nacre alternatingly comprises stiff but brittle mineral layers (typically hexagonal aragonite platelets arranged like "brickwork" in the biological role models) with a thickness of several hundred nanometres (typically ca. 500 nm) and very thin layers of a tough but flexible organic matrix (thickness 30–50 nm). This combination results in an extraordinary high toughness and stiffness as well as fracture-insensitivity of the biological micro-laminate (Jackson et al. 1988; Nudelmann et al. 2007). Due to the dimensions of the layers of the micro-laminate, nacre shows an interesting optical property, namely the famous iridescent nacreous lustre which represents a structural colour. Nacreous lustre can be considered as an emergent property of type two that is caused by interference due to the thickness of the stacked mineral layers that are in the size range of visible light (400–725 nm). The inner shell surface is typically covered from the soft body of the molluscs and the optical quality of nacre is of unknown selective advantage. Therefore, nacreous lustre may represent an emergent property,

Fig. 2 Nacre at the inner surface of the abalone shell. The iridescent nacreous lustre is visible representing an emergent property caused by interference due to the stacked mineral layers

inevitably caused by structural requirements that originated during the evolution of its selectively important mechanical structures or functions.

Nacre has inspired biomimetics research since many years but it was not before the last 20 years that convincing results were realised (e.g. Lu and Mano 2009; Wilkerson et al. 2016). In some biomimetic micro-laminates inspired by nacre, the nacreous lustre representing a second type emergent property can be entirely or partly lost, which is of minor interest as for the potential fields of application of these micro-laminates only the mechanical properties are of interest. In other cases, the nacreous lustre shall be conserved and is of specific interest for applications (e.g. as mechanically stable but also optically attractive surfaces and cases) and may even widen the potential fields of application.

Other biological micro-laminates with excellent mechanical properties and emergent optical properties originating from their hierarchical structures are insect cuticles that also have attracted increasing interest from the field of biomimetics. Due to their fibrous nature, cuticles can best be described as micro-laminated fibre-reinforced composites (Vincent and Wegst 2004). In the cuticle, the orientation of the chitin micro-fibrils gradually changes in successive layers (Bodigand structure) and this stacked helicoidal arrangement not only produces excellent mechanical properties (see also section "Biomimetic Composite Materials"), but often also optical properties emerge such as different structural colours including in some species even the reflection of circularly polarised light (Parker 2000; Kinoshita 2008; Kinoshita and Yoshioka 2005). In contrast to molluscs, in insects the emerging optical properties are as various types of signals of high biological significance and evolutionary importance. Biomimetics has started increasingly to use insect cuticles as role models in materials research and architecture (e.g. Parascho et al. 2014) and also tries to take advantage of the emergent optical properties (Kinoshita 2008; Vincent 2009; Dumanli and Savin 2014).

Biomimetic Façade Shading Systems

The third type of emergence in biomimetics deals with emotional aspects, which are related to the aesthetics ("beauty") and the symbolic character of biomimetic products and of their biological role models. These aspects have been mainly discussed in relation to architecture and design [e.g. in the "Growing as Building—GraB" project in Austria (Imhof and Gruber 2015)]. The idea of using bio-inspiration for a novel eco-friendly, aesthetical and human architecture has attracted increasing interest over the last decades as well from architects and civil engineers as from materials scientists and from researchers working in the field of biomimetics (Gruber 2011; Knippers and Speck 2012; Nachtigall and Pohl 2013; Armstrong 2015). The qualities aesthetics ("beauty") and symbolic character belong to an entirely different realm than the physical–mechanical properties discussed in the previous two chapters. Aesthetics and symbolism are attributed by humans to many biological structures and functions and are abstract concepts. These characteristics represent the third type of emergence and are very difficult if not impossible to grasp and to quantify, especially as they are deeply involved in human life. Therefore, we will focus on two examples: selected movement patterns of plant organs, which are aesthetically assessed by humans and in which the aesthetic of functioning was successfully conserved during the transfer process in the biomimetic façade shading systems flectofold® and flectofin.

Characteristically, in biomimetics research the central objective is the development and production of novel biomimetic materials systems with clearly defined specific (multi-)functionalities (Speck and Speck 2008; Bar-Cohen 2011). Also in biomimetic architecture the transfer of the aesthetic and symbolic values, like the elegance of functioning, is even though highly desired but is often not considered as basic prerequisite in the typical "function-centred approach" of biomimetics. In biomimetic approaches, the predictable transfer of functional principles from the biological concept generator to the biomimetic product is central to obtain a targeted function, and the loss of (parts) of the aesthetic value existing in the biological role model is inevitable or has at least to be accepted, i.e. the emotional aspects often are considered to be of minor importance. The first laboratory demonstrators developed during the biomimetic transfer process sometimes differ entirely as to their appearance from their biological role models. Even though having transferred completely and perfectly the functional principles of the biological concept generators and thus attained the desired function they may have lost their aesthetic value partly or entirely. However, sometimes the developed biomimetic products not only possess the physical function transferred from the biological role models but also show a functional elegance similar to the highly aesthetic biological role models. In architecture, these (still rare) cases can be considered as the "royal road" in biomimetic product development and represent ideal solutions that combine bio-inspired function with the emergent natural beauty of functioning (Speck 2015).

The biomimetic façade shading systems flectofin® and flectofold are hingeless, robust and resilient structures inspired by elastic deformation processes taking place in plants. Both were technically realised on the basis of glass-fibre-reinforced polymers and fine-tuned by bio-inspired local adjustment of the fibre orientation (see also section "Biomimetic Composite Materials"). The façade shading system flectofin® is inspired by the landing perch of the bird-of-paradise flower (*Strelitzia reginae*), which bends downwards and opens due to torsional buckling caused by the weight of a pollinating bird. This opening process in which no localised hinges are involved is based on elastic deformation. The flectofin® was produced as a 2 m long and 0.25 m wide demonstrator as well with a single fin as with double fins acting as compliant flapping structures which are actuated by bending the backbone with external hydraulic pistons (Lienhard et al. 2011; Schleicher et al. 2015; Speck et al. 2015). A strong example proving how this type of bio-inspired research can influence architecture is the One Ocean Thematic Pavilion for EXPO 2012 in Yeosu, South Korea (realised by soma architecture and Knippers-Helbig Advanced Engineering). The slightly curved fins are only 9 mm thick and almost 14 m long (Knippers et al. 2012). A local compression force at top and bottom of single fins leads to a controlled torsional buckling and reversible elastic deformation of the fins causing a very aes-

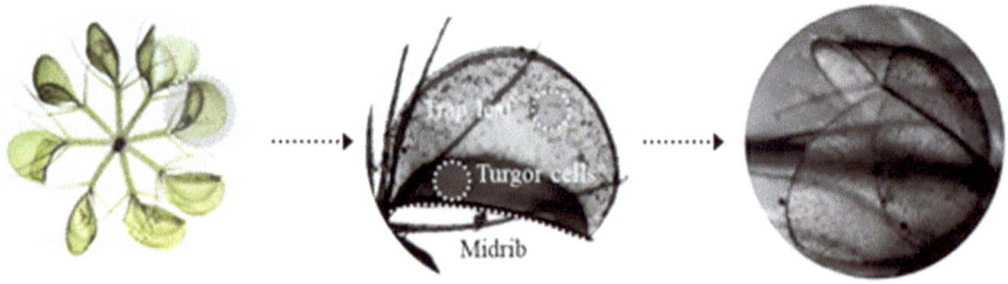

Fig. 3 Snap traps of the waterwheel plant (*Aldrovanda vesiculosa*), concept generator for the mode of function and aesthetic motion pattern of the biomimetic façade shading system flectofold

thetic motion pattern, which represent an example for the third type of emergence in biomimetics.

The biomimetic façade shading system flectofold is inspired by the underwater snap-trap of the waterwheel plant (*Aldrovanda vesiculosa*), which snaps very quickly within 20–100 ms after being triggered by a prey (typically small crustaceans). This fast motion is brought about via motion amplification by a small bending of the midrib of the trap, which causes a fast and large closing motion of the leaf halves forming the two cup-like trap walls (Fig. 3). The motion is actuated by a combination of hydraulic turgor change in the motor cells along the midrib and the release of mechanical pre-stress stored in the trap (Westermeier et al. 2018). The principle of motion amplification was transferred into the biomimetic façade shading system flectofold that consists of a lens-shaped backbone and two side fins attached to the backbone by curved folds (hinge zone) (Fig. 4). Small bending of the backbone actuated by a pneumatic cushion or a cellular pneumatic actuator causes closing or opening of the side fins with a very aesthetic motion pattern, representing another example for the third type of emergence in biomimetics (Körner et al. 2018; Born et al. 2017).

It has to be discussed if aesthetics, or even further reaching, the evolutionary originated "beauty of biological design", could be defined as an inherent emergent (design) characteristic of biological role models. If this is the case, it would be interesting to see whether this characteristic can be purposefully transferred in a biomimetic process to novel technical materials, structures or products, in the same manner as other characteristics, e.g. physical properties or various functions can be transferred (Speck 2015).

Bio-inspired Sustainability (Assessment)

Sustainability represents a mission statement that is desired from the human perspective and especially over the last decades became one of the most and often controversially discussed topics in politics and society. Due to a—at least partly—misin-

Fig. 4 Demonstrator of the biomimetic façade shading system flectofold during an exhibition in the Stuttgart State Museum of Natural History (SMNS) in different states of opening

terpretation of the so-called biomimetic promise it is often expected that biomimetic developments and products are sustainable "per se" (von Gleich et al. 2010, German edition 2007). The misunderstanding seems to be that living nature is seen as the only real sustainable system to which everything is referenced. A closer look in the following remarks will show, however, that this simple conclusion is not admissible. Because if one accepts this interpretation, sustainability would ultimately be an emergent characteristic of all biomimetic products, which is caused solely by inspiration from the living nature on which the development of such products is based.

In the context of the ongoing discussion on sustainability as emergent moral value, it is important to consider that sustainability is a man-made concept with the defined goal of preserving the status quo for the next generations. In contrast, biological evolution is a "blind" process characterised by the dynamics of evolutionary adaptations in a constantly changing environment (Horn et al. 2018). If the biomimetic product would nevertheless contribute to sustainability, this would be an emergent morale value that appeared during the biomimetic transfer and development process. In order to understand the relation between biomimetics and sustainability, it is worthwhile to shortly call back to mind the definition of both terms. As discussed in section "Biomimetic Materials Systems", biomimetics aims for ".... solving technical problems through ... application of knowledge gained from biological models" (VDI 6220 2012). The term sustainability was used first by Hannß Carl von Carlowitz in 1773 in his book on forestry "Sylvicultura oeconomica oder haußwirthliche Nachricht und Naturmäßige Anweisung zur wilden Baum-Zucht". Since then, the term "sustainability" or "sustainable development" has been taken up again and again. The Report of the Brundtland Commission of 1987 contains the only internationally recognised definition of sustainable development: "Sustainable development is development that meets the needs of the present without compromising the ability of future generations to meet their own needs", which is still used today. Sustainable development in general is a challenge to be addressed in societal consensus and learning from nature is linked with the hope of learning from biological solutions that seem to be evolutionarily "optimised", ecologically adapted and low risk. von Gleich et al. (2010/German edition 2007) coined the term "biomimetic promise", which implies that, precisely because of their inspiratory flow from nature to technical products, biomimetic solutions seem to have the specific potential to contribute to sustainable technology development. Considering this definition of sustainability literally, it is obvious that living nature cannot be sustainable, since the sustainability concept requires planned future-oriented goals with an anthropocentric perspective (Antony et al. 2013; Horn et al. 2016; Speck et al. 2017). This means that if sustainability would exist in living nature, evolution would have to be "aiming for a predefined target (sustainability)" solely with a focus on mankind, i.e. evolution would have to be teleological and anthropocentric, which is not the case.

Bearing this in mind the question arises if anything can be learnt at all from living nature in general and from evolution in particular for improving sustainability in the human realm. This seems to be indeed the case if one reduces the demands one has on living nature as a model for sustainability from that of a general role model solving all questions arising, to an idea provider from which it can be learnt for answering specific questions that are to be addressed in the context of sustainable development. These questions can concern specific properties and functions of biomimetic products as well as how the sustainability of biomimetic products is assessed and accompanied during biomimetic development. One aspect, very important for sustainable development, for which inspiration can and already successfully has been taken from living nature is material and energy efficiency. It is well known that efficiency as to material use and energy consumption is of high selective significance and represents an advantage not to be underestimated in evolutionary processes as pointed out by Vincent (2002). In his article with the provocative title "Survival of the cheapest", he showed examples of how in living beings structures with fantastic mechanical and other (partly emergent) characteristics can be formed by hierarchical structuring (see section "Introduction") and genetically controlled self-assembly with economical use of material and energy. This becomes possible, first, as the (few) structurally important macromolecules, which consist predominantly of light atoms (H, C, O, N, P), are processed at ambient temperature and pressure and, second, as their building blocks are usually taken in from the immediate environment of the organisms (Fratzl and Weinkammer 2007; Masselter et al. 2012; Knippers and Speck 2012; Knippers et al. 2016). In addition, these macromolecules and by this the biological materials systems built are biodegradable under ambient conditions and the building blocks can be fed back into the material cycles of nature. If material and energy efficiency are considered as emergent characteristics of the internal construction and formation process of biomaterials, these qualities can be transferred to biomimetic products. To a certain extent, this transfer has already been successfully established in some biomimetic materials systems and products as can be seen as to material efficiency, e.g. in (ultra-)lightweight fibre-reinforced composites used in cars, windmills and especially in air- and space-crafts. If biodegradability at ambient conditions is also rated as emergent in biomaterials, this represents an additional emergent characteristic, which can be used as concept generator for realising improved degradability and recyclability also in bio-inspired materials systems built of technical components and artificial building blocks.

Recent studies test whether the transfer of the above-mentioned emergent characteristics, which are important for sustainable development, can be accompanied, assessed and probably improved by applying novel bio-inspired sustainability assessment methods (Horn et al. 2016, 2018; Speck et al. 2017). An example is the newly developed Bio-inspired Sustainability Assessment (BiSA), which allows for quantitatively assessing the self-imposed claim of sustainability in the building sector by comparing newly developed—including bio-inspired—products and conventional products.

Inspired by the common denominator between biological and technical systems, namely the relationship of function fulfilment and respective resource demand, a suitable basis for assessing sustainability had been found. BiSA focuses on the assessment of environmental, social and economic functions and the corresponding burdens within a consistent, quantitative framework based on life cycle thinking (Horn et al. 2018). In this context, bio-inspired sustainability can be defined as the continuous ability of society to provide specific functions through utilising resources and ensuring their future availability (Fig. 5). This proves again that sustainability is neither an emergent characteristic of living nature nor emerges per se during the biomimetic transfer and development but represents a consciously created societal value, which should help to ensure the survival of future generations of mankind.

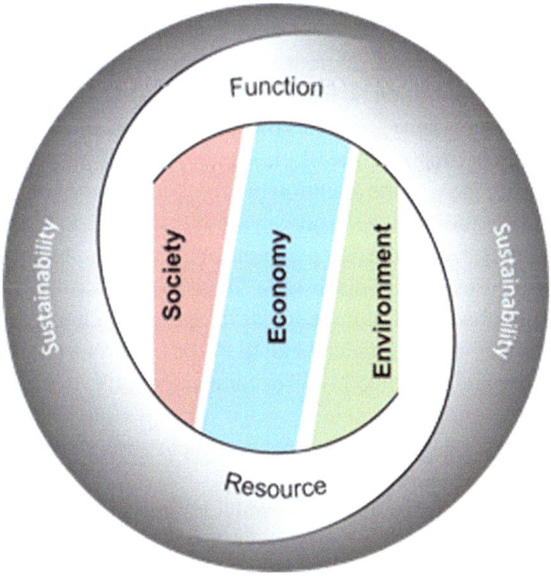

Fig. 5 Bio-inspired sustainability defined as the trade-off between the fulfilment of specific functions and the respective resource demand (from Horn et al. 2018)

Discussion and Perspectives

Emergent characteristics are widespread in nature and technology, but it is often not realised that an important or—in technology—desired physical property or societal value is caused by emergence, i.e. that it is an inevitable result of the interaction of the individual elements of the system. It may be that our "blindness" to emergence is also influenced by our predominantly linear way of thinking and planning: if we add two components, we expect to get the sum of both and not something over and above and for free. In the course of biomimetic transfer and development emerging characteristics that already exist in the biological role model may be conserved (e.g. in fibre composite materials and some micro-laminates), independent if they are or not of ecological or evolutionary significance in the biological concept generator. Conservation of emerging characteristics is typically found if physical properties are transferred from biological role models to biomimetic products. These emergent characteristics may concern the same type of physical property, e.g. combining two components with different mechanical properties results in an improvement of these properties and in emerging novel mechanical properties (section "Biomimetic Composite Materials"). Or the emerging properties concern another type of physical property, e.g. combining two components with different mechanical properties results in an improvement of these mechanical properties, emerging new mechanical properties and in emerging novel optical properties (section "Biomimetic Micro-laminates").

Other emerging effects in context with biomimetic developments arise as biomimetic products are used in the interaction with humans and society and thus human emotions and/or societal values are involved in the assessment of these products. The impact of these emergent values is complex and can be only understood if discussed with reference to their emotional (aesthetics and symbolic character, section "Biomimetic Façade Shading Systems") and normative aspects (contribution to sustainability, section "Bio-inspired Sustainability (Assessment)"). Independent of the complex nature of these emerging values, they represent the ones which contribute most to the attention biomimetics gains today in the fields of architecture and design (aesthetics, beauty, symbolic significance) and in the realm of politics and society, where the potential contribution of biomimetics to sustainable development is controversially discussed ("biomimetic promise"). In anticipation of an increasing influence of biomimetic developments in many areas of technology and society, the analysis and understanding of the importance of different types of emergence in biomimetics (Table 1) will become imperative. It is also interesting in this context that no "biomimetic design language" emerges that makes it possible to identify products that have a biological model by simple observation and thus easily distinguish them from conventional products.

Emergence is also important for multifunctionality, another striking characteristic of many biological materials system that is also of high interest for biomimetics. As well from an evolutionary as—after biomimetic transfer—from a technical point of view it is interesting to research if these (multi-)functions are conflicting or not.

However, this and the question of whether or how these conflicts are solved during evolution (Pareto optimisation) and whether emergence is relevant for solving these "conflicts" go far beyond the scope of this book chapter.

Acknowledgements We thank our colleagues from the Plant Biomechanics Group, from the Excellence Cluster *liv*MatS (EXC 2193) and from the CRC 141 for many helpful discussions. We are grateful to the German Research Foundation for the funding of our biomimetic projects in the framework of the CRC-Transregio 141 "Biological Design and Integrative Structures—Analysis, Simulation and Implementation in Architecture" and the Ministry of Science, Research and the Arts of Baden-Württemberg for additional support within the framework of Sustainability Center Freiburg.

References

Antony F, Grießhammer R, Speck T, Speck O (2013) Natur – (k)ein Vorbild für nachhaltige Entwicklung. In: Kesel AB, Zehren D (eds) Bionik: Patente aus der Natur. Tagungsbeiträge zum 6. Bionik-Kongress in Bremen, Bionik-Innovations-Centrum (B-I-C), Bremen, Germany, pp 164–170

Antony F, Grießhammer R, Speck T, Speck O (2014) Sustainability assessment of a lightweight biomimetic ceiling structure. Bioinspir Biomim 9:016013. https://doi.org/10.1088/1748-3182/9/1/016013

Armstrong R (2015) Vibrant architecture: matter as a codesigner of living structures. De Gruyter Open LTD, Warsaw, Berlin

Bar-Cohen Y (2011) Biomimetics: nature-based innovations. Biomimetic series. CRC Press, 778 pp

Born L, Körner A, Schieber G, Westermeier AS, Poppinga S, Sachse R, Bergmann P, Betz O, Bischoff M, Speck T, Knippers J, Milwich M, Gresser GT (2017) Fiber-reinforced plastics with locally adapted stiffness for bio-inspired hingeless, deployable architectural systems. Key Eng Mater 742:689–696 (Trans. Tech. Publications, Switzerland)

Brundtland Commission (1987) Our common future. Oxford University Press, Oxford

Budd M (1998) Aesthetics. In: Craig E (ed) Routledge encyclopedia of philosophy. London (https://www.rep.routledge.com/) (Accessed 28 May 2018)

Burgert I, Fratzl P (2009) Actuation systems in plants as prototypes for bioinspired devices. Philos Trans R Soc A 367:1541–1557. https://doi.org/10.1098/rsta.2009.0003

Dumanli AG, Savin T (2014) Recent advances in the biomimicry of structural colours. Chem Soc Rev 45:6698–6724. https://doi.org/10.1039/C6CS00129G

Fratzl P, Weinkamer R (2007) Nature's hierarchical materials. Prog Mater Sci 52:1263–1334. https://doi.org/10.1016/j.pmatsci.2007.06.001

Frey E, Masselter T, Speck T (2011) Was ist bionisch? Eine Analyse des Ideenflusses von der Biologie in die Technik an ausgewählten Beispielen „bionischer" Entwicklungen. Naturwiss Rundschau 753(64/3):117–126

Gibson LJ (2012) The hierarchical structure and mechanics of plant materials. J R Soc Interface 9(76):2749–2766. https://doi.org/10.1098/rsif.2012.0341

Gruber P (2011) Biomimetics in architecture: architecture of life and buildings. Springer, Wien

Horn R, Gantner J, Widmer L, Sedlbauer KP, Speck O (2016) Bio-inspired sustainability assessment—a conceptual framework. In: Knippers J, Nickel KG, Speck T (eds) Biomimetic research for architecture and building construction: biological design and integrative structures. Biologically-inspired systems, vol 9. Springer International Publishing, Switzerland, pp 361–377

Horn R, Dahy H, Gantner J, Speck O, Leistner P (2018) Bio-inspired sustainability assessment for building product development—concept and case study. Sustainability 10(1):130–154. https://doi.org/10.3390/su10010130

Imhof B, Gruber P (eds) (2015) Built to grow—blending architecture and biology. Birkhäuser, Basel

Jackson AP, Vincent JFV, Turner RM (1988) The mechanical design of nacre. Proc Roy Soc Lond B Biol 234:415–440. https://doi.org/10.1098/rspb.1988.0056
Kinoshita S (2008) Structural colors in the realm of nature. World Scientific Publishing, Singapore
Kinoshita S, Yoshioka S (eds) (2005) Structural colors in biological systems—principles and applications. Osaka University Press, Osaka
Knippers J, Speck T (2012) Design and construction principles in nature and architecture. Bioinspir Biomim 7:015002. https://doi.org/10.1088/1748-3182/7/1/015002
Knippers J, Scheible F, Jungjohann H, Oppe M (2012) Kinetic media façade consisting of GFRP louvers. In: Proceedings of the 6th international conference on FRP composites in civil engineering (CICE 2012), 8 pp
Knippers J, Nickel KG, Speck T (eds) (2016) Biomimetic research for architecture and building construction: biological design and integrative structures. Biologically-inspired systems, vol 9. Springer International Publishing, Switzerland, 408 pp
Körner A, Born L, Mader A, Sachse R, Saffarian S, Westermeier AS, Poppinga S, Bischoff M, Gresser GT, Milwich M, Speck T, Knippers J (2018) Flectofold—a biomimetic compliant shading device for complex free form facades. Smart Mater Struct 27(1):017001. https://doi.org/10.1088/1361-665X/aa9c2f
Lienhard J, Schleicher S, Poppinga S, Masselter T, Milwich M, Speck T, Knippers J (2011) Flectofin: a hinge-less flapping mechanism inspired by nature. Bioinspir Biomim 6:045001. https://doi.org/10.1088/1748-3182/6/4/045001
Luz GM, Mano JF (2009) Biomimetic design of materials and biomaterials inspired by the structure of nacre. Philos Trans A Math Phys Eng Sci 367:1587–1605. https://doi.org/10.1098/rsta.2009.0007
Masselter T, Speck T (2011) Biomimetic fiber-reinforced compound materials. In: George A (ed) Advances in biomimetics. Intech, Rijeka, Croatia, pp 195–210
Masselter T, Barthlott W, Bauer G, Bertling J, Cichy F, Ditsche-Kuru P, Gallenmüller F, Gude M, Haushahn T, Hermann M, Immink H, Knippers J, Lienhard J, Luchsinger R, Lunz K, Mattheck C, Milwich M, Mölders N, Neinhuis C, Nellesen A, Poppinga S, Rechberger M, Schleicher S, Schmitt C, Schwager H, Seidel R, Speck O, Stegmaier T, Tesari I, Thielen M, Speck T (2012) Biomimetic products. In: Bar-Cohen Y (ed) Biomimetics: nature-based innovation. CRC Press/Taylor & Francis Group, Boca Raton, London, New York, pp 377–429
Menges A, Reichert S (2015) Performative wood: physically programming the responsive architecture of the hygroscope and hygroskin projects. AD Architect Des 85(5):66–73. https://doi.org/10.1002/ad.1956
Milwich M, Speck T, Speck O, Stegmeier T, Planck H (2006) Biomimetics and technical textiles: solving engineering problems with the help of nature's wisdom. Am J Bot 93:1455–1465
Nachtigall W, Pohl G (2013) Bau-Bionik: Natur – Analogien – Technik, 2nd edn. Springer Vieweg, Berlin, Heidelberg
Nudelman F, Chen HH, Goldberg HA, Weiner S, Addadi L (2007) Spiers memorial lecture—lessons from biomineralization: comparing the growth strategies of mollusc shell prismatic and nacreous layers in *Atrina rigida*. Fraday Discuss 136:9–25. https://doi.org/10.1039/B704418F
Parascho S, Knippers J, Dörstelmann M, Prado M, Menges A (2014) Modular fibrous morphologies: computational design, simulation and fabrication of differentiated fibre composite building components. In: Block P, Knippers J, Mitra NJ, Wang W (eds) Advances in architectural geometry. Springer, Vienna, pp 29–45
Parker AR (2000) 515 million years of structural colour. J Opt A Pure Appl Opt 2(6):R15–R28. https://doi.org/10.1088/1464-4258/2/6/201
Poppinga S, Masselter T, Speck T (2013) Faster than their prey: new insights into the rapid movements of active carnivorous plants traps. BioEssays 35:649–657. https://doi.org/10.1002/bies.201200175
Poppinga S, Körner A, Sachse R, Born L, Westermeier A, Hesse L, Knippers J, Bischoff M, Gresser GT, Speck T (2016) Compliant mechanisms in plants and architecture. In: Knippers J, Nickel KG, Speck T (eds) Biomimetic research for architecture and building construction: biological

design and integrative structures. Biologically-inspired systems, vol 9. Springer International Publishing, Switzerland, pp 169–193
Poppinga S, Nestle N, Šandor A, Reible B, Masselter T, Bruchmann B, Speck T (2017) Hygroscopic motions of fossil conifer cones. Sci Rep 7:40302. https://doi.org/10.1038/srep40302
Poppinga S, Zollfrank C, Prucker O, Rühe J, Menges A, Cheng T, Speck T (2018) Towards a new generation of smart biomimetic actuators for architecture. Adv Mater 30(19):1703653. https://doi.org/10.1002/adma.201703653
Reichert S, Menges A, Correa D (2015) Meteorosensitive architecture: biomimetic building skins based on materially embedded and hygroscopically enabled responsiveness. Comput Aid Des 60:50–69. https://doi.org/10.1016/j.cad.2014.02.010
Rüggeberg M, Burgert I (2015) Bio-inspired wooden actuators for large scale applications. PLoS ONE 10(4):e0120718. https://doi.org/10.1371/journal.pone.0120718
Schleicher S, Lienhard J, Poppinga S, Speck T, Knippers J (2015) A methodology for transferring principles in plant movements to elastic systems in architecture. Comput Aid Des 60:105–117. https://doi.org/10.1016/j.cad.2014.01.005
Sepp HR, Embree L (2010) Handbook of phenomenological aesthetics. Springer, Dordrecht, New York
Speck T (2015) Approaches to bio-inspiration in novel architecture. In: Imhof B, Gruber P (eds) Built to grow—blending architecture and biology. Birkhäuser, Basel, pp 145–149
Speck T, Burgert I (2011) Plant stems: functional design and mechanics. Annu Rev Mater Res 41:169–193. https://doi.org/10.1146/annurev-matsci-062910-100425
Speck T, Speck O (2008) Process sequences in biomimetic research. In: Brebbia CA (ed) Design and nature IV. WIT Press, Southampton, pp 3–11
Speck T, Knippers J, Speck O (2015) Self-X-materials and -structures in nature and technology: bio-inspiration as driving force for technical innovation. AD Architect Des 85(5):34–39. https://doi.org/10.1002/ad.1951
Speck O, Speck D, Horn R, Gantner J, Sedlbauer KP (2017) Biomimetic bio-inspired biomorph sustainable? An attempt to classify and clarify biology-derived technical developments. Bioinspir Biomim 12(1):011004. https://doi.org/10.1088/1748-3190/12/1/011004
VDI 6220 (2012) Bionik: Konzeption und Strategie—Abgrenzung zwischen bionischen und konventionellen Verfahren/Produkten;Biomimetics: conception and strategy—differences between biomimetics and conventional methods/products. VDI, Beuth, Berlin
VDI 6226 (2015) Bionik: Architektur, Ingenieurbau, Industriedesign—Grundlagen; Biomimetics: architecture, civil engineering, industrial design—basic principles. Beuth, Berlin
Vincent JF (2002) Survival of the cheapest. Mater Today 5(12):28–41
Vincent JFV (2009) Biomimetics—a review. Proc Inst Mech Eng H - J Eng Med 223(8):919–939.https://doi.org/10.1243/09544119JEIM561
Vincent JFV, Wegst UGK (2004) Design and mechanical properties of insect cuticle. Arthropod Struct Dev 33:187–199. https://doi.org/10.1016/j.asd.2004.05.006
von Carlowitz HC (1773) Sylvicultura oeconomica oder haußwirthliche Nachricht und Naturmäßige Anweisung zur wilden Baum-Zucht. Johan Friedrich Braun, Leipzig
von Gleich A, Pade C, Petschow U, Pissarskoie E (2010) Potentials and trends in biomimetics. Springer, Heidelberg
Wegst UGK, Bai H, Saiz E, Tomsia AP, Ritchie R (2015) Bioinspired structural materials. Nat Mater 14:23–36. https://doi.org/10.1038/nmat4089
Westermeier A, Sachse R, Poppinga S, Vögele P, Adamec L, Speck T, Bischoff M (2018) How the carnivorous waterwheel plant (*Aldrovanda vesiculosa*) snaps. Proc Roy Soc Lond B Bio l285:20180012. https://doi.org/10.1098/rspb.2018.0012
Whitesides GM (2015) Bioinspiration: something for everyone. Interface Focus 5:20150031. https://doi.org/10.1098/rsfs.2015.0031
Wilkerson RP, Gludovatz B, Watts J, Tomsia AP, Hilmas GE, Ritchie RO (2016) A novel approach to developing biomimetic ("Nacre-Like") metal-compliant-phase (Nickel–Alumina) ceramics through coextrusion. Adv Mater 28(45):10061–10067. https://doi.org/10.1002/adma.201602471

Roots of Complexity in the Self-referential Genetic Code

Romeu Cardoso Guimarães

Abstract The genetic code is the correspondence between 'letter' units that cells utilize for translation: triplets of bases in the producers (genes) and amino acids in the products (proteins). The self-referential model indicates that the codes resulted from proto-tRNA dimer-directed protein synthesis. The dimerized proto-tRNAs became codes when the peptides they produced bound back to them and stabilized the correspondence between the units and the protein production system. Anticodons are representative sites of the initial binding oligomers that guided the complementariness at dimerization. The process of producing stabilized associations is a 'dynamic, epigenetic kind of memory'. The associated system is a module for the construction of polymers—genes, in the realm of 'memories in strings.' Memories guarantee stability while plasticity refers to the dynamics, which are the two main and interdependent characters of the living. Further stabilization and partial autonomy come from diversity in proteins at construction of structures and functions for the metabolic flow network. The metabolic system remains dependent on the environment, in a tense relationship with the degradation it provokes. A necessary component of biological complexity is the plasticity in behaviors, which mediates the diversity, adaptations and open-ended evolution. It is constitutive to protein structures and functions. Plastic behaviors are enhanced through the network organization of the system. Interactions that build networks are dependent on the wide range adhesiveness and binding sites of proteins. The model indicates that networks of nucleoprotein interactions are superposed on those of anticodon dimers, while all components are polymers with variable sequences. The complex behaviors of the resulting multi-synthetase complexes are now minimally rationalized.

R. C. Guimarães (✉)
Laboratório de Biodiversidade e Evolução Molecular, Departamento de Genética, Instituto de Ciências Biológicas, Universidade Federal de Minas Gerais, Av Pres Antonio Carlos 6627, Belo Horizonte MG 31270-901, Brazil
e-mail: romeucg@icb.ufmg.br; romeucardosoguimaraes@gmail.com

L. H. Wegner and U. Lüttge (eds.), *Emergence and Modularity in Life Sciences*,
https://doi.org/10.1007/978-3-030-06128-9_6

Graphical Abstract (Left, bottom) **The encoding process** A module is a simple network of tRNA pairs. The first module encodes the anticodon pairs (1) 5′GGG:YCC3′ and (3) GGA:YCU. Complex are the many cycles of evolutionary adjustments between the sequences of the synthetases and of the tRNAs (Right, top) **The ribonucleoprotein network in the multi-synthetase complex of mammals** There are nine enzymes [highlighted pink] and three auxiliary proteins [numbered]. Interactions via anticodon pairs are in straight lines, via protein binding in hand-drawn curved lines. The latter are more abundant upon the subnetworks that are less connected through the tRNA pairs and that join central A:U triplets.

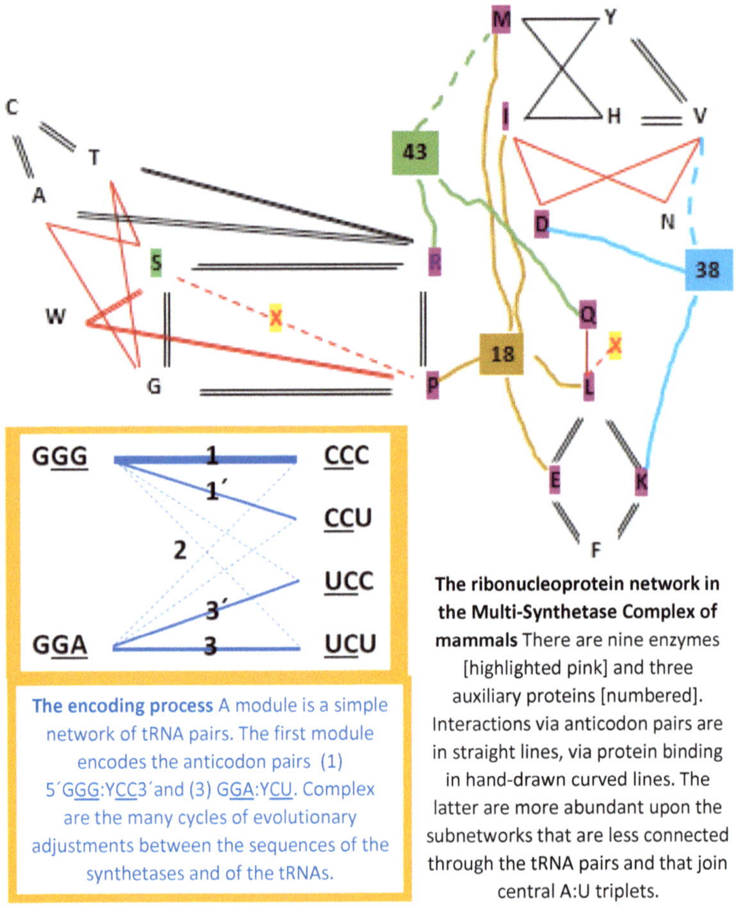

Graphical abstract

Keywords Genetic code · Self-reference · Coherence-decoherence · Memory · Metabolic flow · Plasticity · Networks · tRNA dimers · Multi-aminoacyl-tRNA synthetase complex · Cohesiveness

Introduction The genetic code is the set of correspondences between the aminoacyl-tRNA synthetases (synthetases) and their substrates tRNAs and amino acids. The working products of the correspondences, the aminoacyl-tRNAs, are represented by the tRNA anticodon triplet codes and the amino acids—the meanings—that are carried by the tRNAs. The correspondences have been settled following functional necessities of the cellular system but from the strictly biochemical point of view they may be considered nearly symbolic or arbitrary because there are no strong evidences of chemical relatedness between codes and meanings (Fig. 1).

The anticodons find their complements in the codons of mRNA sequences at translation. The translation process would be more strictly called transliteration in view of its punctual or 'letter-by-letter' (triplet-to-amino acid, 'digital') nature, without any hints at interpretation of messages. This set of 'letter' codes is at the origins of cellular organization, from which precision in the structures and functions can be obtained through the construction of sequences—genes and proteins. These acquire functional conformations that build the metabolic flow, which sustains the organisms, through the adequate body structures (Fig. 2).

Model Understanding the structure and the process that formed the code is still non-consensual, in spite of the half century span from the deciphering of the meaning of the triplets. The self-referential model (Guimarães et al. 2008) indicates that the formation of the code was based on 'protein synthesis directed by dimers of proto-tRNAs.' The dimers are considered mimics of the ribosomes—structures that hold two tRNAs together and facilitate the transferase reaction—but may be also considered among other instances of non-ribosomal protein synthesis (Fung et al. 2016; Goudry et al. 2009; Mocibob et al. 2010; Moutiez et al. 2014). While the couple of tRNAs is laterally associated with ribosomes, the dimers associate proto-tRNAs through pairing of the anticodon loops (Fig. 3). Information on the original data, going back to 1996 (Guimarães 1996), and further references are compiled and reviewed in Guimarães (2013, 2017).

We concentrate here on mechanisms involved with the generation of complexity in biosystems that are centered on the evolutionary construction and diversification of sequences of the biopolymers—proteins and nucleic acids. These combine with the construction of cellular bodies through the criterion of functionality of the metabolic flow that is directed to serve the regeneration of the system of biopolymers. The fundamental structure is of networks that are rich in self-referential loops and provide for partial sustainment of the system—the environmental dependency is irrevocable. It is reported here on the progressively increased complexity of the networks formed by the basic encoding/decoding components, which proceed in three levels. First, the dimers of tRNA anticodes. Second, the connections the dimers make through the aminoacyl-tRNA synthetases that bind concomitantly to the members of the dimers. The synthetases may also get involved with protein-protein binding to each other. An added degree of integration arises from the expanded degeneracy of various synthetases. Finally, a higher level of integration of various components is obtained through the addition of auxiliary proteins that bind them together in the Multi-Synthetase Complexes (MSC). Main components of complexity are the

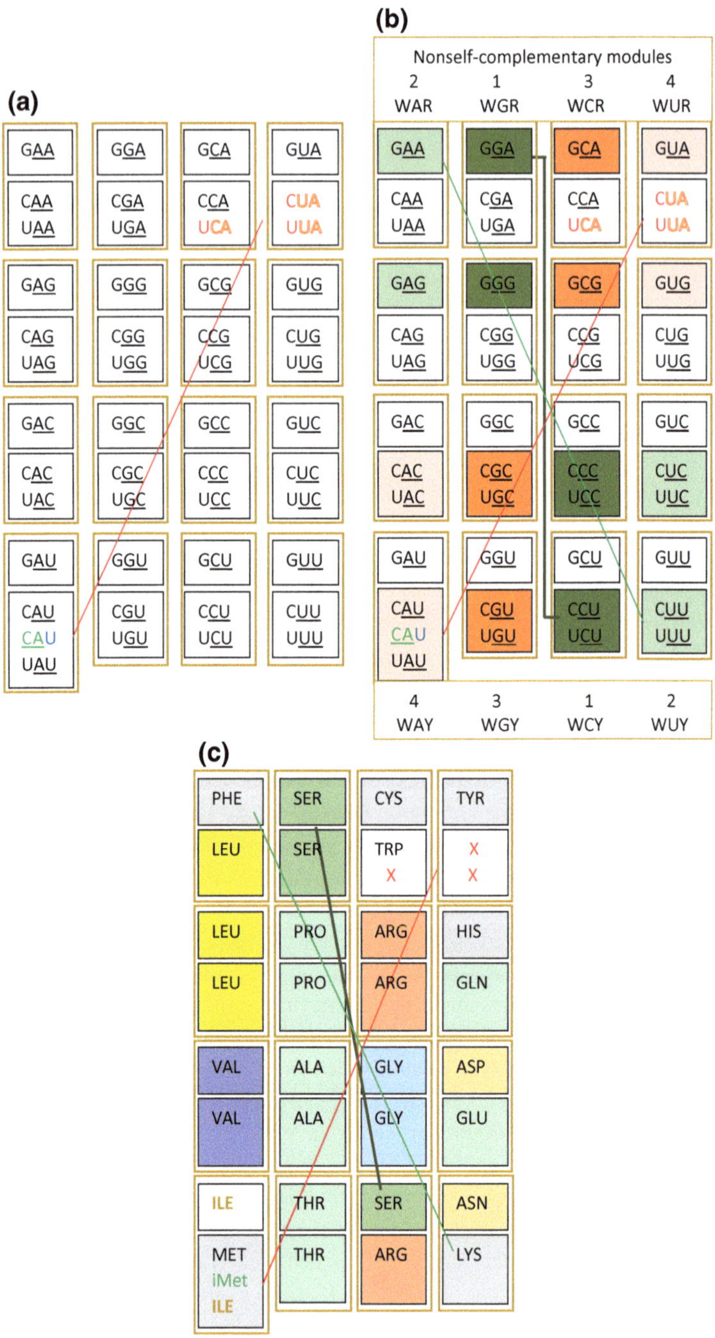
(a)
GAA
CAA
UAA
GGA
CGA
UGA
GCA
CCA
UCA
GUA
CUA
UUA
GAG
CAG
UAG
GGG
CGG
UGG
GCG
CCG
UCG
GUG
CUG
UUG
GAC
CAC
UAC
GGC
CGC
UGC
GCC
CCC
UCC
GUC
CUC
UUC
GAU
CAU
CAU
UAU
GGU
CGU
UGU
GCU
CCU
UCU
GUU
CUU
UUU
(b)
Nonself-complementary modules
2 WAR
1 WGR
3 WCR
4 WUR
4 WAY
3 WGY
1 WCY
2 WUY
(c)
PHE
LEU
LEU
LEU
VAL
VAL
ILE
MET
iMet
ILE
SER
SER
PRO
PRO
ALA
ALA
THR
THR
CYS
TRP
X
ARG
ARG
GLY
GLY
SER
ARG
TYR
X
X
HIS
GLN
ASP
GLU
ASN
LYS

◀**Fig. 1** Genetic anticode triplets and meanings in the matrix format. **a** The matrices differ from the traditional in exchanging the positions of the two right columns and the two bottom rows, therewith conducing to easy visualization of the symmetries produced by the triplet pairs. The principal dinucleotides, underlined, define the 16 boxes. They are constituted by the columns or the central bases, and the rows or the 3′ or last bases of the triplets; the first base, 5′, is the variable or wobble position, W (G, C, U). The direction is given by the biosynthesis of the polymers: start with the leftmost monomer and keep adding new ones at the right side. Codons are 64 but anticodons are 46: three per box plus the initiator less the three terminators. One of the symmetries is highlighted in colors and by the diagonal line uniting the initiator and the terminator codes. This relationship builds a punctuation system that is examined in details in Guimarães (2017). Here, it is only pointed out the complex relationships: both the initiator and the elongator Met triplets have the same composition but different functionalities of the principal dinucleotides, CAU for elongation and **CA**U for initiation; the termination codons correspond to the anticodons (red) that were eliminated, with a Y**YA** constitution, except the CCA that is maintained for Trp. **b** The four modules of nonself-complementary triplet pairs, which are the primary encoding modules, are numbered and highlighted in colors; the networks of pairs are drawn in Fig. 6. Two other diagonals are indicated, which will be commented upon later: the complementary principal dinucleotides $\frac{\rightarrow NGA}{\leftarrow UCN}$ for the same amino acid Ser; the two atypical synthetases that occupy complementary triplets $\frac{GAAPhe}{UUYLys}$. **c** The matrix of the meanings. There are three hexacodonic attributions: LeuRS NAG plus YAA runs along one same column inside the homogeneous principal dinucleotide sector; ArgRS NCG plus YCU runs along one same column and traverses from the homogeneous to the mixed principal dinucleotide sector; SerRS utilizes complementary principal dinucleotides in module 1. One half of the boxes is single-meaning, the other half is multi-meaning, and these are all paired symmetrically: single meanings in the core boxes (NGG Pro: NCC Gly, NCG Arg: NGC Ala); multi-meaning in the boxes at the tips (NAA Phe, Leu: NUU Asn, Lys; NAU Ile, Met, iMet: NUA Tyr, X); single-meaning pairing with multi-meaning, respectively: NGA Ser: NCU Ser, Arg; NAG Leu: NUC Asp, Glu; NAC Val: NUG His, Gln; NGU Thr: NCA Cys, Trp, X

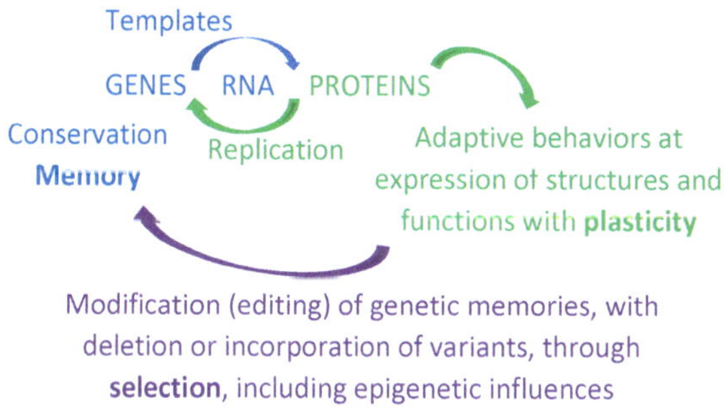

Fig. 2 General structure of the nucleoprotein system of cells. Aspects highlighted are the conservative (blue) and the evolutionary (green), which are contrasting functions in mutuality and interdependence. Conservation of memories is necessary for identity and regeneration of the proteins. Proteins execute replication of the memories and the vast majority of other functions, constructing the body and relating it to the environments, through their own activities and through regulation of the expression of the memories. These activities are mediated by diverse RNA types together with the proteins, which are the epigenetic mechanisms. Some of the epigenetic signals produce hotspots or facilitate genetic activities that are related to the generation of variability. Selection acts upon the variant sets—the 'editing function' (violet)—evaluating the system's fitness in relation to the environments, which results in changes in the populations of individuals with their genomes

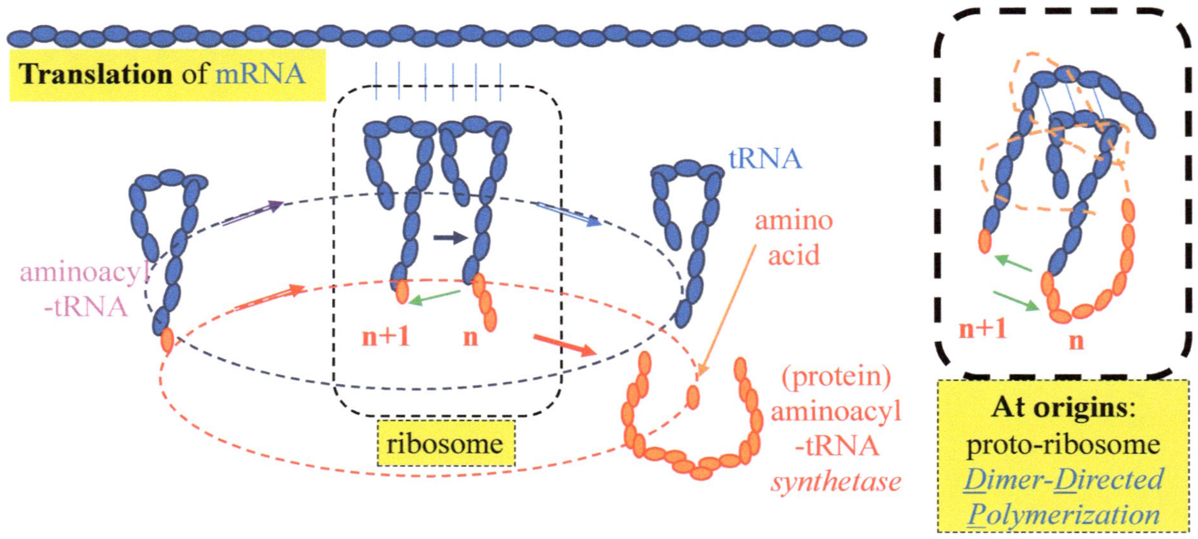

Fig. 3 Ribosomal, mRNA-directed protein synthesis and proto-ribosomal dimer-directed protein synthesis. The first (left) is unidirectional and the transferase reaction (green arrow) occurs between a couple of laterally associated tRNAs. The peptide in the peptidyl-tRNA (n) is added to the entering aminoacyl-tRNA; the peptidyl becomes n + 1, and this peptidyl-tRNA is translocated (blue arrow) to the peptidyl site. The second (right) is bidirectional, and the tRNAs are paired through the anticodon loops. The paired proto-tRNAs have an undecided identity, the anticodons being at the same time codons one for the other. The product protein may be semi-repetitive in constitution, combining some preferential amino acids in consequence of some chemical affinities, but also openness to external availability of amino acids. The dashed curved line elongates the chain of amino acids in the peptide and wraps around the producer dimer in the process of stabilization and coating/protection. Variations in the product peptides that bind to the producers may lead to continuation of the 'superposed or coherent' qualities (in analogy with quantum systems properties) or to differentiation ('decoherence') into singular identities after preferential binding of the products to one or the other proto-tRNAs. It is possible that the Ser codes are reminiscent ('chemical 'fossils') of the coherent state, conserving the complementary principal dinucleotides (NGA:NCU) and an auto-aggregated synthetase

multiple and weak connections that proteins get involved with, therewith building networks with highly plastic and dynamic behaviors.

RNA world The dimer-directed encoding process overcomes and bypasses the problems introduced by the RNA world hypothesis, of having RNA-only protocells that, besides not having a solid foundation on abiotic availability of nucleotides, requires long chains of these, which are known to be fragile and unstable. Instead of starting with long RNA molecules to be translated, only thereafter acquiring a meaning other than the self-directed (RNA makes RNA that makes RNA- - - - -), the dimers may be composed of oligomers; sizes below or around 20 mer are compatible with abiotic synthesis. The constitution of the oligomers is left open in view of the many possibilities that are offered by prebiotic chemistry (e.g., Francis 2015); they might have been polymerized on crystal or clay surfaces. It is only required that they should function similarly to the tRNAs, being able to carry attached molecules, including amino acids, to dimerize (Fig. 4) and to transfer the load from one to the other.

Precursors of anticodon triplets would have been oligomeric complementary sites, like others that participate in intermolecular associations. Present-day triplet structures would have been derived from a 'compression' process imposed upon the mRNA chain and the tRNA L-shape inside the ribosome. In order to accommodate the interacting segment of the mRNA plus the two tRNAs inside the organelle, the anticodon loop and the mRNA developed torsions and curvatures that should reflect the 1 + 2 functional differentiation of the wobble + principal dinucleotide positions. The ribosomal decoding site structure became physically separated into a 1 + 2 non-contiguous construction. The synthetases conserve the preferential interaction with the principal dinucleotide of anticodons at most of the single-meaning boxes. While it is not possible to have the relevant prebiotic samples to work with, biochemical tests may utilize the known tRNAs or some mini-versions of them, as proxies, at the same time attributing known functions to the codes and adding biological qualities to the model components.

Network origins A sketch of the organization of the code based on pairs of anticodons is presented [detailed in Guimarães (2017)], on which basis the formation of biological networks can be visualized. These go from the more rigid and regular kinds of interactions between the triplets of bases in RNAs, which are reminiscent of nearly crystalline structures (Fig. 5), to the more plastic and pleomorphic that are formed by protein interactions, sometimes described through the similarity with sticky gels.

Triplets and dimers The pairs of anticodons have a structure $\frac{\rightarrow 5'\,wobble-\mathbf{central}-3'}{\leftarrow 3'-\mathbf{central}-wobble\,5'}$ that generates small networks due to the choices allowed by the composition of bases in the wobble position of the triplets. The central base pair is of the strict G:C and A:U kind. The lateral pairs are dictated by the 3′ base of the principal dinucleotide, choosing the complement among the possibilities offered by the wobble position and accepting the generic R:Y pairs. This is necessary in view of the elimination of A at the 5′ position. A basis for the self-referential model is the full credit given to this 1 + 2 structure of the anticode triplets, which is only now being introduced into studies of codons (Seligmann and Ganesh 2017).

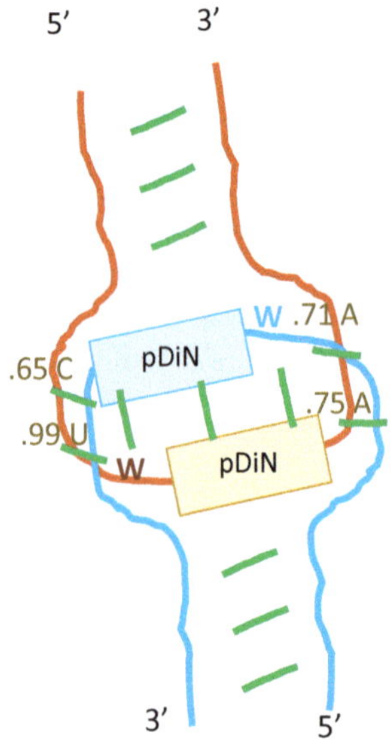

Fig. 4 Sketch of a general structure for a pair of anticodon loops. The central base pair joining the principal dinucleotides is of the standard G:C or A:U kind. The 3′ base of a principal dinucleotide pairs with one of the choices offered by the wobble (W) position in the other strand, according to the generic R:Y rule. In pairs of present-day tRNAs the two bases lateral to the anticodon, in both sides, are indicated to extend the pairing, since they are frequently purines in one side and pyrimidines in the other side (Widmann et al. 2005), namely (in the brown strand) base 32 is 65% C: base 37 is 75% A, base 33 is 99% U: base 38 is 71% A. The structure in (Moras et al. 1986) is not entirely adequate for comparison due to having been obtained from pairs that went through harsh purification procedures and show only the triplet pair. The thermodynamic data (Grosjean and Houssier 1990) indicated stability strength compatible with about seven base pairs. Possible involvement of curvatures in the anticodon loops, such as the U-loop involving U33 and the central purine in the same anticodon loop (Lehmann and Libchaber 2008), is not drawn

Nonself-complementary triplets and modules The networks are of two types, distinguished by the kinds of triplets that suffered the consequences of the 5′A elimination in different ways (Fig. 6). In the matrix of triplets, note the hemi-boxes called nonself-complementary: both lateral bases are of the same kind, both R or both Y. These triplets pair only with others of the same nonself-complementary set. This set of pairs builds two sectors in the matrix that run along the diagonals. In one sector, from the upper left corner to the lower right, the principal dinucleotides are called homogeneous, composed of either two R or two Y bases. One nice consequence of being nonself-complementary is that the triplets display for interactions fully planar surfaces, where each kind of radicals reaches the same height for either run of three R or three Y. The repetitiveness also means simplicity in the set of radicals along the triplet, and symmetry from the center to the sides. It is indicated nevertheless that it is not the symmetric character per se that constitutes a qualitative requirement of the encoding process; it is the non-complementary character of the lateral bases that adds

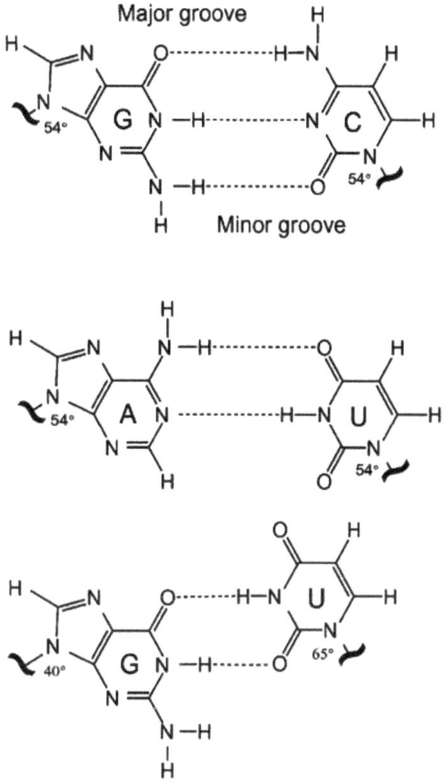

Fig. 5 The common base pairs in RNA. The G:C pair is strong, with three hydrogen bonds. All others are weak, with two hydrogen bonds. The G:C and A:U are standard, conserving the precise angles and distances, which are important for the double-helical strict regularity. The G:U pair is somewhat weaker than the A:U due to some distortion in distances and angles. The A:C pair, not shown, is topologically similar to the G:U but even weaker and rarer. The general rule is purine: pyrimidine, R:Y, (G or A):(C or U), allowing for shuffling of kinds along the sequences. Picture obtained from Google Images—EteRNA WiKi, August 2017

the meaningful quality. In other words, it might be possible to have another molecule in the place of one of the bases that would create asymmetry but if it maintained the avoidance of the complementary pairing it could still be accepted by the encoding system.

The sector of the nonself-complementary triplets that runs from the lower left to the upper right corners of the matrix is called mixed due to the composition of the principal dinucleotides, with one R and one Y. It is, therefore, also more complex structurally than the homogenous sector due to the rugged surface in the single strands, where the R bases are bulkier than the Y bases. The networks of dimers formed by these triplets become, after the 5′A elimination, asymmetric due to having two 5′G triplets pairing with four 5′Y triplets, and this structure is fully repetitive the whole matrix traverse. Four modules are generated, introducing a quality of importance for the encoding process: What has been learned and developed in the first module may be applied with expediency to the others, through duplications. The process of evolution that is facilitated by duplication followed by diversification is common and may be applied to various kinds of structures (Diss et al. 2017; Donoghue et al. 2005; Iranzo et al. 2016), possibly made easier in the case of modular repeats such as indicated by the self-referential model.

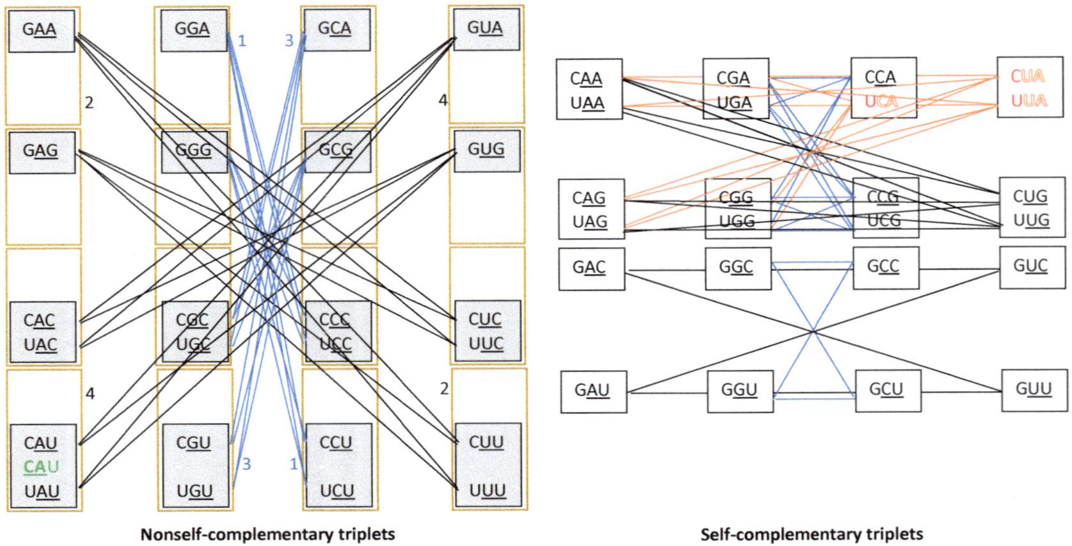

Fig. 6 Networks of anticodon pairs. The pairing rules are: central standard G:C and A:U. Laterals generic R:Y dictated by the 3′ base, choosing the base to pair among the variety offered by the wobble position $\frac{\rightarrow 5' WNN3'}{\leftarrow 3' NNW5'}$. The anticode matrix is separated in the nonself-complementary (left) and self-complementary (right) types of triplets. The first kind finds complements among themselves diagonally and forms four modules (1–4) with identical topology, asymmetric due to combining two triplets from the upper two rows (5′G-central-3′R) and four from the lower two rows (5′Y-central-3′Y), totaling 8 pairs per module. The four modules compose two sectors, with homogeneous (NRR and NYY) or mixed principal dinucleotides (NRY and NYR). The self-complementary triplets from the upper two rows (5′Y-central-3′R) are untouched by the 5′A elimination and conserve the 4 × 4 pairs symmetrical topology. The topology is identical for the central G:C and central A:U modules that find pairs combining the two sectors, horizontally along the upper two rows. In the lower two rows, the self-complementary triplets are all 5′G-central-3′Y, there being only four triplets in each group of central base kind; the two networks have symmetric 2 × 2 pairs, again combining the two sectors. Module numbers indicate the order of encoding: homogeneous principal dinucleotides before the mixed and inside a sector, the central G:C before the central A:U

Encoding letters The process envisaged for the encoding is the generation of a circularly structured association system that keeps practicing cycles of the transferase reaction. It is composed by (1) producers, which are the dimers of oligomers, and the (2) products—proteins (oligomers to polymers), among other possibilities, depending on the kinds of substrates utilized for the synthesis. (3) When the products acquire the adequate composition for not being lost to the environment and for binding back to their producers, therewith coating and protecting them from degradation, stabilization of the ensemble is reached. (4) Stability means habituation, where the system develops a longer duration and, provided that its original function is not impaired, it will keep producing more of what it got used to do, according to the natural availability of reactants. The end result of the cycles of synthesis and of the association is (5) a stable producer-product correspondence, which is a letter code. The evolved producer acquires the property of memory for the product; the cycle is identical in structure to the epigenetic processes (Ptashne 2013). Composition of the products of the dimer-directed syntheses might be originally biased with respect to the constitution of the dimers and of the available monomers that they carry, due to chemical affinities and abundances, therefore not homogeneous but also not entirely dictated by external availability. The cycling process would contribute to the enforcement of some aspects of the interactants and lead to mutual adjustments.

Encoding practical The nonself-complementary modules fit convincingly the requirements for encoding structures. Their simplified and asymmetric character facilitates the process. A high-stability dimer is encoded first (1). This is composed by triplets belonging to the two middle rows of the matrix. Data on the estimated stability of triplets are in Guimarães (2012). (2) Considering that the triplets that form the most stable dimer are practically sequestered one with the other, a consequence is that all other dimers that could be formed with any of them become scarce. A further consequence is that (3) another set of triplets is left free for dimerizing among themselves and at a high concentration, which facilitates their utilization for the second encoding in that module. This set is composed by the triplets in the upper and lower rows. At the encoding, the synthetase joins the degeneracies in the wobble position together in the same principal dinucleotide (1′, 3′) (Fig. 7).

The principle governing the choices among the four modules for establishing a chronological succession in the encoding process is to obey first the structural simplicity in the interacting sites of the partners. The homogeneous sector is encoded first, where the triplets and the synthetase active sites would be structurally more repetitive and less complex. The mixed sector is encoded afterwards. Inside a sector, precedence is given to encode modules with central G:C pairs first, with central A:U later, which says that the intrinsic dimer stability facilitates the encoding process. This was found entirely in accordance with the starting metabolic pathway, which is the Glycine-Serine Cycle, and also with the late installation of the specific punctuation system.

Diversity and combinatorics The availability of a letter code is a main and essential attribute of living beings, which allowed the construction of specific structures and functions through the enchaining of the letters into linear sequences of polymers; these fold in the 3D space into precise functional conformations. An apparently end-

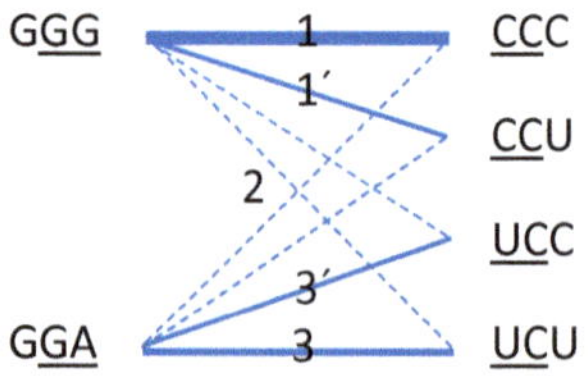

Fig. 7 Encoding the two pairs of boxes in the asymmetric nonself-complementary triplet modules through cycles of dimer-directed protein synthesis. Among the products of each dimer, there are the aminoacyl-tRNA synthetases that materialize the encoding. Module 1 is an example. The high ΔG pair (1) is stable enough to facilitate protein synthesis from which a stabilized precursor-product correspondence is generated. This pair is composed by triplets belonging to the two middle rows of the matrix. All other pairs the triplets of pair 1 would be involved with become of low concentration (2), leaving the other pair (3) at high concentration, from which another correspondence is fixed. This pair is composed by the triplets in the upper and lower rows. Synthetase recognition of tRNA evolves from—the initial state **a**—a collection of distributed sites along the protein and the tRNA sequences, which may or may not involve the anticodon, to **b** involve specificity toward the principal dinucleotide of the anticodon. At these stages, high degeneracy is the rule so that the other triplets in the module (1′, 3′) follow their cognate principal dinucleotides

less array of sequences is possible to be generated, giving support to the enormous diversity of the biosphere, all possibly based on combinatorics, among other processes, with some intriguing similarity to human languages, at least metaphorically or as an appealing analogy. The evolutionary panorama of large diversity of living beings that form highly complex bodies and ecosystems would be adequately described by the already well-settled assertion of an endless or open process—open-ended evolution, which is the biological counterpart to the infinites of logics and mathematics.

Encoding sequences The encoded letters would be able to generate sequences or chains of codes through ligation and other choices among the variety of molecular evolution processes, therewith forming genetic sequences—chains of the tRNAs or of segments of their sequences. These would have the quality of being meaningful from the beginning, therewith reducing the problem of having nonsense or stop segments inside the coding sequences. The first kinds of selection criteria (Fig. 2) would be plain and simple stability of the products, thereafter their ability to bind the producers, their protection from degradation, and the capacity of not harming their activities. In this way, the small precursor-product system may keep producing more of itself, which is akin to reproduction. The process has similarities to other models called auto-catalytic sets or systems (Hordijk and Steel 2017).

Early sequences It is envisaged that at some early stage in evolution of the system its composition would be biased according to the affinities presented by the monomers that were obtained from external sources. If, e.g., they would be similar to the products of Miller type experiments, they would contain mostly organic acids, including keto acids, which may be aminated to become less reactive and more stable in the form of amino acids, among other compounds. Organic materials from meteorites indicate the presence of amino acids in the parts-per-million range, of nucleobases in the parts-per-billion range, which could give estimates on what could have been the composition of some early oligomers. At some later stages, this could have given rise to the variety of non-ribosomal protein synthesis, including the proto-tRNA dimer-directed protein synthesis and the formation of RNP systems, such as the synthetase-tRNA and the ribosomal. It is appealing, e.g., the known high participation of Glycine in the RNA-binding sites of today's proteins, whose affinity might have been relevant for selection in favor of RNA along the process.

Evolution It is adequate to highlight three key words in the context of early biological evolution. (1) Stability is the crux in a process that should ideally acquire the (2) self-stimulation capacity (positive feedback, self-feeding). Stability should be partial so that the structures do not 'freeze' in one form but keep open to change and evolution, which is the attribute of (3) plasticity. In later steps of the process, we might identify two characters that summarize and identify biosystems, in interdependent circularity and in spite of the danger of reductionism but with the quality of simplicity: (a) stability and conservativeness, which are main attributes of the genes, nucleic acids and replication; (b) evolutionary potential and realization, openness to change and adaptation, which might be pinpointed to protein plasticity (Colussi et al. 2014; Kenkel and Matz 2016; Murren et al. 2015).

The process of formation of the code is pre-Darwinian, in the self-organization realm. It may be estimated that it would have taken hundreds of million years, in the interval from the origin of the Earth to the paleontological dating of cellular microfossils, this at $\leq$3.5 Gya. This is the period of maturation of the LUCA lineages and their confluence into the populations that share the nearly universal code. We stress, inside the self-organization paradigm, the self-referential mechanism, which is intended to be at the same time wider and softer than the auto-catalytic; it would be closer to the more systemic 'auto-catalytic sets' (Hordijk and Steel 2017). The specificity in the case of the encoding indicates the supposed dominance of characters of the products, which are the proteins made of simple monomers, in shaping the outcomes in the system under construction. It is the quality in the product (such as stability and the binding ability) that chooses among variants of the producers which will be adequate for the ensemble to fit together in a system. In the case of encoding, it is suggested that peptides that were stable in themselves and adequate for binding to the proto-tRNAs, directed the development of the RNA structure, that is considered of biotic origin.

A result of the chronology of amino acid fixation in the code that highlights the precedence of the plastic character over the internally organized protein structures is the composition of initial set of amino acids, which is more adequate to build intrinsically disordered regions of proteins. The evolutionary path indicated is from disorder to order, meaning that disordered regions are original and open to develop order and informational patterns at the interaction events, in mutuality and in accordance with the kinds of interactants (Guimarães 2015).

Metabolism Besides delineating modes of experimentally testing the dimer-directed protein synthesis activity, the significance of the self-referential encoding process for cellular systems was immediately evaluated through overlaying the structure of the sets of dimers upon various sets of properties of amino acid residues in

Table 1 Overview of the chronology of encodings. Annotations can be detailed in the homogeneous principal dinucleotide sector, where there are constraints dictated by the constitution of the Gly-Ser Cycle of assimilatory metabolism. This supports the five first encodings, all in single-meaning boxes; Leu is the only synthetase class I. Additional encodings in this sector (total ten) are proposed to have been added at maturation of the central metabolic pathways, starting with gluconeogenesis, then glycolysis and the pentose-phosphate shunt (necessary for the biosynthesis of Phe) plus the Krebs Cycle. The latter includes the precursor to the Glu family of amino acids (Pro, Arg), while Lys may be obtained from both Asp and Glu sources. The richer amino acid repertoire allows synthetase specificity to develop, including the generation of multi-meaning boxes. Encodings in the mixed sector are not metabolically constrained. The sector was initiated by the ArgRS expansion from the YCU hemi-box to the NCG box. Note the complete substitution of Gly by Pro in the NGG box. Other concessions are partial, from the first amino acid (in the left) to the new attributions, separated by comma(s)

Homogeneous sector

1a	GGG → CCC; ↘ CCU	2a	GAG → CUC; ↘ CUU
1b	GGA → UCU; ↗ UCC	2b	GAA → UUU; ↗ UUC

First encodings

1 a	GGG/CCY	→ NGG/CCN	→ Gly/Gly	2 a	GAG/CUY	→ NAG/CUN	→ Leu/Asp
1 b	GGA/UCY	→ NGA/UCN	→ Ser/Ser	2 b	GAA/UUY	→ NAA/UUN	→ Leu/Asn

Present encodings

1 a	NGG/CCN	Pro/Gly	2 a	NAG/CUN	Leu/Asp, Glu
1 b	NGA/UCN	Ser/Ser, Arg	2 b	NAA/UUN	Leu, Phe/Asn, Lys

Mixed sector

3a	CGC ← GCG; UGC ↙	4a	CAC ← GUG; UAC ↙
3b	UGU ← ACG; CGU ↖	4b	UAU ← AUG; CAU ↖

3 a	GCN/NGC	Arg/Ala	4 a	NUG/CAN	His, Gln/Val
3 b	NCA/UGN	Cys, Trp, X/Thr	4 b	NUA/UAN	Tyr, X/Ile, Met, iMet

proteins and of properties of proteins. The chronology of amino acid encoding that was generated found support in a biosynthesis pathway that makes sense as candidate among the first in the metabolic network, the Glycine-Serine Cycle: (1) It is the simplest among the central metabolic pathways—it starts with C1 and its most complex components reach only the C4 level. (2) It is the only one containing amino acids, while the others contain precursors to amino acids. (3) It contains the two direct precursors of gluconeogenesis. The chronology of amino acid incorporation into the code is presented in Table 1.

The flow It is not adequate to ask which came first, the code or the metabolic pathways. It is not possible to have one without the other. It is considered that generation of triplets would be simple, from replication, but only after the monomers can be available from biosynthesis, and amino acids are precursors to the nucleobases. Amino acid biosynthesis is also difficult, besides having to adjust to—therefore being directed by—physiologic necessities, so being the main constraint on the process. On the easier side, it is considered that the most relevant amino acids (Gly and Ser) come directly from C1 sources. The solution has to reside in coevolutionary processes among various members of the system, which should include these two components plus the upstream substrates or sources for metabolism and the downstream destinations for its products. This metabolic ensemble composes a flow system where the flow itself is the organizer since it is immersed in the preexistent general universal flow of masses and energy. In nested circular structures, (a) the flow that is made by the metabolic system is the measure through which the quality of the system is evaluated; (b) the flow produced is checked in relation to and in accordance with the environmental flow. Metabolism, which is the biological dynamics, only adds a new segment with its own preferential sources—C1 organics, and sinks—starting with the uptake and synthesis of amino acids, thereafter with their sequestration in proteins.

Superposition and the pair of Serine codes A problem with the dimer-directed encoding, as a model for the proto-ribosome, would arise from the equal value of the oligomers that dimerize complementarily, while the rule in the set of correspondences for translation is the individual encoding of distinct tRNAs. This indicates a biochemical situation, in the dimer, that is analogous to the phenomenon of superposition of states in quantum objects (Park et al. 2017; Schlosshauer 2014; Zurek 1991, 2002). It is said that the superposition corresponds to the undecidedness between the states or the coherence of one with the other. The quantum objects will adopt the classical behavior, or give rise and transform into them, after a process of decoherence that is triggered by interactions with some kinds of perturbations that work as if separating the components of the quantum object into classical singular states or classical objects.

There are choices to be investigated for identification of the interacting perturbation in the case of the dimers. They could be the product peptides that bind preferentially to one of the proto-tRNAs of the dimers, or the possible different products from one dimer that could bind differently to the two members, and among other possibilities, they could be different proto-RNAs that would interfere with the pairing of the dimers at the proto-anticodon sites. The latter could be assuming the role of the present-day mRNAs. While in the dimer the anticodons are at the same time codons for each other—states superposed; the singular anticodon identity is defined when an external RNA substitutes the codon function of the other member of the dimer.

The self-referential model finds in the case of the serine codes, to my knowledge an enigma that has never before received any minimally suggestive interpretation, a remnant of the ancient and original situation of dimer-directed protein synthesis where both members of the dimer were adopted by one same synthetase enzyme.

SerRS is one of the few synthetases that maintain the original situation of not requiring to interact with the anticodon bases. The anticodon principal dinucleotides are complementary $\frac{WGA}{UCW}$, and both the tRNAs and the enzyme are exceptional. The tRNAs have a very long variable arm, and the synthetase is auto-aggregated, with two tRNA binding sites (Gruic-Sovulj et al. 2002), which seems to be a unique situation. The usual interpretation for homodimers is the allosteric where the binding of a substrate to one of the sites activates the other in a synergistic mode.

Meanings in hemi-boxes and the atypical pair $\frac{GAA}{UUY}$ $\frac{PheRS}{LysRS}$ A long known regularity in the code attributions, namely the distribution of meanings according to hemi-boxes (of course, in the cases of multi-meaning boxes), now receives a functional explanation, not just the R versus Y description, that is the partition of the set of codes in a box into the self-complementary and nonself-complementary types. The encoding process starts upon the nonself-complementary high ΔG pair of triplets, but the synthetase specificity follows a succession of degrees of precision in discrimination with respect to the participation of the wobble position of anticodons.

The anticodon may not participate in recognition; that is, it does not interact directly with the synthetases (SerRS, LeuRS, AlaRS). When it participates in recognition, it may do so only through interactions with the principal dinucleotide, generating a single-meaning box (the three above plus ArgRS, GlyRS, ThrRS, ProRS, ValRS); all kinds of wobble bases have the same value and meaning, whether generating self- or nonself-complementary triplets. Multi-meaning boxes require that the synthetases interact specifically with the base in the wobble position. A rule that describes the occupation of the wobble bases would be: (a) the nonself-complementary is the original or first triplet occupied, but it may be 5′G or 5′Y; (b) this encoding passes through the single-meaning stage where the trace of the triplet of origin is erased (the wobble is any base); (c) when a new meaning is to be encoded in a box, the original or initial meaning of the box retains the specificity for the 5′G triplet and (d) concedes the 5′Y triplets to the new synthetase; this may or may not coincide with the primary encoder in the box. The reason for step (c) would be the strong and specific pair the 5′G will form at pairing with a 3′C.

The rule is followed by another character that the second meaning in a multi-meaning box corresponds to class I or punctuation. Such rule of concession of YNN hemi-boxes to the second meanings is followed by six of the eight multi-meaning boxes: WCA Cys/Trp and X, WUA Tyr/X, WUG His/Gln, WUC Asp/Glu, WCU Ser/Arg and WAU Ile/Met, iMet. The other two follow a pair of atypical enzymes, which are in accordance with the self-referential model. The LysRS class II of some organisms is atypical with respect to the rule above, besides the exceptionality of LysRS being of different classes in different organisms. The case of the PheRS/LeuRS split box does not contribute to the accounting above since the model says that LeuRS was originally octacodonic WAR (Table 1); the concession to PheRS was followed by recession of the LeuRS, maintaining the W$\underline{AG}$ + Y$\underline{AA}$ contiguity, while it is the G$\underline{AA}$ PheRS that developed the atypical character of being a class II enzyme acylating in the class I mode.

This explanation is also partly historical and contingent: The amino acids Phe and Lys are the only large of class II enzymes and of extreme hydropathies; they should have been taken up by class I enzymes, but these were lacking at the time of fixation of the codes, generating the atypical behaviors. The coincidence of the couple being settled precisely upon a pair of triplets adds confidence to the proposition. The moment of these encodings, at the transition between the sectors, should have been a critical period in the system, possibly of enrichments in the nucleic acid subsystems, as indicated by the rise of sugars, which are necessary for the biosynthesis of Phe, and the rise of the basic amino acids Lys and Arg. This is the last in the homogeneous sector and the first in the mixed sector.

Self-complementary triplets and modules are not good for encoding The triplets containing bases of different kinds in the lateral positions, one purine and one pyrimidine, would not be the choice for the encodings in view of their experimentally observed formation of auto-dimers, especially when they contain the small pyrimidines at the central position; the prime example is the very stable auto-dimer of the tRNA$^{ASP\text{-}G\underline{UC}}$, in spite of the central mismatch $\frac{GUC}{CUG}$. There follows lower concentration of the bona fide hetero-dimers $\frac{GUC\,Asp}{CAG\,Val}$. This rationale should explain why these self-complementary modules would not be chosen for encoding, especially in the presence of the competing nonself-complementary (Xia et al. 1998).

The modules formed by the self-complementary triplets are of two kinds (Fig. 6). Triplets initiated by a Y base are not affected by the 5′A elimination so that the topology is the original symmetrical 4 × 4. Those initiated by an R base are more reduced than the nonself-complementary, and the topology of the network is symmetrical 2 × 2. Each topology is repeated twice, according to the G:C or A:U central pair. It is estimated that a decision process based on symmetrical networks relying only upon differential abundances of elements and differential thermal stability could eventually happen but would take too long in face of the expediency expected for the nonself-complementary networks.

While in the nonself-complementary modules the pairs unite triplets along diagonals of the matrices, following the axes of the sectors, the triplets in the self-complementary modules belong to boxes in the same rows of the matrix, with horizontal connections. Therewith comes the main attribute of the latter modules: They are integrators, connecting the sectors, but this job is accomplished by the special mode of evolution of the synthetase specificity for the triplets that starts with full box degeneracy. Were it not for this stage of the synthetase degeneracy (encoding directed to the principal dinucleotide alone) where the distinction between the self- and nonself-complementary triplets is erased, these two kinds of triplets would belong to disjoint modules from the beginning. We are now proceeding the algebraic treatment of the self- and nonself-complementary submatrices (Fig. 6), plus their summed set (Fig. 1a), in order to check for formal properties and for their possibly being at the roots of the thermodynamical distinction that was observed by Xia et al. (1998).

Integration of the system via proteins—the code degeneracy The self-complementary modules are in charge of integrating the other modules into the RNP system; that is, they participate together with proteins in the integrative process. Here enters a new instance of self-reference where (a) the encoding utilizing a fraction of the triplets produces proteins, the synthetases. (b) Each synthetase, at the development of specificity directed to the principal dinucleotide, will incorporate into its meaning the other triplets with the same principal dinucleotide, which is the development of a full box degeneracy, each box containing the self- and nonself-complementary triplets joined into a coherent set (Fig. 8). Expansion of this process will progressively dissipate the whole space of triplets, all of them being recruited for participation in the code system.

Further consequences of the presence of proteins derive from their ability to bind a large variety of other molecules, since they are the sticky or cohesive components of biosystems, in charge of holding all others together into a whole, usually acquiring the shape of a globule when immersed in a watery solution. In the present case, we follow the formation of a network of proteins, the synthetases, superposed upon the tRNA network. This combined RNP network would be near the roots of cellular organization, where RNAs and proteins functionally meet, and also near the roots of biological complexity. It is convenient and didactic to study this case, in view of its small size and apparent simplicity.

Protein plasticity in complexity One component of the complexity comes from the very wide range of diversity in the modes that protein sites—oligomeric motifs—can accomplish the binding function. The combinatorial possibilities at composing the sites allow for fine-tuning so that it is possible to utilize the almost digital properties of single amino acids to reach the analogic properties in the site sequences. Definitions of complexity, especially when referring to biological objects, are problematic in themselves because they would have to take into account the great variety of components and of behaviors in the systems the definition refers to, generally ending with non-satisfactory assertions and lack of consensus.

Another aspect to be considered is that complex objects or entities are systems presenting behaviors that may change along time or may accept some degree of non-constancy in composition, therewith increasing the difficulty of capturing in a definition these ranges of variations. Such considerations may suggest that it should not be expected to reach one consensual proposition, but it should be accepted that approaches to the examination of complex systems would inevitably be many and diverse, each of them adequate under its own limits and purposes, and that a composite picture [e.g., Souza and Lüttge (2015)] would be always under construction and revision.

For the biological setting, the proposition I could reach says: (1) Living beings are metabolic flow systems that self-construct on the basis of memories and adapt/evolve on the basis of constitutive plasticity. (2) Life is the ontogenetic and evolutionary process instantiated by living beings (Guimarães 2017). These attempts at definition place complexity, as a quality, in the realm of behaviors, and its material basis would be constitutive to biological entities, here adopting the near synonym plasticity, which is of more general use in biochemistry and not so 'heavy' with the load of associated connotations as the first term. Plasticity is less intense in biomolecules other than proteins, such as DNA, intermediate in RNA.

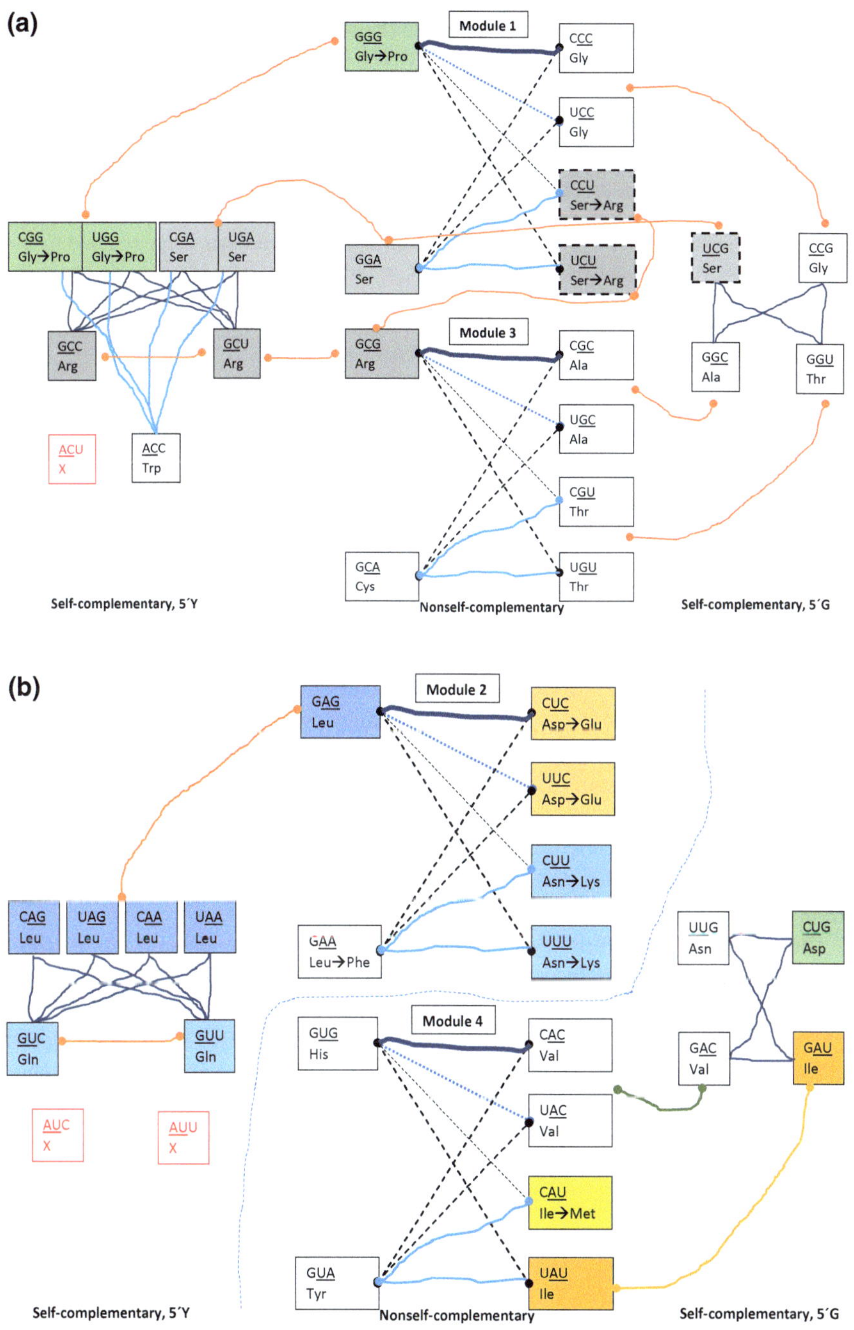
(a)
Module 1
GGG
Gly→Pro
CCC
Gly
UCC
Gly
CCU
Ser→Arg
UCU
Ser→Arg
CGG
Gly→Pro
UGG
Gly→Pro
CGA
Ser
UGA
Ser
GGA
Ser
UCG
Ser
CCG
Gly
GCC
Arg
GCU
Arg
GCG
Arg
Module 3
CGC
Ala
GGC
Ala
GGU
Thr
UGC
Ala
ACU
X
ACC
Trp
CGU
Thr
GCA
Cys
UGU
Thr
Self-complementary, 5´Y
Nonself-complementary
Self-complementary, 5´G
(b)
Module 2
GAG
Leu
CUC
Asp→Glu
UUC
Asp→Glu
CUU
Asn→Lys
CAG
Leu
UAG
Leu
CAA
Leu
UAA
Leu
GAA
Leu→Phe
UUU
Asn→Lys
UUG
Asn
CUG
Asp
GUG
His
Module 4
CAC
Val
GUC
Gln
GUU
Gln
GAC
Val
GAU
Ile
UAC
Val
AUC
X
AUU
X
CAU
Ile→Met
GUA
Tyr
UAU
Ile
Self-complementary, 5´Y
Nonself-complementary
Self-complementary, 5´G

◂**Fig. 8** Network graphs of connections between synthetases that are facilitated by the pairwise interactions between their correspondent tRNAs. **a** The central G:C and **b** the central A:U subnetworks. The structure of the subnetworks of tRNA pairs/dimers is sketched; details are in Table 3. Note that the dimer connections are self-contained and separated into the nonself- and self-complementary kinds, while the integration is developed gradually via properties of the synthetases. In these graphs, the integration comes from the degeneracy properties. Specificity of a synthetase toward tRNAs may be strict single (Σ 9 cases: Phe, Cys, Trp, Tyr, His, Asp, Met, iMet, Asn), or minimally degenerate, to the couple of pyrimidines, which does not add integration beyond the triplet kind of the pyrimidines (Σ 3 cases: Gln, Glu, Lys). The synthetase specificity becomes integrative of different kinds of triplets in the other cases: the tetracodonic or full box Pro, Gly, Ala, Val, Thr (Σ 5), the Ile GAU and UAU, and the three hexacodonics—the simpler case of Leu that is NAG plus YAA, with limited 3′R ambiguity; the peculiar case of Ser that conserves the original complementary principal dinucleotides while concedes YCU to Arg and recedes to GCU; ArgRS is also most peculiar in bridging the two sectors through a wider 3′ ambiguity, going from a homogeneous principal dinucleotide YCU to the mixed NCG. The NUC box is shared by these two hexacodonic synthetases

Network plasticity—the case of the Multi-Synthetase Complex The most interesting aspects in protein plasticity come from the wide range in the diversity of interactions. Besides the twenty-few encoded amino acids, there are the posttranslational modifications that enlarge widely the repertoire of elements. This should be reflected strongly in the other material substrate for plasticity that is the networks. These are originated mostly in consequence of the presence of components—nodes—with three or more interactive sites or functions, which is a common feature of biopolymers.

It has been shown above that the tRNA dimers compose a few types of small networks that are lowly connected, one possible strong limitation being the constraint imposed by the strict central base pairing rule. These are partially joined by the synthetase wide degeneracy, when specificity is based on sequence features that are distributed along the tRNA and synthetase molecules without recognition of the anticodon or when it is directed to the principal dinucleotide of the anticodon. These are the cases of the hexa- and tetracodonic attributions (Fig. 8).

Dimers place different synthetases in contact and propitiate integration through binding Beyond these integrative advancements based on synthetase degeneracy, a new level of integration comes with the possibility offered by the dimers, when they associate tRNAs belonging to different synthetases. The proteins placed in close contact may develop binding sites through evolution of adjustments in these sites, in case the association, which was initially driven by the tRNAs, proves beneficial to the system (Cho et al. 2015; Fang and Guo 2017). This process may be placed in the context of internalization or endogenization of the benefits of an external influence into the genetic sequences: selection in favor of variants, in the interacting sites, whose effects will mimic the work of the external factor and may now do without it.

Some quantitative observations are consistent with this possibility. Data on the distribution of dimers, the whole matrix of anticodes traverse, were overlaid upon the data on the constitution of the multi-synthetase complexes (MSCs). The MSCs that have been observed across the evolutionary scale build an apparent succession of increasing size, expressed as the number of different enzyme specificities that associate into a complex (Table 2). It is noticed that data on plants are lacking from

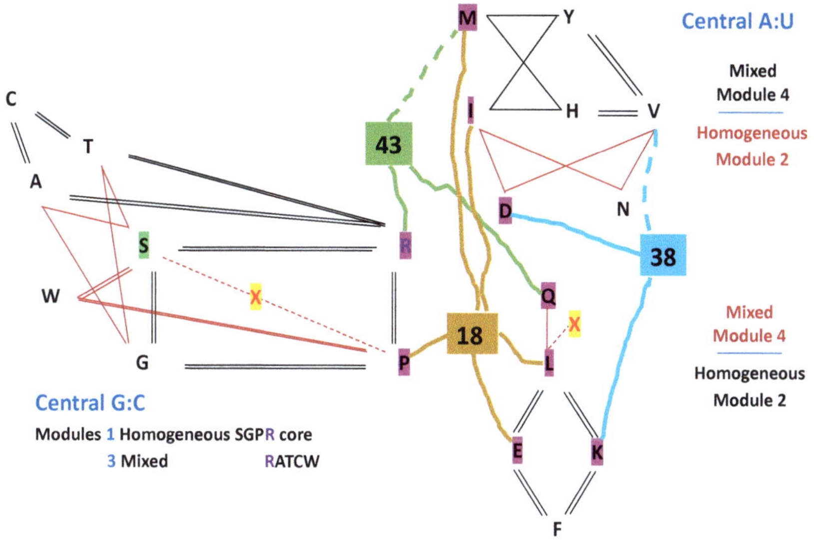

Fig. 9 Network graph of the connections between synthetases mediated by tRNA pairs and in the Multi-Synthetase Complex. Two sources of the interactions are superposed: those facilitated by the pairing/dimerization of their correspondent tRNAs (straight lines) and those observed experimentally in the isolated Multi-Synthetase Complexes (MSC; curved hand-drawn thick lines). These are colored green for the subcomplexes mediated by protein 43, brown for those mediated by p18 and blue for p38. The straight black lines indicate connections through nonself-complementary triplet pairs, in red the connections through self-complementary pairs. Data on the MSC were taken from the two most complex, that are from Bilaterian animals (see Table 3): the mammalian type includes four species of mammals, plus *Drosophila melanogaster* and *Artemia salina*, while *Caenorhabditis elegans* is a variant: Mamm 43RQ, 38KD, 18LIEPM; Cele 43RQM, 38 kV, LIE. The LIE group of Cele is an example of synthetases that bind to the complex without the intervention of auxiliary proteins. Note the establishment of connection between the two central A:U subnetworks by p38, in different ways in the two MSCs. Protein 43 connects the central G:C with one or both of the central A:U subnetworks in the different kinds of MSC. Protein 18 connects the three subnetworks but it is not present in the MSC of the worm, where connections are only between the two central A:U and obtained by properties of the synthetases

the compilation. The peak sizes are in the Bilateria animal groups, with a variant containing eight enzymes and two auxiliary proteins in the worm *C. elegans*, and another of wider distribution (Crustacea, Insecta, Mammals) composing nine enzymes with three auxiliary proteins (Havrylenko et al. 2011; Havrylenko and Mirande 2015).

Our counts of the connections between enzyme specificities that would be propitiated by the dimers are detailed in Table 3, where observations on the enzymes that compose the mammalian type of MSC are marked in blue, as well as the auto-aggregated SerRS marked in green. These observations are sketched in a network graph format in Fig. 9.

All synthetases that compose the different MSCs were counted, reaching a sum of 31 occurrences. Among these, it has already been pointed out the excess (3/4) of

Table 2 Composition of Multi-Synthetase Complexes along the evolutionary scale. An intermediate step of encoding is added between the homogeneous and the mixed sector (modules 1 and 2, +) to indicate the maturation of the central metabolic pathways and full occupation of the homogeneous sector, beyond the Gly-Ser Cycle. Arg belongs to both modules 1 and 3, but was added at the end of the homogeneous sector. It is indicated that the aggregation involves mostly synthetases for the Glu family of amino acids, class I enzymes, and pairs of anticodons with central A:U, which reflects absence of systems involving amino acids from Module 3. Oldest auxiliary protein entering the composition of the complex is p43, youngest p18 (Havrylenko and Mirande 2015)

Modules		Archaebacteria *Methanothermobacter thermoautotrophicum*		Apicomplexa *Toxoplasma gondii*	Euglenozoa *Trypanosoma brucei*	Fungi *Saccharomyces cerevisiae*	Nematoda *Caenorhabditis elegans*	Insecta *Drosophila melanogaster* Crustacea *Artemia salina* Mammalia Four spp				
1 Gly Ser			Ser									
2 Leu Asp Asn		Leu					Leu			Leu	Asp	
1, 2, +: Phe, Glu Pro Lys Arg		Pro Lys	Arg	Glu	Pro	Glu	Glu	Lys	Arg	Glu Pro	Lys	Arg
3 (Arg) Ala Cys Trp Thr												
4 Tyr His Gln Val Ile Met				Tyr Gln Met	Gln Met	Met	Ile	Val	Gln Met	Ile Met		Gln
Synthetase class	II Σ8	PK	S		P			K		P	DK	
	I Σ23	L	R	EYQM	QM	EM	LEI	V	RQM	LEIM		RQ
Central base	G, C Σ7	P	SR		P				R	P		R
	A, U Σ24	LK		EYQM	QM	EM	LEI	KV	QM	LEIM	DK	Q
Auxiliary	p43			p43	YbaK p43 p43c	p43			p43			p43
	p38							p38			p38	
	p18									p18		

Table 3 Connections between synthetases facilitated by the pairing of their correspondent tRNAs and mediated by ribonucleoprotein aggregation. **A** Number of anticodon pairs formed by the specificity in the left column with other specificities. The auto-aggregated SerRS is highlighted green. The synthetases belonging to the MSC of the mammalian type are highlighted blue. These detailed data are summarized in the graph sketched in Fig. 9. The isolation of the subnetworks is overcome by the cohesive properties of the synthetases and the auxiliary proteins. Synthetase specificity is indicated to have started wide, now shown by the hexacodonics, then entering partial reduction when directed to the principal dinucleotides—forming the tetracodonics, up to the finer tuning of the di- and mono-specificities. **B** Only a few hints can be extracted from an evaluation of the information on the tRNA dimers and on the synthetases aggregated into the MSC. **B1**. It is possible that the presently seen structure is still under evolution, but indications may be that the integration by aggregation should be partial. Would this mean: don't over-integrate but leave some specificities free for independent and autonomous work and regulation? While we could not extract regularities referring to specific qualitative amino acid properties, it is possible to infer some quantifications for further evaluation. The central G:C set of pairs is hyperconnected so that it conceded the least number of enzymes to the aggregate and its highest connected specificity SerRS was excluded from hetero-aggregation via auto-aggregation. The two central A:U subnetworks are lowly connected through tRNA pairs and became more strongly integrated into the system via synthetase aggregation

A	In nonself-complementary modules	Σ pairs	In self-complementary modules	Σ pairs	Σ Σ	Synthetases, X
Central G:C, NSC Modules 1 (homogeneous pDiN) and 3 (mixed pDiN)						
Ser	Arg 2, Gly 2	4	Arg 4, Trp 2, Ala 1, Thr 1	8	12	5, $X^{\underline{UCA}}$
Arg	Ser 2, Pro 2, Ala 2, Thr 2	8	Ser 4, Pro 4	8	16	4
Ala	Cys 2, Arg 2	4	Ser 1, Gly 1	2	6	4
Thr	Cys 2, Arg 2	4	Ser 1, Gly 1	2	6	4
Gly	Pro 2, Ser 2	4	Ala 1, Thr 1	2	6	4
Pro	Arg 2, Gly 2	4	Arg 4	4	8	2, $X^{\underline{UCA}}$
Cys	Ala 2. Thr 2	4			4	2
Trp			Pro 2, Ser 2	4	4	2
8	7 occurrences	32	Trp new out of 7	30	62	
Central A:U, NSC Module 2 (homogeneous pDiN)						
Leu	Glu 2, Lys 2	4	Gln 8	8	12	3, $X^{\underline{YUA}}$
Gln			Leu 8	8	8	1
Glu	Leu 2, Phe 2	4			4	2
Lys	Leu 2, Phe 2	4			4	2
Phe	Glu 2, Lys 2	4			4	2
5	4	16	Gln new out of 2	16	32	

(continued)

Table 3 (continued)

A	In nonself-complementary modules	Σ pairs	In self-complementary modules	Σ pairs	Σ Σ	Synthetases, X
Central A:U, NSC Module 4 (mixed pDiN)						
Ile	His 1, Tyr 1	4	Asn 1, Val 1	2	6	4
Val	His 2, Tyr 2	4	Asp 1, Ile 1	2	6	4
Tyr	Ile 1, Val 2, Met 1	4			4	3
His	Ile 1, Val 2, Met 1	4			4	4
Met	His 1, Tyr 1	2			2	2
Asp			Asn 1, Val 1	2	2	2
Asn			Asp 1, Ile 1	2	2	2
7	5	18	Asp, Asn new out of 4	8	26	
20	16 occurrences	66	4 new out of 13	54	120	

B How to Distribute Synthetases to Compose the Multi-synthetase Complex

Global connectivity through anticodon pairs		Central G:C Modules 1 + 3 Homogeneous principal dinucleotide core GPSR plus Mixed RATCW	Central A:U Mixed Module 4 VMIYH plus Homogeneous 2: DN	Central A:U Homogeneous Module 2 LEFK plus Mixed 4: Q
Three-node cycles		SGA, SGT, SRA, SRT	–	–
Four-node cycles		SRPG, SRPW, GACT, SACT, RACT	VYIH, VYMH, MYIH, VNID	LEFK
Number of nodes connected to a node	3 + 21	A T S G R P C W -	Y V H I M D N -	L E K F Q
Anticodon pairs 120/20 synthetases = 6		62/8 = 7.75	26/7 = 3.71	32/5 = 6.4
Synthetases in MSC 9/20 = 0.45		2/8 = 0.25	3/7 = 0.43	4/5 = 0.8
Synthetases not in MSC		Ser auto-aggregated Cys Trp Gly Ala Thr	Val Tyr His Asn	Phe

class I enzymes (just 8 class II). It is now realized that this corresponds to the excess of central A and central U attributions (just 7 central G and central C). A possible rationale, merely quantitative, would be that the aggregation into the MSC would be of help to the synthetase activity via stabilization inside the MSC. It is not possible to detail suggestions on the mechanisms since there is only the hint given by the known low ΔG of the central A:U dimers. The stabilization would be of help also in view of the relative isolation of the two small central A:U subnetworks, relative to the tight integration of the central G:C single network.

The scarcity of the central G:C attributions is not easy to explain, even making room for suspicion on the organization of the mixed principal dinucleotide sector. The enzymes lacking in the MSCs correspond mostly to module 3. There are five among the eight synthetases of the central G and C kind missing from the MSCs, GACWT, the last four belonging to module 3 (50%), while there are only three missing among the 12 of the central A and U kind (25%). Module 3 is also the only one not showing an independent evidence of the organization based on anticodon pairs (Fig. 1). On the contrary, the termination codes are distributed along the NYA hemi-row, which have given room to the proposition of the 'windmill' organization, with different topologies for the modules between the two sectors (17).

Such kinds of aggregate cytoplasmic organizations are usually interpreted as having the function of bypassing the need for transcriptional or translational regulation, therewith providing for quick and strong responses to necessities of the material stored in the aggregates. Some of the material stored may be used as such, which is the case of tRNAs and synthetases when they are recruited for the translational function. The model says that tRNAs would not be naked but associated with proteins: They would leave the MSC as aminoacyl-tRNA already in association with EF1A, and after the translational function, they leave the ribosomes in association with synthetases and get into the MSC again. The MSC aggregate may therefore have a dynamic and variable constitution, so that it is not expected to present precise stoichiometry.

Other functions of synthetases that participate in MSC are non-translational and much varied, a large part of them related to fragments of their proteolytic processing. It is said that these would depend on the release of the synthetases from the MSC so that they could be activated, meaning that the enzymes integrated into the MSC would be in a precursor form with respect to the non-translational function (Fang and Guo 2017; Ognjenović and Simonović 2017). The same might be said of the tRNAs, whose fragments may acquire different functions (Balatti et al. 2017; Keam et al. 2017; Millán et al. 2016; Schimmel 2017). While not being able to rationalize in detail the whole set of functional qualities compiled until now, a generic homeostatic participation of the MSC materials is proposed. It is typical of complex systems to offer challenges to explanatory endeavors.

Acknowledgements The friendly collaboration and support of Gustavo Maia Souza and Alfredo Pereira Junior, personally and along our journey in the self-organization group of UNICAMP.

Conflicts of interest No conflicts of interest are involved with the present communication.

References

Balatti V, Nigita G, Veneziano D, Drusco A, Stein GS, Messier TL, Farina NH, Lian JB, Tomasello L, Liu CG, Palamarchuk A, Hart JR, Bell C, Carosi M, Pescarmona E, Perracchio L, Diodoro M, Russo A, Antenucci A, Visca P, Ciardi A, Harris CC, Vogt PK, Pekarsky Y, Croce CM (2017) tsRNA signatures in cancer. PNAS 114(30):8071–8076. www.pnas.org/cgi/doi/10.1073/pnas.1706908114

Cho HY, Maeng SJ, Cho HJ, Choi YS, Chung JM, Lee S, Kim HK, Kim JH, Eom CY, Kim YG, Guo M, Jung HS, Kang BS, Kim S (2015) Assembly of multi-tRNA synthetase complex via heterotetrameric glutathione transferase-homology domains. J Biol Chem 290:29313–29328. https://doi.org/10.1074/jbc.M115.690867

Colussi TM, Constantino DA, Hammond JA, Ruehle GM, Nix JC, Kieft JS (2014) The structural basis of tRNA mimicry and conformational plasticity by a viral RNA. Nature 511:366–369

Diss G, Arsenault IG, Coté AMD, Vignaud H, Ascencio DI, Berger CM, Landry CR et al (2017) Gene duplication can impart fragility, not robustness, in the yeast protein interaction network. Science 355:630–634

Donoghue PCJ, Mark A, Purnell MA (2005) Genome duplication, extinction and vertebrate evolution. Trends Ecol Evol 20:312–319. https://doi.org/10.1016/j.tree.2005.04.008

Fang P, Guo M (2017) Structural characterization of human aminoacyl-tRNA synthetases for translational and nontranslational functions. Methods 113:83–90. https://doi.org/10.1016/j.ymeth.2016.11.014

Francis BR (2015) The hypothesis that the genetic code originated in coupled synthesis of proteins and the evolutionary predecessors of nucleic acids in primitive cells. Life 5:467–505

Fung AWS, Payoe R, Fahlman RD (2016) Perspectives and insights into the competition for aminoacyl-tRNAs between the translational machinery and for tRNA-dependent non-ribosomal peptide bond formation. Life 6:2

Goudry M, Saugnet L, Belin P, Thai R, Amoureux R, Tellier C, Tuphile K, Jacqet M, Braud S, Courçon M et al (2009) Cyclodipeptide synthases are a family of tRNA-dependent peptide bond-forming enzymes. Nat Chem Biol 5:414–420

Grosjean H, Houssier C (1990) Codon recognition: evaluation of the effects of modified bases in the anticodon loop of tRNA using the temperature-jump relaxation method. In: Gehrke CW, Kuo KCT (eds) Chromatography and modification of nucleotides. Elsevier, Amsterdam, pp A255–A295

Gruic-Sovulj I, Landeka I, Söll D, Weygand-Durasevic I (2002) tRNA-dependent amino acid discrimination by yeast seryl-tRNA synthetase. Eur J Biochem 269:5271–5279

Guimarães RC (1996) Anti-complementary order in the genetic coding system. Int Conf Orig Life 26:435–436

Guimarães RC (2012) Mutuality in discrete and compositional information: perspectives for synthetic genetic codes. Cogn Comput 4:115–139

Guimarães RC (2013) Formation of the genetic dode—review and update as of November 2012. http://www.icb.ufmg.br/labs/lbem/pdf/GMRTgeneticodeNov12.pdf. All original publications. https://www.researchgate.net/profile/Romeu_Guimaraes (both sites accessed on August 2017)

Guimarães RC (2015) Emergence of information patterns: in the quantum and biochemical realms. Quantum Biosyst 6:148–159

Guimarães RC (2017) Self-referential encoding on modules of anticodon pairs—roots of the biological flow system. Life 7:16. https://doi.org/10.3390/life7020016. www.mdpi.com/journal/life

Guimarães RC, Moreira CHC, Farias ST (2008) A self-referential model for the formation of the genetic code. Theory Biosci 127:249–270

Havrylenko S, Legouis R, Negrutskii B, Mirande M (2011) Caenorhabditis elegans evolves a new architecture for the Multi-aminoacyl-tRNA synthetase complex. J Biol Chem 286(32):28476–28487. https://doi.org/10.1074/jbc.M111.254037

Havrylenko S, Mirande M (2015) Aminoacyl-tRNA synthetase complexes in evolution. Int J Mol Sci 16:6571–6594. https://doi.org/10.3390/ijms16036571

Hordijk W, Steel M (2017) Chasing the tail: the emergence of autocatalytic networks. Biosystems 152:1–10
Iranzo J, Puigb P, Lobkovsky AE, Wolf YI, Koonin EV (2016) Inevitability of genetic parasites. Genome Biol Evol 8(9):2856–2869. https://doi.org/10.1093/gbe/evw193
Keam SP, Sobala A, St H, Hutvagner G (2017) tRNA-Derived RNA fragments associate with human multisynthetase complex (msc) and modulate ribosomal protein translation. J Proteome Res 16:413–420. https://doi.org/10.1021/acs.jproteome.6b00267
Kenkel CD, Matz MV (2016) Gene expression plasticity as a mechanism of coral adaptation to a variable environment. Nat Ecol Evol 1:0014
Lehmann J, Libchaber A (2008) Degeneracy of the genetic code and stability of the base pair at the second position of the anticodon. RNA 14:1264–1269
Millán PF, Schelcher C, Chihade J, Masquida B, Giegé P, Sauter C (2016) Transfer RNA: From pioneering crystallographic studies to contemporary tRNA biology. Arch Biochem Biophys 602:95–105. https://doi.org/10.1016/j.abb.2016.03.005
Mocibob M, Ivic N, Bilokapic S, Maier T, Luic M, Ban N, Durasevic IW (2010) Homologs of aminoacyl-tRNA synthetases acylate carrier proteins and provide a link between ribosomal and nonribosomal peptide synthesis. Proc Natl Acad Sci USA 107:14585–14590
Moras D, Dock AC, Dumas P, Westhof E, Romby P, Ebel J-P, Giegé R (1986) Anticodon-anticodon interaction induces conformational changes in tRNA: yeast tRNAASP, a model for tRNA-mRNA recognition. Proc Natl Acad Sci USA 83:932–936
Moutiez M, Schmidt E, Seguin J, Thai R, Favry E, Belin P, Mechulan Y, Goudry M (2014) Unravelling the mechanism of non-ribosomal peptide synthesis by cyclopeptide synthases. Nat Commun 5:5141–5146
Murren CJ, Auld JR, Callahan H, Ghalambor CK, Handelsman CA, Heskel MA, Kingsolver JG, Maclean HJ, Masel J, Maughan H et al (2015) Constraints on the evolution of phenotypic plasticity: Limits and costs of phenotype and plasticity. Heredity 115:293–301
Ognjenović J, Simonović M (2017) Human aminoacyl-tRNA synthetases in diseases of the nervous system. RNA Biol 23:1–12. https://doi.org/10.1080/15476286.2017.1330245
Park JW, Yan ZZ, Loh H, Will SA, Zwierlein MW (2017) Second-scale nuclear spin coherence time of ultracold $^{23}Na^{40}K$ molecules. Science 357:372–375
Ptashne M (2013) Epigenetics: core misconcept. Proc Nat Acad Sci USA 110:7101–7103
Schimmel P (2017) The emerging complexity of the tRNA world: mammalian tRNAs beyond protein synthesis. Nat Rev Mol Cell Biol. https://doi.org/10.1038/nrm.2017.77
Schlosshauer M (2014) The quantum-to-classical transition and decoherence. In: Aspelmeyer M, Calarco T, Eisert J, Schmidt-Kaler F (eds) Handbook of quantum information. Springer, Berlin/Heidelberg
Seligmann H, Ganesh W (2017) Genetic code optimization for cotranslational protein folding: codon directional asymmetry correlates with antiparallel beta-sheets, tRNA synthetase classes. Comput Struct Biotechnol J 15:412–424. https://doi.org/10.1016/j.csbj.2017.08.001
Souza GM, Lüttge U (2015) Stability as a phenomenon emergent from plasticity–complexity—diversity in eco-physiology. In: Lüttge U, Beyschlag W (eds) Progress in Botany 76, 211–239. Springer International Publishing Switzerland. https://doi.org/10.1007/978-3-319-08807-5_9
Widmann J, Di Giulio M, Yarus M, Knight R (2005) tRNA creation by hairpin duplication. J Mol Evol 61:524–530
Xia T, SantaLucia JJr, Burkard ME, Kierzek R, Schroeder SJ, Jiao X, Cox C, Turner DH (1998) Thermodynamic parameters for an expanded nearest-neighbor model for formation of RNA duplexes with Watson-Crick base pairs. Biochemistry 37:14719–14735
Zurek WH (1991) Decoherence and the transition from quantum to classical. Phys Today, 36–44 Oct
Zurek WH (2002) Decoherence and the transition from quantum to classical—revisited. Los Alamos Sci 27:2–25

Brains Emerging: On Modularity and Self-organisation of Neural Development In Vivo and In Vitro

Paul Gottlob Layer

Abstract Molecular developmental biology has expanded our conceptions of gene actions, underpinning that embryonic development is not only governed by a set of specific genes, but as much by space–time conditions of its developing modules (*determinate vs. regulative development*; or, *nature vs. nurture* discussion). Typically, formation of cellular spheres, their transformation into planar epithelia, followed by tube formations and laminations are modular steps leading to the development of nervous tissues. Thereby, actions of organising centres, morphogenetic movements (in- and evaginations), inductive events between epithelia, tissue polarity reversal, widening of epithelia, and all these occurring orderly in space and time, are driving forces of emergent laminar neural tissues, e.g. the vertebrate retina. Analyses of self-organisational formation of retina-like 3D structures from dispersed cells (so-called *retinal spheroids,* also called *retinal organoids*) under defined cell culture conditions (in vitro) demonstrate that not only particular genetic networks, but—at least as important—the applied culture conditions (in vitro constraints) define phenotypes of emergent tissues. Such in vitro approaches allow assigning emerging tissue formation to ground-laying genetic networks separately from contributions by conditional constraints.

Introduction: Biologic Determinism Revisited

Preformation and *epigenesis* as mutually exclusive ideas have over centuries dictated the quest for understanding of how organisms come into living. Epigenesis (not to be mistaken for *epigenetics*), as was first formulated by Aristotle, postulates new formation of the entire organism in each generation from scratch, i.e. envisions concepts of development. On the other side, ideas of preformation hold that the final organism is already somehow preformed in the egg (or, alternatively the sperm head; Malphigi 1672; see in Jahn 2000; Gilbert 2016), which then has only to be unrolled during embryonic growth. Preformationism, which has never vanished in biology

P. G. Layer (✉)
Darmstadt, Germany
e-mail: layer@bio.tu-darmstadt.de

L. H. Wegner and U. Lüttge (eds.), *Emergence and Modularity in Life Sciences*,
https://doi.org/10.1007/978-3-030-06128-9_7

completely, belongs to the category of determinism, while concepts of epigenesis rely on processes of emergence.

As biologists in the nineteenth century tried to advance their science to a more "exact science", determinism became a common position of eminent figures in biology. Ernst Haeckel presented hundreds of newly discovered protozoa in his famous plates not only as shiny colourful beauties, but also in perfect geometrical symmetry, certainly trying to make the point that a mathematical precision was behind their making (Haeckel 1904, 1998). Haeckel, certainly a shiny figure himself in many respects, was reductionist, monist and determinist. August Weismann, after having detected the early separation of germ and somatic cell lines in embryos, concluded that certain distinct (chemical) "determinants" would predetermine the fate of all cell types, and that only germ cells contained all determinants for the entire future body of a next-generation organism ("mosaic development"). Accordingly, each and every feature (morphologic, physiologic, etc.) would be completely determined by its respective determinants. Supporting this concept, Wilhelm Roux in 1887 had achieved half frog larvae (hemi-embryos), after having killed experimentally one cell of the two cell-staged frog embryos (an experiment which was hampered by methodological flaws). Hans Driesch, in trying to provide support of Roux' findings, managed to separate a four cell-stage sea urchin embryo into its four cells. To his surprise, four little but quite normal sea urchin larvae developed in his culture dish. What became to be called *developmental regulation*, was at the same time the discovery of stem cell totipotency. By then, embryologists had revealed good reasons to conceive development of an organism not as a mere unrolling of a prefixed programme.

During the same period, however, deterministic concepts in biology received strong support through great progress of the upcoming genetic era. Works of Beadle and Tatum in the early forties on the ascomycete *Neurospora grassa* (co)-founded the so-called *dogma of molecular biology* (see Strauss 2016), which stated that one gene codes for one (and only one) protein, and that each protein subserves one distinct function (e.g. enzymatic, structural, etc.). Although these early geneticists themselves were quite cautious in interpreting their findings one-dimensionally, genes then became more and more considered as completely autonomous, autocratic players ("determinants" in Weismann's words), each one sitting on top of a hierarchical cascade.

The development of Neo-Darwinism during the first half of the last century as a standard theory of evolution was much influenced by this concept. It led Ernst Mayr and colleagues to their famous saying "nothing comes between genotype and phenotype"; in fact stating that in order to understand evolution we do not have to bother with development and/or morphologies of embryos (phenotypes), but only with the genomes of adult organisms (capable of reproduction). What presumptuous, exclusive misconceptions, which have come to be called *gene-centrism* and *adultocentrism*: biologic determinism at its best! As a rather new subbranch of *Developmental Biology* now *EvoDevo* (idiom. for *Evo*lutionary *Develo*pmental Biology) has developed, which for the first time provides reasonable clues to mechanisms of macroevolutionary change (Gilbert 2016).

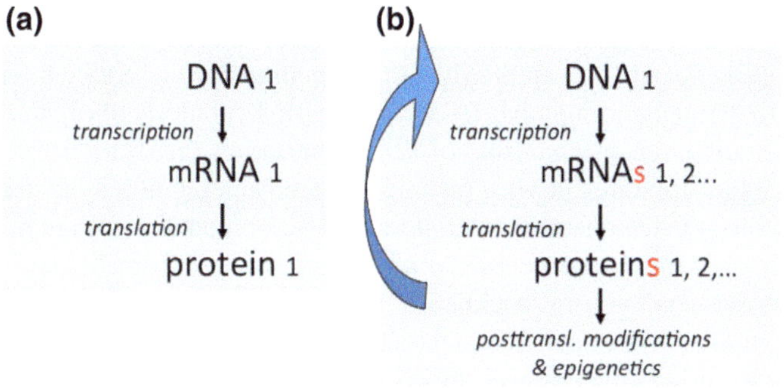

Fig. 1 Classic (**a**) and modern (**b**) concepts of gene realizations. According to (**b**), one gene ("DNA_1") can code for many different proteins, and proteins can feedback on gene activities. Further see text

Time was waiting for the rise of molecular developmental biology from the seventies onwards to achieve a new concept of development. As more and more model organisms were studied, minds of researchers were opened. Actions of genes became conceived as embedded within widely distributed networks, regulated by complex signalling cascades (Fig. 1). Thereby, feedback mechanisms between proteins and genes (transcription factors) can lead to prominent autocatalytic amplifications, or, as well, to silencing of particular genes (inhibition). *Time* and *space* of gene expression became decisive aspects of their actions, revealing the insight that one particular gene can affect many different things. Strict determinism in biology lost its appeal.

Concepts of biological emergence take a decisive anti-deterministic stand; they decline exclusive *gene-centrism*, and favour concepts of "nature *and* nurture". Emergence has been defined as the appearance of a new property in a system at a higher level of organisation, which is not explained by properties of a lower, more fundamental level. Such new properties are not predictable by, and not reducible to the more fundamental properties. Emergence deals with dynamic processes, e.g. processes of appearance (and disappearance), by the insight that "…something comes out from something …" (Fromm 2005; see other contributions in this book). Typically, *weak emergence* is distinguished from *strong emergence* (Chalmers and Jackson 2001). Thereby, "weak" means that the emerging properties are unexpected based on the lower-level properties, while "strong" defines new properties which—even in principle—are non-deducible and unpredictable from the given lower-level properties.

Clearly, the field of *Developmental Biology* is governed by emerging properties. As in all fields, features of emergence in biology are difficult to grasp. Nonetheless, are there means to characterise such processes for a developing organism? What are distinguishable levels of development of an animal? What are building modules, which level is lower, and which is above, if these levels are interrelated by complex feedback mechanisms? What means self-organisation, is it predictable; if not, why not? Such are the questions which are tackled in this chapter, which is divided into three parts.

1. A description of general aspects of normal (e.g. in vivo) animal development from a fertilised egg until—exemplarily—the formation of a vertebrate brain, thereby trying to define building modules and morphological levels of organisation.
2. Considerations on mechanisms of self-organisation (generation) of organised tissue/organ structures in vitro (as nowadays emanating into stem cell regeneration biology), demonstrating that normal developmental paths are not the only possible ones to achieve a certain goal ("many roads to Rome") and
3. A discussion on "genetic backbones" of modules in relation to "environmental constraints" (physical, chemical and ecological) that could drive emergent processes during development, independently from a particular causative gene action.

Modules Governing Normal Development

For long periods in the prehistory of life on our planet, life existed only in the form of unicellular organisms (3.5–1.8 Gya, giga years ago, or, billion years ago). The so-called prokaryotic cell was a "simple" molecular bag, having—as one of its notable features—no real nucleus. A major change occurred with the invention of an entirely new form of cell. Besides other essential novel organelles, the *eukaryotic cell* was equipped with a complete nucleus containing the genetic information and a double-layered outer cell membrane (plus a cell wall in the plant cell). Illustrious, spectacularly shaped unicellular organisms, called Protista, began to populate our planet (1.8–1.4 Gya). Only now the scene was set for the evolution of higher life, which—as we should have understood by now—certainly never was, and still is not possible without continuous mutual interactions with the prokaryotic world (McFall-Ngai et al. 2013).

Cells Forming Spheres

At some later point of evolution (1.4 Gya), particular eukaryotic cells developed a tendency to form small cell clusters, as a first sign of development of multicellular organisms. As still nowadays can be observed with green algae new species emerged step by step that would form larger and larger cell aggregates (here not considering that some prokaryotes also can associate to large biofilms). There are multiple hypotheses how multicellularity was achieved during evolution (Grosberg and Strathmann 2007), one of them suggesting colony-forming signals from bacteria onto eukaryotic cells (Alegado et al. 2012). Such colonies could still disaggregate under certain circumstances, and each individual cell would multiply by normal cell division (mitosis). Eventually, much larger, more organised species emerged as—for instance—are represented by some green algae of the order *Volvocales*. Presenting themselves under the microscope as splendid translucent spheres, they steadily rotate in their water habitat; that is why they became named "Volvox" (from *latin* "volvere", to roll, rotate). Their individual cells were not identical any longer, but began to show signs of specialisation (e.g. flagella for motion), revealing the evolutionary onset of differentiation. Besides so-called somatic cells, they also produced reproductive cells. Their progeny was kept inside the spherical body, there forming *spheres within spheres*, until the outer body would release them and the original parent sphere would disintegrate and die. Along with the invention of multicellularity, cell and tissue differentiation, sexual reproduction and cell death had entered the living world. Hence, aggregation of cells into more and more regular spheres characterised this period.

This is not the place to engulf further into the spectacle of early evolution, but only to point out that the first multicellular shape within which cells organised themselves during *phylogeny* was the *cellular sphere*. Amazingly, a similar sequence of early events happens during the development of nearly each and any individual animal, during their *ontogeny*. After fertilisation of the egg, fast cell divisions amplify cell numbers (*cleavage divisions*), thereby forming a spherical ball of cells, a *blastula*. As in phylogeny, the sphere is the earliest and simplest multicellular structure in each individual´s life. Such an assembly of cells could be considered the simplest developmental module, with which new capacities/functionalities can and will emerge (e.g. communication between cells; see below). Sphere formation is an ever-recurring theme in biology: for instance, during the development of kidneys, liver, lungs and testes; in brain formation, cellular spheres will form brain nuclei or ganglia (e.g. dorsal root ganglia, DRG). Not to forget, as tissues disintegrate during cancerogenesis, tumours grow in the shape of spheres.

From Hollow Spheres to Planar Tissues

As we follow the developmental paths in different animals, patterns of development become more difficult to generalise. As blastulae in model animals like sea urchin or frogs grow bigger, a fluid-filled space emerges in their interior (blastocoel). Nearly in all animals, the following process of *gastrulation* represents a real cellular revolution. Spherical blastulae become quite abruptly transformed by an invagination of their outer parts (note: shapes of blastulae and types of *morphogenetic movements* differ greatly, depending on species). A distinction between inner and outer parts emerges with entoderm and ectoderm representing the first two *germ layers*. In most animal branches, the mesoderm as an intermediate germ layer pushes itself in between the other two (in fact, the—future—mesoderm appears to exert an initiating and driving force during gastrulation). Notably, along with these transformations creating three novel modules, cells transit from a more globular to a layered arrangement. Concomitantly, in some animal groups (Coelomata) a secondary fluid-filled bodily space forms the so-called *coelom* (abdominal cavity, dt. *sekundäre Leibeshöhle*). That is, from now onwards cells are not assembled any longer within a spherical volume, but they have become organised within planar cell layers, which marks the beginning of tissue formation.

The Epithelium, the Most Basic Tissue

In histology, several types of tissues are distinguished (epithelium, blood, fat, nerve, muscle and bones/supportive). The only one that is relevant here is the epithelium. Epithelia are widespread in all animal bodies, covering outer and inner bodily surfaces, like skin, gut and capillaries, in embryonic and mature organisms alike. In an epithelium, many cells of a particular type are arranged "side-by-side", forming (in its simplest form) a one cell-wide layer in planar register. Along with their integration into a compound tissue, cells attain the same cell shape (e.g., cylindrical, cuboidal, etc.). Driven partially by active as well as passive forces, formation of epithelia represents an emergent process. The cell plane as a whole is polarised by a basal and an apical side, representing its inner and outer surface, respectively. The basal side is endowed with an extracellular matrix for optimal contact; the apical side presents protrusions (e.g. cilia, microvilli) for secretion, transport of fluids, etc. Several types of cell-to-cell junctions connect neighbouring cells, to stabilise the whole tissue and allow communication between all cells of the tissue. Each epithelium will subserve specialised functions, such as mechanical protection, containment of fluids and gases, ingestion or glandular secretion. Planar epithelia of diverse morphologies (simple, stratified and pseudostratified) will form tubes as essential parts of intestines, lungs, blood circulations (called *endothelia*) and heart. Each one tissue type represents an organismic building block, a module, which only as such (not the individual cells) can fulfil its distinct function(s).

Brain and Eyes Emerging from the Body Surface Epithelium

The initial step of neurogenesis is nothing but formation of an epithelial tube, derived from the ectoderm, a process called *neurulation*. Shortly following gastrulation, a mesodermal rod-like structure, the *chorda dorsalis*, is formed along the length of the embryo and becomes an *organising centre* for the steps coming. Chemical factors secreted from the chorda induce the overlying ectoderm to form an inwardly oriented, longitudinal groove. The groove closes dorsally to form a tube and separates from the overlying ectoderm. Then, the tube enlarges and differentiates in rostral–caudal direction, e.g. the future head is always farther developed than trunk and tail regions. Notably, some features that could be marginalised as "inevitable side products" will be indispensable for development of the nervous system. A population of cells that "accidentally" escapes during the process of tube closure, called *neural crest cells*, will migrate on defined paths out into the body space. The *neural crest* represents a major building module to form—besides other parts—the entire

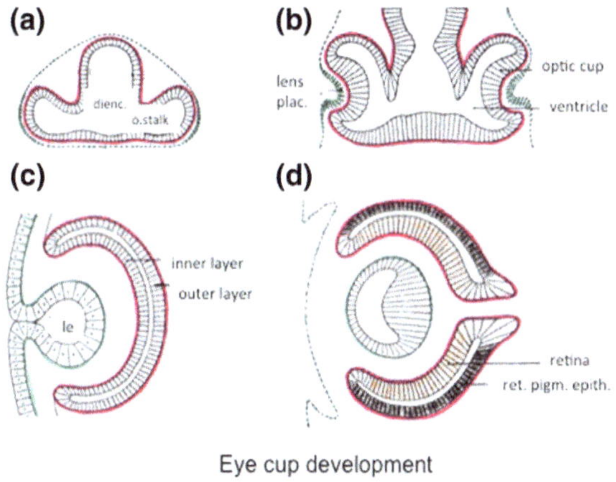

Fig. 2 Schematics of vertebrate eye cup formation. **a** Stage of optic stalk evagination from diencephalon. **b** Invaginating neuroepithelium after contact with ectoderm; lens placode is induced; **c** an inner and an outer layer of the neuroepithelium form the eye-cup; lens vesicle has enlarged; **d** inner layer forms retina, outer layer forms pigmented retinal epithelium (RPE), lens differentiates

peripheral nervous system. Due to invagination of the ectoderm during neural tube formation, its inside-out polarity becomes reversed, e.g. the basal side will become the outside of the neural tube (see Fig. 2, and further below on eye development). As the tube extends in length and thickness, space restrictions within the future head will cause tube flexures, bends and partial rotations (note: this result is an excellent example for a mechanic rather than genetic causation). Along with it, the rostral (front) end of the tube is constricted into first three, then five brain vesicles (front-, mid- and hindbrain vesicles, or *Latin*, tel-, mes- and rhombencephalon), representing the first subdivisions of the rostral tube. All brain vesicles will be further subdivided into *neuromeres*. These become most evident in the hindbrain (rhombencephalon) as a series of numbered *rhombomeres* (Fig. 3). The number one rhombomere will later develop into the cerebellum. Following differentiation of the tube towards more caudal parts, the future trunk and tail regions will be segmented. Thereby, a close interplay between neural tube structures and mesodermal tissue (e.g. somites), muscular and skeletal anlagen is strictly controlled by a rostro-caudal clockwork (not further detailed here). Modularity of brain development is overtly demonstrated by these longitudinal subdivisions of the frontal neural tube since from each and every neuromere a distinct part of the future brain will develop (Lumsden and Keynes 1989; Layer and Alber 1990; Puelles 2001).

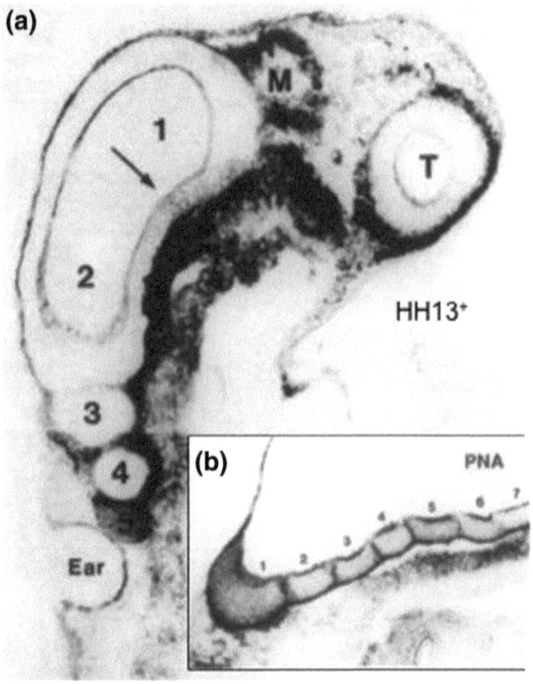

Fig. 3 Emergence of molecular boundaries in hindbrain of chicken embryos. **a** Sagittal section of a 2 day-old ($HH13^{+}$) chicken head and **b** a more horizontal section of a 3 day-old hindbrain, both stained by PNA lectin (black). Rhombomeres of hindbrain are numbered 1–7. Note diffuse emergence of staining between R1 and R2 at $HH13^{+}$ (arrow in **a**). By HH17 (**b**), all boundaries in between rhombomeres 1–7 are strongly stained. Further see text. Pictures taken from Layer and Alber (1990)

Neural Tube Evagination, Invagination and Widening to Form an Eye

The eye, in particular, retina and pigmented epithelium (RPE) are derived from the neural tube also. From the first brain vesiclc, the neuroepithelium evaginates laterally to eventually touch the ectodermal surface (Fig. 2; eye formation). Being stopped at a point that marks the origin of the lens, the so-called optic stalk once again invaginates to form a double-layered optic cup; the outer layer will soon turn into the black RPE, the inner will differentiate into the retina. Similar to movements during gastrulation, evagination and invagination of epithelial tissues lay the grounds for eye-cup formation.

The neural tube presents some unique epithelial features that found later formation of neuronal cell layers and networks during brain development (*lamination* or *stratification* of brain regions). As cells heavily divide within the neural tube, individual cell bodies shift back and forth between inside (apical) and the outside (basal) side, while their radial processes remain anchored to both epithelial surfaces. Each transversal (radial) position of a cell body correlates with a specific state within the cell cycle. Due to these *interkinetic migrations*, the neuroepithelium is wider than other unistratified epithelia. Under a microscope, it appears as if it would be stratified; therefore, it is called *pseudostratified neuroepithelium*. After a dividing cell undergoes its last mitosis, one of the emerging two daughter cells will continue to divide, while the other cell, which has now become "postmitotic", will migrate to the outer surface and begin to differentiate, e.g. it will send out a neuronal process. Consequently, a mantle layer forms on the outside of the tube, which marks the beginning of cell layer formation (lamination and stratification; see Weikert et al. 1990). In different areas of the future brain, lamination will follow different schemes (e.g. inside-out scheme in cortex, lamination of cerebellum or retina, etc., see below). Now, future network formation will set in: neuritic outgrowth, path and target finding of neurites to/into distant brain areas (e.g., eye/retina to tectum), thereby establishing connections between neurons of different layers and areas. Synapse formation, refining of connections by their use, according to *fire-and-wire* mechanisms (see Glossar), only are some of further emerging steps of a maturing complex brain (here not further discussed).

Retinogenesis is comparable in all vertebrates, forming three nuclear (ONL, INL and GCL, see Abbrev.) and two plexiform layers (OPL, and IPL); of course, in detail, there are many species-specific differences not dealt with here (Fig. 4). In the forming eye-cup, the inner layer widens, since interkinetic cell migrations are prominent in the future retina. The first cells begin to differentiate at the inner border of the retina (e.g. basal side). The retina differentiates gradually from central to the eye periphery near the lens. As a rule, big cells are born before small cells, e.g. ganglion cells and photoreceptors, then amacrine and horizontal cells, and finally bipolar and radial glial cells (for different retinal cell types, see below and legend to Fig. 4). Vertebrate photoreceptors, which are considered the most complex cells in nature, become located at the outer interface next to the RPE. Their well-being during development and adult functioning depends heavily on mutual relationships with the RPE. During the first phase, photoreceptors in some species target directly on to ganglion cells, the terminal retinal cell type which will send an axon to the brain. Only as the network further matures, entrance (PRs) and exit cells (GCs) will become interconnected through interneurons. As amacrine ("without process"), horizontal and bipolar cells are born, they become located in an intermediate "inner" nuclear layer (INL). All neurons become wired together at the level of two synaptic layers, called inner and outer plexiform layers: first the inner plexiform layer (IPL) will emerge, followed by the outer OPL. Precursors of radial glial cells (Müller cells) spanning through the entire retina, stabilise the tissue during development (Reichenbach and Bringmann 2013). Being last to differentiate, they retain hidden features of stem cells, rendering them with capacities for retinal homeostasis and regeneration.

In summary, formation of cellular spheres, their transformation into planar epithelia, followed by tube formations are decisive steps leading to the development of nervous systems, which—as is dealt with in section "Decoding Self Organisation of Brain Tissue Formation (Genetic Backbone Versus Non-genetic Constraints)"—can be conceived as developmental modules. Thereby, morphogenetic movements (e- and invaginations), mechanic forces, inductive events between epithelia, polarity reversal, widening of epithelia are driving forces of emergent laminated neural tissues, like the retina.

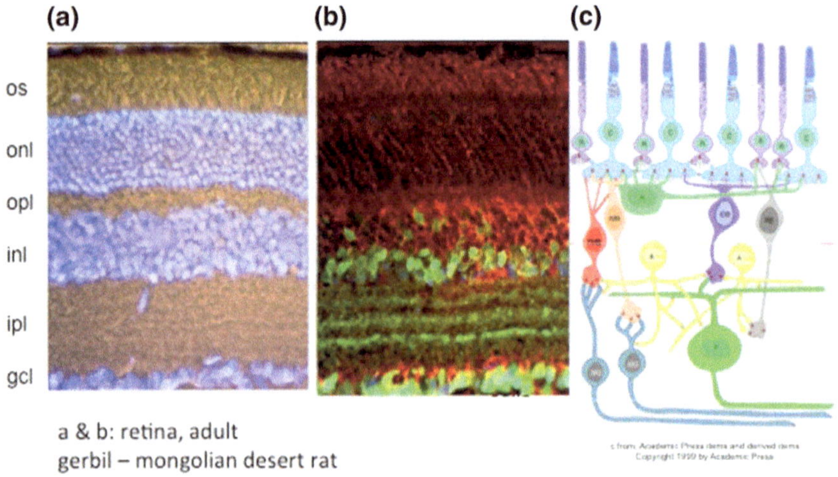

Fig. 4 Stratified (laminar) structure of vertebrate retinae, as represented by DAPI- (**a**), and Pax6-stained (green in **b**) retina sections of an adult Gerbil. Note three layers of cell bodies (ONL, INL, GCL in **a**), and synaptic sublaminae formed by $Pax6^+$ neurites from neurons in INL and GCL. **c** Network scheme of vertebrate retinae, consisting of five major neuronal cell types (photoreceptors, horizontal, bipolar, amacrine and ganglion cells), interconnected in OPL and IPL; radial Müller glial cell is not shown

Self-organisation of Neural Tissues In Vitro from Stem Cells

When development of a tissue or organ is being studied in its normal in vivo environment, effects due to cell-autonomous factors often cannot be clearly distinguished from external factors. Thence, causes of self-organisation or emergence of tissues remain ambiguous or occluded. One way to overcome this drawback relies on performing tissue culture experiments. With standard procedures, cells isolated from a specific tissue (e.g. embryonic, brain part, diseased organ, etc.) are raised in a tissue culture dish, whereby the cell environment (atmosphere, media supplements, temperature, etc.) can be fully controlled. Depending on chosen culture conditions, cells will settle on the surface of the dish and proliferate. Cell division stops as soon as a more or less densely populated cell carpet is formed, and cells begin to differentiate. For instance, conditions of neurite outgrowth from embryonic neurons and of synapse formation between them can be studied at ease. In such two-dimensional (2D), or "flat" cell cultures, however, a cellular compound resembling a normal tissue formation is never achieved (except for some clustering of cells, in particular so with malignant cancer cells).

Emergence of Tissues In Vitro: Cell Reaggregation and Sphere Formation

As at the phylogenetic base of multicellular organisms (see above), formation of cellular spheres from the fertilised egg represents the most basic module of each individual development. In this respect, the postulate of *a recapitulation of phylogeny in ontogeny* fits well (ascribed to Haeckel, but in fact, was already formulated earlier by Johann Friedrich Meckel and Fritz Müller; see Jahn 2000, p. 373). Hence not surprisingly, 3D cell cultures provide a superior approach over 2D cultures to demonstrate and probe self-organisational cellular capacities to form distinct tissues. In applying 3D cell culture techniques, fully dissociated stem cells from embryonic organ anlagen, or from some other source are constantly kept under rotation during their culturing (suspension cultures). Thereby, dispersed cells quickly reaggregate and form more or less regular cellular spheres. Under defined and optimal in vitro conditions, they can form tissue-specific structures. Besides improved nutritional and oxygen supplementation of cells, a major advantage of using 3D over 2D cultures are enhanced interactions between aggregating cells, which are promoted through constant movements of dispersed cells.

Self-organisation of a Chicken Retina from Precursor Cells

To form an organised "histotypic" tissue in vitro needs more than initial reaggregation and sorting-out processes. To this end, the chicken embryonic retina had proven an ideal study model already in the forties, not only because the retina is easily reachable within the eye, but also because retinal cells can be instantly distinguished from black cells of the retinal pigmented epithelium (RPE). Earlier work had revealed that RPE cells sort out in the centre of mixed retina/RPE reaggregates. Since RPE and retinal cells mutually influence each other (reviewed in Layer and Willbold 1994; Layer et al. 2010), in the early eighties we added RPE cells to retinal 3D cultures of the chick. Immediately, we could detect highly ordered spherical structures (Fig. 5; Vollmer et al. 1984). The histology of *stratospheroids* reveals an almost complete threefold retinal lamination, much comparable with the normal retinal lamination (Fig. 5c). This experiment demonstrated for the first time in history that formation of a nearly complete neuronal tissue can be experimentally reconstituted through self-organisation from stem cells in vitro (we called these structures *retinal stratospheroids*). Before their formation can be analysed in more detail, a more basic type of retinal reaggregate, which we have called *rosetted spehroids*, needs to be explained (Figs. 5b and 6).

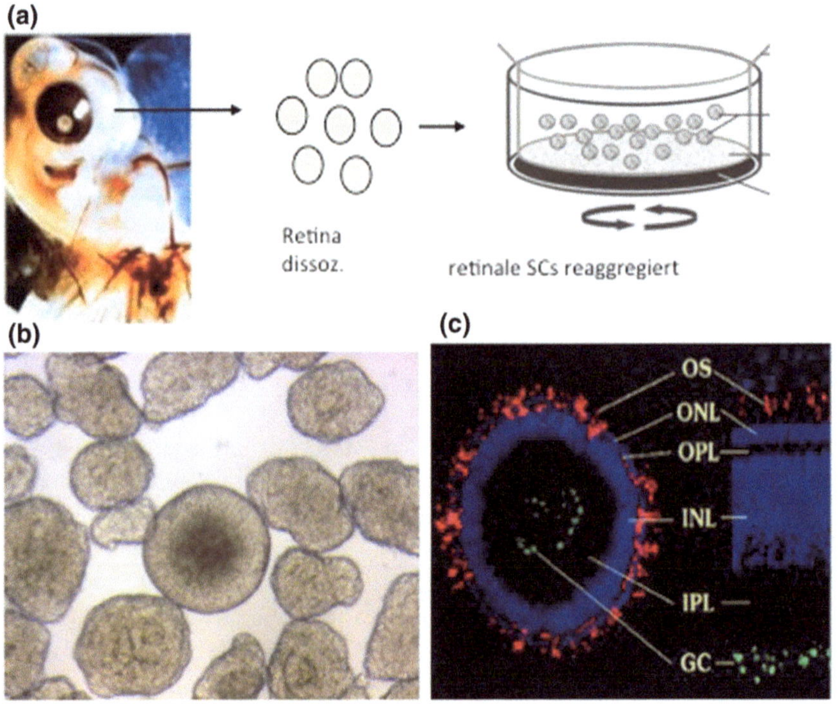

Fig. 5 Production (**a**, **b**) and histologic structure of correctly stratified retinal spheroids (called *stratospheroids*, **c**; see one in centre of **b**) from retinal precursor stem cells of the chicken embryo. **a** The retina is isolated from the eye and dissociated into single cells. Cultured under constant rotation, cells reaggregate into more or less regular cellular spheres (**a**, **b**). The potato-shaped spheres in (**b**) are *rosetted spheroids* (see Fig. 6)

Spheres Within Spheres: Rosettes and Clonal Cell Columns as Modules

As cells have been sorted out within spheres, their initial random distribution has much diminished. As a next step of tissue organisation, sorting-out is directly associated with emergence of rosettes (note: with murine cells, different processes apply; see below "many roads to Rome"). Groups of few segregated cells form several small cell rosettes within a much larger spheroid (within hours for chick cells). Thereby, rosettes are dividing stem cells that have—in principle—formed a small circular, but already epithelial compound (Fig. 6a, equiv. to spheres within a sphere; cf. *Volvox*). Through cell division newborn mitotic cells are integrated laterally into this rosette, which thereby enlarges; internally, a fluid-filled space inflates. At the same time, clones of postmitotic cells are produced from precursor cells within the rosette (Fig. 6a, b). These daughter cells are stacked upon each other to present transversally oriented cell columns, which are stabilised by processes of radial glial

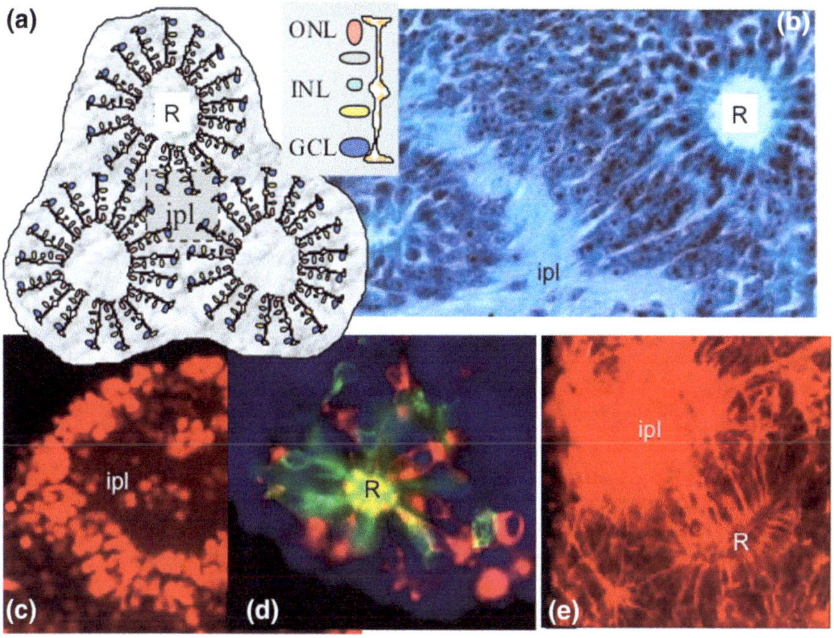

Fig. 6 Rosette ("R") and cell-column formation in *rosetted retinal spheroids*. **a** Schematic of internal structure of rosetted spheroids; note that photoreceptors point inside the rosette; insert in middle represents one cell column, consisting of all major cell types as a basic construction module. **b** HE-stained section of a rosette; coherent cell columns are evident. **c** Shows Pax6-stained amacrine cells of INL and GCL; **d** rod (rot) and cone photoreceptors (green) are located in rosette; **e** radial glial cells emanate from rosette towards IPL-like space

precursor cells. Columnar cell clones become neatly stacked one-by-one, thereby surrounding each one rosette (see Fig. 6a, b). Cells within columns then differentiate into various retinal cell types, e.g. photoreceptors, amacrine, horizontal and bipolar cells. Therefore, by the two processes of *rosette enlargement* and *column formation* (*lateralisation* and *radialisation* of rosettes), modules of laminar retinal tissues have emerged within a larger spheroid.

From Spherical Compounds to Planar Tissue: Fusion and Tissue Inversions

How can transformation from a *rosetted* into a *planar* arrangement of cells be achieved? At the outset of retinal spheroid formation, development of *strato-spheroids* follows a similar path as that of *rosetted spheroids*. As their modular units (rosettes, see above) have reached a certain size, several of them will fuse. Often, these larger structures present an inverted laminar arrangement, e.g. future photoreceptors tend to be found internally and amacrine cells on the outside (note: in vitro ganglion cells quickly will die, due to absence of growth factors). Only after a complete reversal of the entire spheroid, a correctly layered retinal sphere, the retinal *stratospheroid* will be achieved. Thus, formation of rosettes and of cell columns represent spatial in vitro preconditions for further cell-layer differentiation, followed by the establishment of interconnecting networks (synaptic layers of IPL and OPL; not further discussed here). These different retinal spheroid models became the most instrumental to learn about self-organisational tissue formation from isolated cells (see below in section "Decoding Self Organisation of Brain Tissue Formation (Genetic Backbone Versus Non-genetic Constraints)").

Brains Emerging In Vitro—Brain Organoids Have a Great Future

Having been neglected for a long time, only with the recent rise of stem cell biology the advantages of three-dimensional suspension cultures were again fully recognised. In particular, the availability of human *induced pluripotent stem cells* (iPSCs), highly structured retinal spheroids derived from human iPSCs now can be produced, called *organoids* (Meyer et al. 2009; Eiraku et al. 2011; Lancaster et al. 2013; Zhong et al. 2014). Organoids from hiPSCs resembling human *gastrulae*, so-called *Gastruloids*, are spectacular since they can form a *primitive streak* (area of gastrulation and onset of neurulation). After some authors considered these structures as "synthetic human embryos" (*sheefs*), a public dispute came up as to whether sheefs may become endowed with a human mind and consciousness. At any rate, organoids from retina or from other organs clearly have a great future in regenerative and transplantation medicine (Huch et al. 2017). The present hype on human organoids is based on two envisioned fields of applications: 1. human organoids could possibly be used for transplantation purposes to replace diseased organs, e.g. to cure blinded people. For some organs, e.g. skin, pancreas and liver, applications may become feasible soon, while for others there are still huge obstacles to be mastered (brain, retina, etc.). Successful first trials are ongoing. 2. At least as important, human organoids are already much applied as test models to analyse causes and possible cures of certain diseases. For instance, causes for congenital microcephaly disorders were analysed in cerebral organoids (Lancaster et al. 2013). Their applications will provide pharmacological and toxicological assay systems, which will help to drastically replace animal experiments. In fact, patient-specific (autologous) assays should become feasible, which would allow to test drugs and their side effects directly on a patient's in vitro tissue. Thus, 3D stem cell cultures form the basis of modern *Tissue Engineering* (Huch et al. 2017). Its present progress would not have been possible without extensive basic analytical research on construction principles of spheroids from different embryonic tissues, which will be described below.

Decoding Self Organisation of Brain Tissue Formation (Genetic Backbone Versus Non-genetic Constraints)

Section "Modules Governing Normal Development" has briefly outlined the development of animals by sequential processes from a fertilised egg to the cellular, then to histological (tissue) and organismic levels. Using retinal in vitro tissue regeneration as an example, section "Self Organisation of Neural Tissues In Vitro from Stem Cells" documented that a population of dispersed stem cells can find ways to rearrange, multiply and eventually form a tissue that is highly comparable to its in vivo counterpart, a result apparently favouring autonomy of retinal tissue formation. However, particular details of in vitro retinal development were clearly dependent on specific features of the provided culture conditions. Can these findings help to analytically resolve to what extent emergent features contribute to brain development?

Each developmental step is regulated by underlying complex genetic-molecular networks. At the same time, each completed step brings with it novel environmental conditions, which in turn exert *constraints* on possible future (genetic) steps. On all organisational levels, from molecular up to organismic (including most decisive interactions with microbioms; see excellent review by McFall-Ngai et al. (2013), and ecological), such constraints bring about situations of needs or even stress that necessitate some reaction(s). Constraints upon genetic activities can be of purely physical nature (e.g. traction, pressure, gravitation, shape, sorting-out, temperature and pH) or chemical nature (cytokines, paracrine factors, hormones and nutritional status). Constraints can also originate from restricted time windows, limited spatial options, evolutionary relicts and more. Recent *EvoDevo* research defines these constraints as heterochronic, heterotopic and phyletic, respectively (Gilbert 2016). The following section attempts to decipher how much of retinal development can be assigned to genetic determination (is predictable), and how much to non-genetic constraints (not reducible and not predictable)?

Common Genetic Backbone In Vivo and In Vitro

Progress of modern molecular biology brought tremendous novel insights into the *nature versus nurture* dispute; whereby "nature" refers to the genetic backbone of a system, while "nurture" points to non-genetic (environmental) actions upon it. In fact, understanding of modular developments—as analysed above histologically—has now achieved molecular and genetic bases. To mention just a few examples: a spatial gradient of a fibroblast growth factor (FGF) and a counter-gradient formed by retinoic acid together balance segmentation of the neural tube in rostro-caudal dimension. Then, codes of *Hox (master)genes* define the identities of hindbrain *rhombomeres*, as well as those of cell layers and cell types in several brain areas (example eye development, see Meyer et al. 2009). Notably, the so-called *Wnt signalling pathway* is one of the most relevant molecular regulators of early development. Briefly, a cell-external *Wnt protein* binds to its cell-surface receptor. Receptor activation then initiates an intracellular molecular cascade, eventually regulating the expression of particular nuclear genes. This cascade is involved in a multitude of developmental processes (e.g. cell movements, axis specification and regionalisation of tissues), including the organisation of planar epithelia. In case of retinal spheroids, the molecular basis of tissue reversal remained obscure for a long time; although we had detected that it can be induced by RPE and also by Müller glial cells. Several groups including ours searched for a lamina-inducing factor in retinal spheroids. Some growth factors, such as FGF, PEDF and GDNF (see Abbrev.) affected the ratio of rods to cones in both types of spheroids; however, they did not promote a laminar retinal structure. Eventually, a Japanese group found that Wnt-2b could induce the transformation of chicken rosetted into laminar stratospheroids (Nakagawa et al. 2003). Supporting this finding, supplementation of retinal cells from the Mongolian desert mouse (Gerbil) with Wnt-3b led to production of the first mammalian retinal *stratospheroids* (Rieke et al. 2018). Up to date, several reports have concluded that genetic networks that regulate retinal development in vitro and in vivo are basically comparable.

Sequence of Gene Activations Is Preserved In Vitro

Importantly, developmental genes have to be activated in the embryo at the right time at the right place. Accordingly, a spatiotemporally appropriate expression of the retinal genetic backbone is indispensable for normal retinal, as well as for retinal spheroid development. Indeed, proliferation and differentiation of cells occur in vitro on a comparable time scale as in vivo, eventually leading to a nearly complete laminar network, presenting all cell types including complex synaptic layers. Within spheroids, the various cell types differentiate quite normally, including expression of specific neuronal genes. As in vivo, in vitro formation of complex retinal connections is established, whereby an inner plexiform layer (IPL) precedes that of an outer (OPL). For instance, IPL sublamination in vitro is detectable in 5–6 days-old rosetted spheroids, corresponding well to completion of lamination around E12 in the normal chick retina. Recent seminal work by David Gamm and colleagues (Madison, WI) has documented that genetic networks that rule normal eye development from the state of a neural tube epithelium until reaching a differentiated retina plus a black RPE compare quite well with in vitro retinal spheroids. Most interestingly, at the earliest onset of aggregate formation of embryonic stem cells (ESCs) or, of induced pluripotent stem cells (iPSCs), *Oct* 4 and *Nanog* genes were expressed. These are genes which characterise the blastula/blastocyst stage, e.g. the earliest spherical multicellular structure following fertilisation. About one week later, genes characteristic of formation of the eye field within the telencephalic brain vesicle, e.g. Pax6, Rx and a.o., and only a couple days later genes characteristic of retina or RPE differentiation became expressed (Meyer et al. 2009). These findings convey important information: irrespective of in vivo or in vitro environments, all development relies on activities of particular genetic networks (with a stress on *networks*, not on *genes*). The fact that most differentiation events occur on a similar time scale as in vivo strongly indicates that differentiation in vitro underlies similar, or even identical regulatory genetic networks. On one side, such networks can be considered as molecular modules (for instance, the Wnt signalling pathway); on the other side they are quite often flexible and/or mutually overlapping (whereby one particular gene can be involved in different modules performing different functions) or can even be exchanged by others. For instance, during eye-stalk formation the Pax6 gene is involved in a different genetic network than it is during later differentiation of amacrine cells, when this gene fulfils a completely different function within another network. Thus, the same gene can be involved in very different events. Often, it remains uncertain what gene is on top, which one is at the bottom of a molecular network, which gene acts above (master-gene), which protein "downstream", which gene regulates which protein, and which protein acts back on which gene (feedback effects, cf. Fig. 1). But noticeably, gene activities are never non-essential, or dispensable.

Non-genetic Constraints on Tissue Self Organisation

Many features of retinal normal and in vitro development are strongly dependent on non-genetic constraints and self-organisational processes. Even at the subcellular level during the cell cycle, a high local chromatin order within cell nuclei is achieved through self-organisation (Cremer et al. 2014). Also, small chromosomal regions become autonomously arranged according to their chromatin class (van de Werken et al. 2017). Two examples for physical constraints during normal eye development are as follows: (i) as the eye stalk protrudes laterally (Fig. 2), it eventually will contact the outer surface ectoderm, which induces the lens placode, and also—due to expanding growth—pressures the neuroepithelium to bend inwards and thus form the two-layered optic cup; [note that in vitro produced "eye-cups" also bend inwards, which may be due to mechanic instability of an enlarging hollow sphere; cf. conflicting interpretation by Eiraku et al. (2011)]. (ii) As a further consequence, the two tissue layers will now touch each other with their apical sides. The opposition of two apical epithelial surfaces provokes a rare situation, leading to mutual inductive events between future retina and RPE, which in turn will determine differentiation of both photoreceptors and RPE.

A Brief History of Spheroids: Self-organisation in Spheres by Sorting-Out

A brief look into the long history of 3D cultures helps to get a better conception of self-organisation and emergence of tissues from individual cells, in particular, in understanding that tissues can be reconstituted by purely physical means in a culture dish. When kept in suspension, dispersed cells enjoy an additional spatial degree of freedom which allows them during and after their primary aggregation (also called *self assembly*) to find the best suitable locations within a growing cellular sphere. 3D cell culturing has begun with "shaking cultures" ("Schüttelkulturen") at the end of the nineteenth century by using sponges, sea urchins and newt larvae, swiftly unravelling basic concepts of cell biology. As an outstanding example, Henry van Peters Wilson dissociated sponges completely into isolated cells, transferred them into glass dishes and shook them softly in salt water, to then follow how they grew into cell clusters ("*reaggregates*"). To Wilson's surprise, his reaggregates eventually self-organised into complete viable sponges (Wilson 1905; Fig. 7). Even more surprising, when he used cells from two different sponge species (which were marked by colours), differently stained cells were either found within separate reaggregates, or they were amassed in distinct areas within one reaggregate. If differently stained cells originated from the same sponge species, but from different individual animals, cells were distributed statistically within reaggregates. What became well-known as phenomenon of "sorting-out" was—at the same token—the striking discovery of cell-cell recognition (distinction of self versus non-self). Townes and Holtfreter documented pronounced sorting-out of epidermal cells from neural plate cells of the amphibian embryo, whereby their relative position within the aggregate resembled that within the embryo (review in Layer and Willbold 1994). Moreover, an advanced tissue-specific differentiation was indicated. Based on the same technique, regeneration of complete hydras from isolated cells became an outstanding animal model, revealing significant genetic, molecular and histologic knowledge of stem cell and regeneration biology of hydrozoa (Gierer 2012).

Malcolm Steinberg provided a theoretical explanation of the sorting-out phenomenon, based solely on physicochemical properties of cells (Steinberg 2007). Accordingly, different cell types in a mixture were assumed to segregate as a consequence of differential strength of intercellular adhesion (*differential adhesion hypothesis*). Indeed, cells in a given tissue compound depend largely on their respective cell surfaces and extracellular matrices. Accordingly, emergence of tissue properties primarily depends on purely physicochemical conditions, and not so much on *one* particular gene. Such short distance forces will mediate cell cohesiveness (adhesion), optimal integration of cells into a given space, growth directions of their processes, etc. It is of note that individual contributions to the whole emergent process will be numerous (e.g., including mechanical forces; see Franze 2013); they cannot be deciphered in detail or estimated by precise numbers. The effects even can turn out anti-intuitively. For instance, minute irregularities of similar cell shapes can have positive pattern-forming power (Lenz and Witten 2017). Together with forces acting on distance (e.g. diffusible growth factors, cytokines), attraction and retraction between cells, cell migration and final placement all contribute to tissue self organisation. *In summa,* combined physical forces can direct primary steps of tissue formation in an artificial "in vitro space".

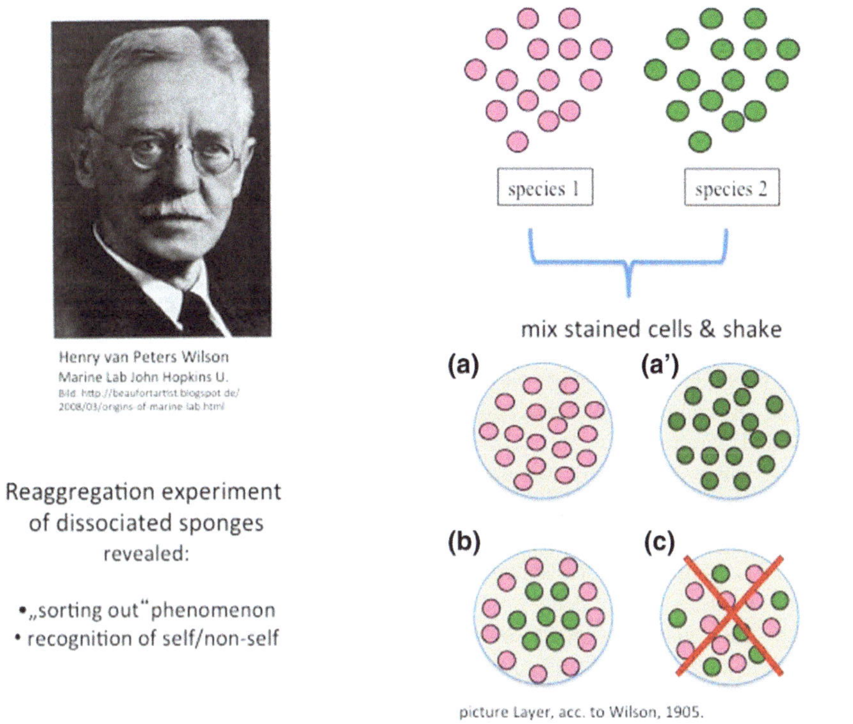

Fig. 7 Discovery of cell communication and sorting out in reaggregation experiments of dispersed sponges (Wilson 1905). After reaggregation of dispersed cells from two different sponge species, cells from the two species were found either in different aggregates (**a**), or within segregated areas of the same aggregate (**b**), but were not distributed randomly (**c**)

Emergent Borders Are Decisive to Structure Tissues and Organs

In a culture dish, separation of similar cells can be directly followed under a microscope (provided that they are somehow labelled). Their segregation leads to "islands", i.e. to regions of similar cells within a larger sphere. However, the process of physical sorting-out is not as obvious during normal development of tissues, yet in principle it also takes place. In fact, it represents a basic process during formation of morphologic/functional subunits. For instance, during subdivision of the early neural tube a series of *rhombomeres* of the early hindbrain become separated by strict (structural) border lines, which can be visualised by appropriate marker molecules (Lumsden and Keynes 1989; Puelles 2001). At onset, some of these markers emerge faintly and are spread quite broadly, to then concentrate more and more towards a focussed border (Layer and Alber 1990; cf. Fig. 3). Eventually, mechanically forced constrictions coincident with these borders further strengthen separation of brain subareas. That thereby sorting-out is involved has been again demonstrated in vitro by mixing and sorting of cells from individual rhombomeres (Götz et al. 1996). Hence, emergence of tissue borders is supported by physical (incl. mechanical, cf. also Franze 2013) means, and without doubt is indispensible for normal embryonic development.

Many Roads to Rome—Plasticity of Tissue Formation

The formation of several distinct types of chicken retinal spheroids highly depends on environmental factors. Retinal spheroids in their most basic form are characterised by internal rosettes and plexiform synaptic regions (*rosetted spheroids*; Fig. 6; their modular structure). Similar rosetted spheroids could be produced from embryonic mouse and rat retinae (e.g., by C. Barnstable, P. Linser, T. Reh; see Layer and Willbold 1994). However, it was most stunning that when retinal spheroids were produced from the Mongolian desert mouse (gerbil), they were not initiated from rosettes, but tissue organisation began at the level of formation of an inner plexiform layer (IPL; Bytyqi et al. 2007). Similarly, retinal spheroids from *Brachydanio rerio* (zebrafish) achieve a laminar structure without being initiated much by rosettes (Eldred et al. 2017). These findings are highly relevant in terms of retinal tissue self-organisation: albeit the basic laminar structure of avian, rodent and fish retinae is very similar (three-layered structure of all vertebrate retinae, see above), to rebuild them from dissociated cells can follow very different paths ("many roads lead to Rome"). Apparently, dispersed cells from different vertebrate origins in a culture dish seem to be determined by an inherent intention of "we are going to build a vertebrate retina" somehow, clearly indicative of a "meta-level" of information above the genetic code that is driving and safeguarding development. The physical nature of this "blueprint" remains widely unclear. At any rate, what becomes instantly clear when working with 3D cultures is that in vitro tissue formation depends to a large extent on culture conditions, e.g. on paracrine factors, on species and many more. Hence, not only particular genes drive formation of a layered neural network tissue, each one performing one specific function (nature versus nurture discussion; indeterminate versus cell-autonomous development), but non-genetic constraints are as decisive.

Conclusions

The idiom of "something comes out of something"—well exemplifying emergence thought—is represented by no other research field more directly than by organismic development (saying this is nearly a tautology). At a first sight, however, normal development appears to follow a determinate one-way road, whereby typically not individual genes, but genetic networks regulate what will happen at a certain place and a certain time in a growing organism. At each given spatio-temporal point in development, distinct *environmental situations* will prevail to cause novel constraints on the genetic backbone. However, as revealed by retinal spheroids, development depends much on environmental conditions. The sequel of any particular "space-time point" under in vivo conditions is only predictable because *the respective constraints themselves are reliably reproduced* during each individual course of normal development. When released from constraints during in vitro development, then development of a system (tissue, organ, organism) is liberated from its determinative power. In summary, we conclude that…

- Normal development of organisms (in vivo DoO) is governed by ground-laying developmental genes.
- In vivo DoO appears as if it were determinate, since the result is predictable.
- However, when analysed under in vitro conditions, emergent principles of DoO are readily revealed, rendering DoO as highly regulative and non-predictable.
- During DoO not individual genes, but rather gene–protein networks represent molecular toolboxes which can be used in changing combinations.
- DoO can resort to such tools for regulating formation of recurring modules, such as cellular spheres, planar epithelia, constricted tissue borders and more.
- In vitro analyses of developmental modules of a tissue, more specifically, of their genetic backbone and environmental constraints (as exemplified here for retina) are essential to understand normal as well as aberrant (diseased) development of a tissue (promoting applicability in stem cell-based regenerative medicine).
- Therefore, earlier prevailing deterministic positions in embryology have been much restricted by insights of modern developmental biology.

Acknowledgements My teachers E. E. Bruchmann (Hohenheim), F. Hucho (Konstanz), E. Shooter (Stanford), H. Meinhardt and A. Gierer (Tübingen) have ignited my passion for science and paved my way into developmental biology research. I thank my students and colleagues G. Bachmann, A. Bytyqi, A. Daus, F. Frohns, M. Reinicke, M. Rieke, A. Robitzki, A. Rothermel, L. Sperling, G. Thangaraj, G. Vollmer and E. Willbold, who have—in spite of difficult infrastructures—promoted our spheroid research with great stamina and enthusiasm. I thank Lynda Wright (Madison, WI) for her careful reading and comments. Editorial assistance by the Chief Editors U. Lüttge and L. H. Wegner is greatly acknowledged.

Glossary and Abbreviations

- Blastocoel—fluid-filled hollow space of blastula;
- Blastula—cell ball (sphere) formed through cleavage divisions;
- Cleavage—rapid cell divisions after fertilisation;
- Coelom—fluid-filled space surrounded by mesodermal epithelium;
- Constraints—limitations of development through environmental (non-genetic) conditions;
- Differential adhesion hypothesis, see *sorting-out*;
- Ectoderm—outer germ layer;
- Endothelium—epithelium forming blood vessels;

- Endoderm (entoderm)—inner germ layer;
- Epithelium—planar tissue covering internal and external surfaces, e.g., skin, gut, etc.;
- *fire-and-wire* mechanism—refinement and stabilisation of neuronal connectivities by their repeated usage;
- Gastrulation—proces by which three germ layers are established in animals;
- Growth factors (cytokines):

  - FGF, fibroblast growth factor;
  - PEDF, pigment epithelium-derived factor;
  - GDNF, glial derived neurotrophic factor;

- Lamination, see *stratification*;
- Mesoderm—middle germ layer in between ecto- and entoderm;
- Morphogenetic movements—classification of cell migratory mechanisms, e.g., during development, such as e- and invagination, ingression, epiboly, etc.;
- Müller glial cell—radial glial cell of retina, spanning its entire width;
- Neural crest—cell population in most vertebrates emigrating dorsally from closing neural tube, which will found peripheral nervous system (and more);
- Neuromeres—early regional subdivisions of frontal neural tube;
- Ontogeny—course/process of development of an individual organism;
- Organising centre—cells or tissue parts, from which particular steps of development are initiated;
- Organoid—from stem cells in vitro regenerated organ-like tissue;
- Phylogeny—course/process of appearance of all phyla (stems) of organisms (phylogenetic tree) over the entire evolutionary period;
- Primitive streak—tissue structure in developing birds and mammals indicating the onset/course of gastrulation;
- Pseudostratified neuroepithelium—monolayered cellular status of neural tube, which due to its width appears to be stratified, but it is not;
- Retinal cell layers:

  - GCL, ganglion cell layer;
  - INL, ONL, inner and outer nuclear layer;
  - IPL, OPL, inner and outer plexiform layer;

- Retinal cell types:

  - AC, amacrine cell—large axon-less cell positioned at inner border of INL, connecting BPs and GCs in IPL;
  - BP, bipolar cell—interneuron in INL, connecting PRs and HCs in OPL, and with ACs and GCs in IPL;
  - HC, horizontal cell—large cell positioned at outer border of INL, connecting PRs with BPs;
  - PR, photoreceptor cell; comes either as rod or several types of cones;

- Rhombomeres—segmental subdivisions of hindbrain;

- Reaggregate—ball (sphere) of adhering cells formed by reaggregation from dispersed cells;
- RPE—retinal pigmented epithelium;
- Sheefs—"synthetic human entities with embryo-like features": a human organoid made from hiPSCs which presents a primitive streak (see, *gastrulation*);
- Sorting-out—process by which different reaggregating cells kept under rotation/in motion associate with similar, and separate from different partner cells; see, differential adhesion hypothesis;
- Spheroids, reaggregated from embryonic chicken retinae,
  - rosetted retinal spheroid—reaggregated cell sphere from dispersed embryonic chicken retinal cells, spatially organised by internal cell rosettes;
  - stratospheroid—dto., achieving a (nearly) complete retina-specific lamination (retinal organoid);
- Stem cells—cell with inherent proliferative ability, which in vitro can be amplified and then directed into one or more distinct differentiated cell type(s);
  - ESCs—embryonic stem cell;
  - iPSCs—induced pluripotent stem cell;
  - hiPSCs—human iPSCs;
- Stratification—arrangement of distinct cell types within cell layers, e.g., in brain and retina;
- Tissue Engineering—artificial (in vitro) reconstruction of tissues from stem cells applying engineering technologies;
- Wnt protein—cell-external ligand protein for the Wnt signalling pathway, a major communication pathway between cells during development and disease (*Wnt* stands for "wingless-related integration site").

References

Alegado RA, Brown LW, Cao S, Dermenjian RK, Zuzow R, Fairclough SR et al (2012) A bacterial sulfonolipid triggers multicellular development in the closest living relatives of animals. eLife 1:e00013

Bytyqi AH, Bachmann G, Rieke M, Paraoanu LE, Layer PG (2007) Cell-by-cell reconstruction in reaggregates from neonatal gerbil retina begins from the inner retina and is promoted by retinal pigmented epithelium. Eur J Neurosci 26:1560–1574

Chalmers DJ, Jackson F (2001) Conceptual analysis and reductive explanation. Philos Rev 110:315–361

Cremer T, Cremer C, Lichter P (2014) Recollections of a scientific journey published in human genetics: from chromosome territories to interphase cytogenetics and comparative genome hybridization. Hum Genet 133:403–416

Eldred MK, Charlton-Perkins M, Muresan L, Harris WA (2017) Self-organising aggregates of zebrafish retinal cells for investigating mechanisms of neural lamination. Development 144:1097–1106. https://doi.org/10.1242/dev.142760

Eiraku M et al (2011) Self-organizing optic-cup morphogenesis in three-dimensional culture. Nature 472:51–56
Franze K (2013) The mechanical control of nervous system development. Development 140:3069–3077. https://doi.org/10.1242/dev.079145
Fromm J (2005) Types and forms of emergence. Cornell University Library. arXiv:nlin/0506028
Gierer A (2012) The hydra model—a model for what? Int J Dev Biol 56:437–445
Gilbert SF (2016) Developmental biology, 11th edn. Sinauer Ass, MA, USA
Götz M, Wizenmann A, Reinhardt S, Lumsden A, Price J (1996) Selective adhesion of cells from different telencephalic regions. Neuron 16:551–564
Grosberg RK, Strathmann RR (2007) The evolution of multicellularity: a minor major transition. Annu Rev Ecol Evol Syst 38:621–654
Haeckel E (1904, 1998). Kunstformen der Natur. Neudruck der Erstausgabe in Faksimile. Leipzig, Wien, Bibliogr Inst. ISBN 3-7913-1979-5
Huch M, Knoblich JA, Lutolf MP, Martinez-Arias A (2017) The hope and the hype of organoid research. Development 144:938–941. https://doi.org/10.1242/dev.150201
Jahn I (2000) Geschichte der Biologie, 3rd edn. Spektrum Akad. Verl. Heidelberg, Berlin
Lancaster MA, Renner M, Martin CA, Wenzel D, Bicknell LS, Hurles ME, Homfray T, Penninger JM, Jackson AP, Knoblich JA (2013) Cerebral organoids model human brain development and microcephaly. Nature 501:373–381
Layer PG, Alber R (1990) Patterning of chick brain vesicles as revealed by peanut agglutinin and cholinesterases. Development 109:613–624
Layer PG, Willbold E (1994) Regeneration of the avian retina by retinospheroid technology. Prog Ret Res 1994(13):197–230
Layer PG, Araki M, Vogel-Höpker A (2010) New concepts for reconstruction of retinal and pigment epithelial tissues. Exp Rev Ophthalmol 5:523–544
Lenz M, Witten TA (2017) Geometrical frustration yields fibre formation in self-assembly. Nat Phys 13:1100–1104. https://doi.org/10.1038/nphys4184
Lumsden A, Keynes R (1989) Segmental patterns of neuronal development in the chicken hindbrain. Nature 337:424–428
McFall-Ngai M, Hadfield MG, Bosch TC, Carey HV, Domazet-Loso T, Douglas AE et al (2013) Animals in a bacterial world, a new imperative for the life sciences. Proc Natl Acad Sci USA 110:3229–3236. https://doi.org/10.1073/pnas.1218525110
Meyer MS et al (2009) Modeling early retinal development with human embryonic and induced pluripotent stem cells. Proc Natl Acad Sci USA 106:16698–16703
Nakagawa S, Takada S, Takada R, Takeichi M (2003) Identification of the laminar inducing factor: Wnt-signal from the anterior rim induces correct laminar formation of the neural retina in vitro. Dev Biol 260:414–425
Puelles L (2001) Brain segmentation and forebrain development in amniotes. Brain Res Bull 55:695–710
Reichenbach A, Bringmann A (2013) New functions of Müller cells. Glia 61:651–678
Rieke M, Bytyqi A, Frohns F, Layer PG (2018). Reconstructing mammalian retinal tissue: Wnt3a regulates laminar polarity in retinal spheroids from neonatal Mongolian rats, while RPE promotes cell differentiation. Int J Stem Cell Res Therapy. https://doi.org/10.23937/2469-570x/1410051
Steinberg MS (2007) Differential adhesion in morphogenesis: a modern view. Curr Opin Genet Dev 17:281–286
Strauss BS (2016) Beadle and Tatum and the origins of molecular biology. Nat Rev Mol Cell Biol 17:266. https://doi.org/10.1038/nrm.2016.42
van de Werken HJG, Haan JC, Feodorova Y, Bijos D, Weuts A et al (2017) Small chromosomal regions position themselves autonomously according to their chromatin class. Genome Res 27:922–933. https://doi.org/10.1101/gr.213751.116
Vollmer G, Layer PG, Gierer A (1984) Reaggregation of embryonic chick retina cells: pigment epithelial cells induce a high order of stratification. Neurosci Letts 48:191–196

Weikert T, Rathjen FG, Layer PG (1990) Developmental maps of acetylcholinesterase and G4-antigen of the early chicken brain: long distance tracts originate from AChE-producing cell bodies. J Neurobiol 21:482–498

Wilson HV (1905) On some phenomena of coalescence and regeneration in sponges. J Exp Zool 5:245–258. https://doi.org/10.1002/jez.1400050204

Zhong X et al (2014) Generation of three-dimensional retinal tissue with functional photoreceptors from human iPSCs. Nat Commun 5:4047. https://doi.org/10.1038/ncomms5047

Plants: Unitary Organisms Emerging From Integration and Self-organization of Modules

Ulrich Lüttge

Abstract The organs roots, stems and leaves of plants have evolved under the challenges of the terrestrial environment where they are firmly rooted. Emergence of plants as unitary integrated organisms occurs by integration of these parts, organs or modules. Below this scaling level of morphology and the organs, there are modules on finer scaling levels, tissues at the level of anatomy, organelles and compartments at the level of cell biology, macromolecules at the molecular level and so on. Modules are connected and integrated forming the knots in networks. Networks at a finer scale can integrate, condense and self-organize to form knots in networks at the next coarser scale. In this way, hierarchies of networks are built up leading to the emergence of whole plants. Requirements for the integration are (i) signals carrying information, (ii) receptors for the signals, (iii) transduction of signals within systems and networks, (iv) cross-talk between different types of signals by their translation into each other and (v) configuration of information as instruction for reactions. Integration is organized during development where tight structural and functional correlations are built up under the systemic control of development. Self-organization uses both correlative inhibition and correlative stimulation, with homoiogenetic induction of self-resemblance and heterogenetic induction of unlike-self, respectively. Examples of integration are source-sink relations of water, nutrients and photosynthetic products, induction of flowering, orientation in space under gravity and light, environmental relations under herbivory and salinity. Plants are unitary organisms without a neuronal system. The whole is more than the sum of its parts (Aristotle 384–322 BC).

Parts and Modules of Plants

Figure 1 shows the scheme of a higher land plant between a giant brown alga of the genus *Laminaria* and the reconstruction from a fossil of one of the earliest land plants *Rhynia* in the fern allies Psilophyta. The major parts of higher land plants are

U. Lüttge (✉)
Department of Biology, Technical University of Darmstadt, Darmstadt, Germany
e-mail: luettge@bio.tu-darmstadt.de

L. H. Wegner and U. Lüttge (eds.), *Emergence and Modularity in Life Sciences*,
https://doi.org/10.1007/978-3-030-06128-9_8

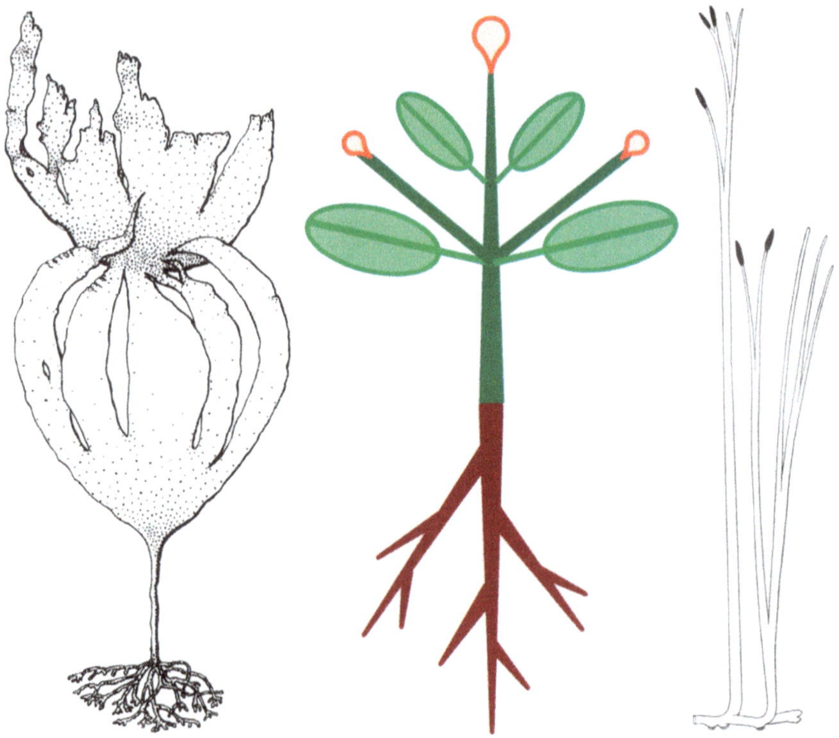

Fig. 1 Scheme of a higher land plant with its organs roots, stems and leaves between the brown alga *Laminaria* with rhizoids, a cauloid and a phylloid (left: After H. Schenck, from Kadereit et al. 2014) and the reconstruction of *Rhynia* from a fossil (right: From Zimmermann 1930)

the organs of roots, stems and leaves. We do not single out flowers here, because they are built up of metamorphoses of leaves. We may consider these organs as "modules," where we apply a very straightforward definition of modules as being parts or components of a system. This is a simple practical definition (Bolker 2000). The term module refers to "repeated, often semiautonomous … units" (de Kroon et al. 2005). The possession of roots, shoots and leaves appears as a general basic building principle of the gestalt of larger multi-cellular plants. The brown algae lineage is not directly linked to the lineage of higher land plants. The large bodies of brown algae or kelp have root-like ("rhizoid"), stem-like ("cauloid") and leaf-like ("phylloid") organs. They evolved parallel to and independent of roots, stems and leaves. It is a case of convergence (Morris 2003). However, *Rhynia* on the direct line to all higher land plants did not have roots and leaves yet. It consisted of slender about 1 m tall dichotomically branched shoots and had horizontal rhizomes, derivatives of shoots, in the soil.

Self-organization leads to the formation of patterns (Johnson and Lam 2010). In this way, development can be a result of self-organization, where patterns originate

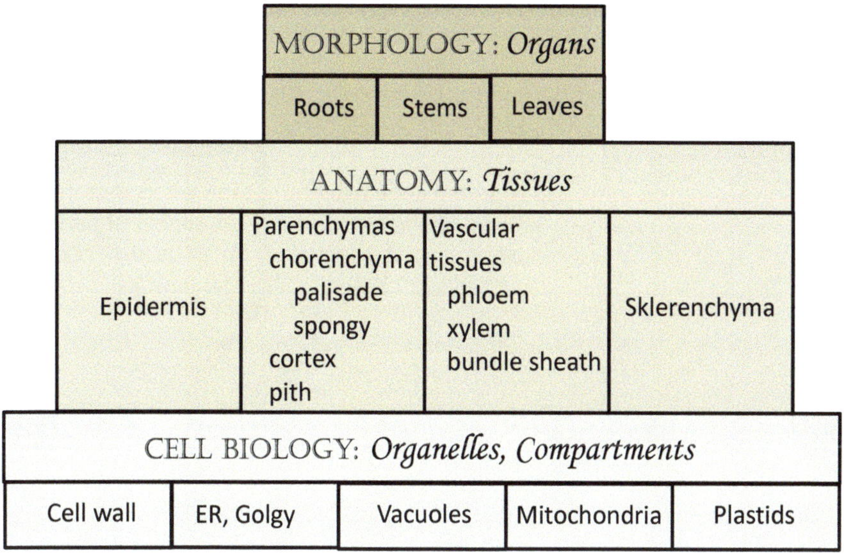

Fig. 2 Modules of plant structures at three different scaling levels

from divisions and differentiation of primordial cells. In morphogenesis, interaction of components of patterns creates new organized structures. The EVO-DEVO concept relates evolution and development. They are linked by feedback loops. There is "*evolution of development*" (Gould 2002). In return, development provides the material for natural selection. When Charles Darwin (1809–1882) developed the theory of evolution, many of his arguments were founded on development and growth (Friedman and Diggle 2011). He considered development and embryology as the most important aspects of natural history. Self-organization marks the frame within which evolution can select, and it is a tool of selection (Johnson and Lam 2010).

In addition to the modules or organs of the whole plant at the scaling level of morphology, Fig. 2 lists modules on two finer scaling levels, namely tissues at the level of anatomy and organelles/compartments at the level of cell biology.

Modularity Versus Emergence

Modularity

Modularity separates components of systems. The components are modules, and each module is regarded by itself. The view of modularity denies that plants with their "vegetative" life lacking integrating neuronal systems are unitary organisms (Fig. 3; Lüttge 2013). It considers plants as being the sum of their parts, where, for

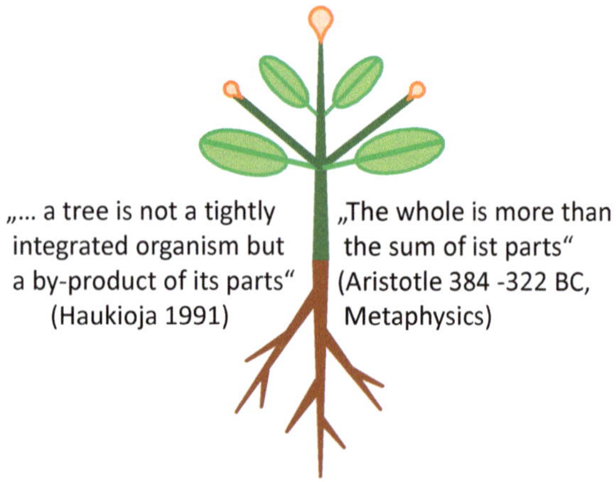

Fig. 3 The points of view of modularity (left) and emergence (right)

example, a tree is just a "by-product of its parts" (Haukioja 1991). The study of separated individual modules is important, and it is essential. Their properties need to be known. Conceptually, the isolation of modules, however, is reductionism. It reduces the degrees of freedom given by interactions. Practically, it is an approach of specialization.

We know structural as well as dynamic functional modules, such as given processes or pathways (Bolker 2000). Going through levels of scaling, we realize that modules at finer scales are themselves composed of modules. Figure 2 shows this for three structural scaling levels of plants. The module "organ" is composed of the modules of tissues, and the module "tissue" is built up of the modules organelles or compartments.

Therefore, even when taken to lack external integration, individual modules clearly show internal integration (Bolker 2000). This leads us to distinguishing a hierarchy of modules over scaling levels. From Fig. 2, we can deduce the circles of Fig. 4. In the latter figure, each circle size would correspond to modules at a specific scaling level. The largest circle would represent the whole plant, which can be a module in vegetation or in an ecosystem, not to be treated in this chapter (but see chapters by Beck 2019 and zu Castell et al. 2019 in this book). The second largest circles would be the organs. Each one as a whole can be taken as a module at the level of morphology. The third largest circles would be modules of organs, i.e., the individual tissues at the level of anatomy. Again taken as individual modules, tissues would be composed of the fourth largest or smallest circles drawn, i.e., the modules of organelles and compartments. The consideration of such hierarchies could be continued for still finer scaling levels, i.e., compartments and organelles are built up of membranes and complex molecular structures, which again would consist of complex macromolecular components and so on.

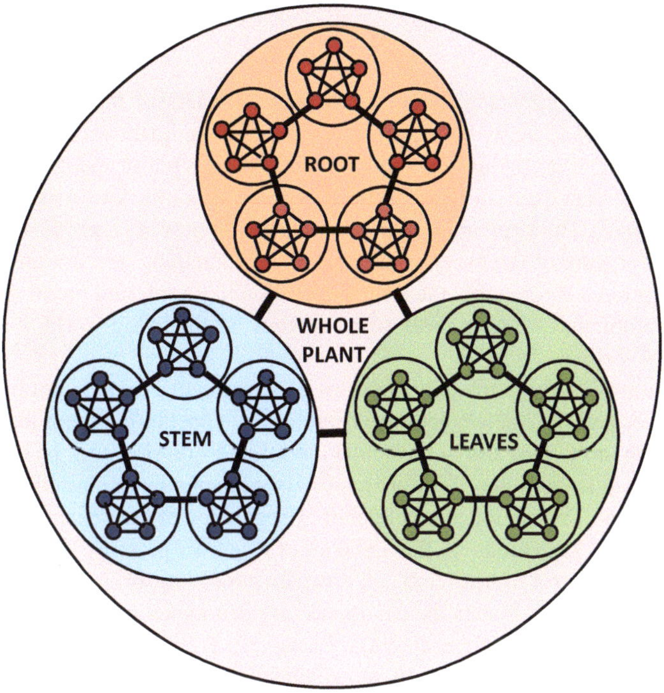

Fig. 4 Multi-scale hierarchical network models linking molecular networks to whole plant. Molecular networks are not drawn in the smallest circles. From smallest to largest circles: cells → tissues → organs → whole plant

Hierarchy of Module-Networks

Above we have already stumbled over "integration," i.e., the internal integration within modules (Bolker 2000). Going through the scaling levels of modules, this proved to be inevitable. When a number of modules at a finer level of scaling become a new module at the next higher level of scaling, we cannot avoid considering them as integrated. They form networks. When each of the networks at a finer scaling level condenses to give a module representing a knot in a network at the next coarser scaling level, it can be easily seen that we arrive at a hierarchy of networks (Souza et al. 2016) over the hierarchy of scaling levels (Fig. 4). Multi-scale network models linking molecular networks to whole-plant development present an integrated view (Lucas et al. 2011; Weston et al. 2012). From this, it is only a very small step toward understanding emergence. Studying individual modules, i.e., reductionism and specialization, is indispensible. We must know structure and function of modules well. However, with this alone, we would not comprehend life (Lüttge 2012). Life is an emergent category (see chapter by zu Castell et al. 2019). Per se, the modularity view of a plant stays short of understanding its life.

Emergence

Interactions of modules and their integration in clusters of modules can lead to a "fulguration" of systems with completely new innate qualities which cannot be predicted by even the closest examination of the modules per se and in isolation. The whole is more than the sum of its parts. This notion goes back to Aristotle (Fig. 3; Wikipedia 2012). The term fulguration was used by Lorenz (1977, pp. 47–48) to grasp the dramatic nature of the event of appearance of something new coming up like a flash of lightning. Regrettably the term fulguration has not found broad application in the literature, and we now speak of emergence. In the living world, it occurs on all levels of scaling.

For reasons of only graphical complexity, the molecular level is not included in Fig. 4. Genes and the nucleic acid macromolecules constituting them may be considered as the most basic and fundamental modules of life. However, by themselves, genes remain quiet modules. They are not life (see chapter by zu Castell et al. 2019). Life is emergent. The organization of genes by regulation of their expression leads to the emergence of gene products, the first ones of which are proteins. The hierarchy of network-based control of upstream and downstream events, or in other words feedback and feed forward, fosters the emergence of phenotypes and creates phenotypic novelties (Müller 2007; Marr et al. 2008; Lüttge 2012; Hütt 2013).

At the level of the gene products of proteins, we can witness most vividly how self-organization and emergence are linked. Self-organization creates fundamentally novel systems, which emerge from the interaction of the protein modules participating. Lipid and protein molecules can self-organize spontaneously to build up biological membranes with distinct properties and functions. Different individual proteins can be self-organized to form multi-subunit macromolecular structures functioning as enzymes. We may also call this self-assembly. Examples are the ATPases in the membranes of mitochondria synthesizing ATP driven by a proton electro-chemical gradient or the proton transporting ATPases building up a proton electro-chemical gradient at membranes by hydrolyzing ATP, e.g., the H^+-ATPase in the tonoplast membrane of plant vacuoles. Ribulose bis-phosphate carboxylase-oxygenase mediating CO_2 fixation in photosynthesis is built up of 8 small and 8 large protein subunits. The pyruvate and α-oxo-glutarate decarboxylases of respiration are also multi-subunit enzyme complexes. There are many other examples. In neither of these cases can their specific enzymatic function be predicted and derived from the particular nature of the individual subunits. Figures 2 and 4 may assist in illustrating the principle of emergence from self-organization with interaction and integration of modules on increasingly coarser levels of scaling: macromolecules → cells → tissues → organs → whole plant.

Mechanisms of Interaction and Integration

Proteins

The primary structure of proteins is given by the sequence of amino acids. The primary chains can undergo self-organization performing some simple folding of given amino acid strings. Parts of the primary chains are aligned, and this is stabilized by different kinds of bonds between amino acid side chains along them. This results in the secondary structure of proteins. These structures can undergo very complex further folding and coiling leading to a conformation called tertiary structure. Such self-organization often is not strictly spontaneous anymore and is regulated by various actions, such as binding of effectors and substrates, by interaction with other proteins, by incorporation into membranes. Special proteins catalyzing and regulating the correct folding are the chaperons (Buchanan et al. 2000; Alberts et al. 2002; Lüttge et al. 2010). When their conformations fit, tertiary structures of proteins can then self-organize into the quaternary structures of multi-protein complexes (section "Emergence").

Requirements at Higher Levels

At higher scaling levels from cells to entire organisms like whole plants, there are a number of requirements for interaction and integration, namely (1) signaling, (2) receptors for signals, (3) transduction of the information inherent in signals, (4) cross-talk by translation of different types of signals into each other and (5) realization of the information by reactions of metabolism and growth. Figure 5 shows the sequence of events integrating modules for the emergence of higher unitary entities such as whole plants.

Signals can be external or internal cues. They carry information. Receptors are needed to fix this information of signals when they arrive at cells in a molecular structure. Often these receptors are specific proteins located in membranes or chromoproteins if the signals are light. Over shorter distances, i.e., within cells, transduction of the information from the receptors to regulation of protein activities and especially of gene expression is a very complex cellular process involving steps of protein phosphorylation catalyzed by kinases and mediated by waves of cellular calcium levels (Buchanan et al. 2000; Alberts et al. 2002; Lüttge et al. 2010). Over longer distances, transduction requires the transport of signals whose nature is mainly electric, hydraulic or chemical. Receptors are needed again at the locations where the signals arrive in cells. It is noteworthy that these three types of signals can be translated into each other before and after transduction (Lüttge 2013, 2016; Shabala et al. 2016). For example, electrical and hydraulic signals are more suitable for long-distance transduction and can be translated into chemical signals eliciting chemical reactions of metabolism and growth. In this way, the information is taken

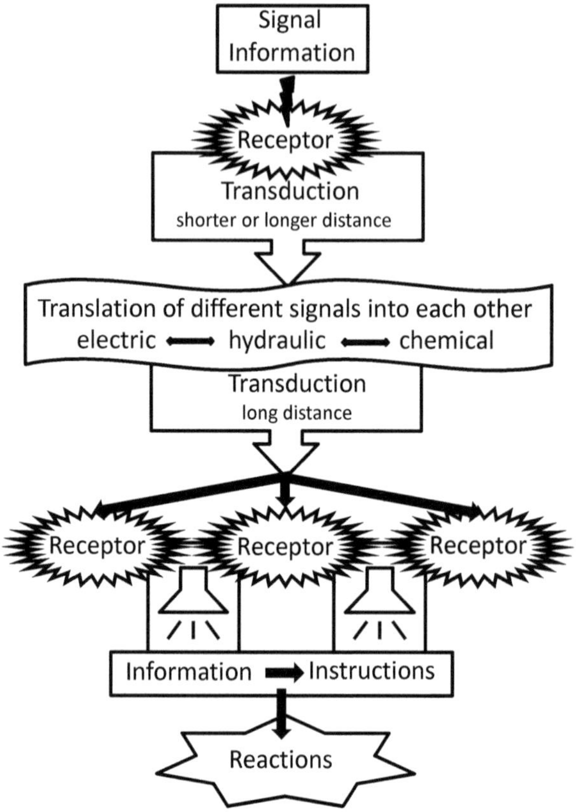

Fig. 5 Flow scheme of the sequence of events integrating modules for the emergence of higher unitary entities such as whole plants

as instructions to perform reactions in the integrated unitary organism. The following sections "Source-Sink Relations in Whole Plants" and "Developmental Relations" shall give some examples to illustrate how this general theoretical scheme is realized in whole plants.

Source-Sink Relations in Whole Plants

A major concept of whole-plant physiology regards source-sink relations mainly of water and various resources, such as mineral nutrients and organic compounds (Fig. 6). They illustrate the integration of roots and shoots for the emergence of whole plants as unitary organisms.

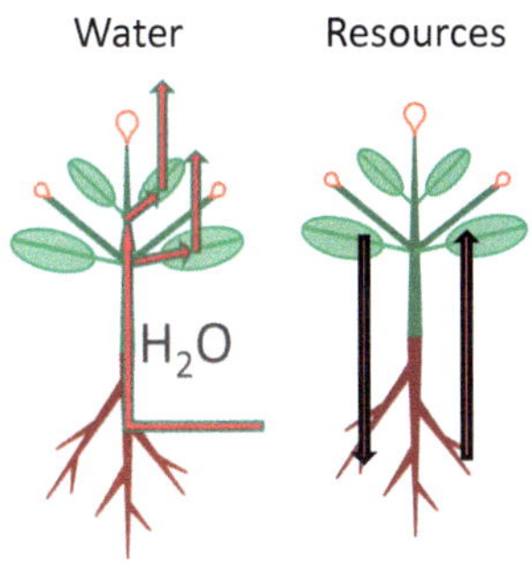

Fig. 6 Integrated allocation of water and resources in the whole plant

Water

The source of water is mainly the environment, i.e., the soil for rooted plants. The distribution of water within plants follows gradients of water potential from high water potential (source) to low water potential (sink). Sinks can be cells and tissues with high solute concentrations and hence low water potentials. The major sink for water is the dry atmosphere. The water potential gradient between the substratum, i.e., soil or a nutrient solution, and the atmosphere drives the flow of water flowing through whole plants, i.e., transpiration. Transpiration is regulated by the opening and closing of small pores, so-called stomata, in the aerial parts, especially the leaves of rooted plants. This process provides a very clear example of the whole-plant integration of the performance of roots and shoots.

Root-shoot signaling of soil drying is one of the most conspicuous features in whole-plant integration. A chemical agent, the phytohormone abscisic acid (ABA), accumulates and elicits stomatal closure when transpiration needs to be reduced to save water under stressful conditions. When the availability of water gets limiting, it is the soil water status rather than leaf water status which primarily controls transpiration of leaves (Turner et al. 1985; Davies et al. 1990; Wartinger et al. 1990; Zhang and Davies 1990). In the leaves, stomata can be warned early about that so that they may close even before the leaves experience water deficit (Davies and Zhang 1991; Gil et al. 2008). Experimental approaches to demonstrate this use split root systems or fence sitter plants. Plants are grown so that their roots develop in two different separated chambers, which looks like someone sitting on a fence and letting her or his legs hang down on either side (Fig. 7; Davies and Zhang 1991; Gersani et al. 1998). This allows it experimentally to subject different parts of the root system to different conditions. When half of the root system was subject to low water potential (Ψ), this elicited drought signals to the leaves, while the other half of the root system was kept at high Ψ and maintained their water supply. Thus, stomata began to close on transduction of a misleading signal, although leaves did not suffer from reduced supply. It was early recognized that a major signaling substance involved was ABA. Other chemical signals participate (Davies et al. 1994) including cytokinin phytohormones (Bano et al. 1993; Shabala et al. 2009), and also inorganic ions and

Fig. 7 Split root system of a fence sitter plant

xylem sap pH (Schurr and Schulze 1996; Wilkinson and Davies 1997; Wilkinson et al. 1998), as well as hydraulic signals (Tardieu and Davies 1993; Tardieu et al. 1993). Hydraulic signals of soil drying can be translated into the chemical phytohormone signals which in turn may be translated into electric signals such as action potentials (APs), variation potentials (VPs) and system potentials (SPs) (Comstock 2002; Davies 2004; Lüttge 2013). Hydraulic signals such as pressure changes and hydraulic waves are rapidly transferred via the xylem (Matyssek et al. 1991; Cermak et al. 1993; Tang and Boyer 2003; Grams et al. 2007). Electrical action potentials are most rapidly propagated in whole plants along the veins via the phloem (Fromm and Eschrich 1993; Fromm and Fei 1998; Volkov 2000; Comstock 2002; Lautner et al. 2005; Fromm and Lautner 2007; Grams et al. 2009; Shabala et al. 2009; Lüttge 2013). Thus, chemical signals in the roots translate into hydraulic and electrical signals and after transduction to the leaves back to the chemical signal of ABA to elicit stomatal closure.

Nutrients

Whole-plant physiology boomed in the 1970s with studies of mineral nutrition (Pitman 1975; Sutcliffe 1976a, b; see Lüttge 2013). The important facts were that nitrogen and sulfur are available in the root medium in the oxidized forms of NO_3^- and SO_4^{2-}. These anions are taken up by the roots, but N and S are required in the metabolism of plants in reduced forms, and the reducing power within the roots is insufficient. It is photosynthetic electron transport in the leaves which provides enough reducing power of nicotinamide adenine dinucleotide phosphate ($NADPH + H^+$) for the reduction processes. If supply of NO_3^- is low, reduction to NH_4^+ can be managed by the plastids, the leucoplasts, in the roots, but at high concentrations, NO_3^- needs to be transported into the leaves to be reduced in the chloroplasts (Wallace and Pate 1967; Kirkby and Knight 1977). Similarly, SO_4^{2-} reduction capacity is weak in roots, and SO_4^{2-} reduction needs to be driven by photosynthetic electron transport in the leaves (Rennenberg et al. 1979; Clarkson et al. 1983; Lüttge et al. 2010).

Nitrate nutrition requires a metabolic circle connecting and integrating roots and shoots. When NO_3^- is taken up by the roots, electrical charge balance of the ions is needed, which can be satisfied either by the release of bicarbonate HCO_3^- or by the uptake of K^+. KNO_3 is then transported in the xylem through the stems to the shoots and leaves. There the reduction of NO_3^- to NH_3 leaves behind a hydroxyl ion OH^-, which is neutralized by fixation of CO_2 and formation of malic acid via phosphoenolpyruvate carboxylase. The K^+ is returned to the roots by transport in the phloem and accompanied by malate anions for charge balance. In the roots, K^+ is available again for transport back to the leaves together with NO_3^-. Malate is decarboxylated. The pyruvate produced is funneled into respiratory energy metabolism, and the HCO_3^- obtained can be used for charge balance of NO_3^- uptake (BenZioni et al. 1971; Frost et al. 1978; Lüttge and Higinbotham 1979; Peuke et al. 1996; Lüttge et al. 2010; Lüttge 2013). Similar loops are envisaged for SO_4^{2-} nutrition.

There is systemic signaling. The Thornley model implies that nutrient deficiency favors photosynthate partitioning to the roots to improve nutrient acquisition capacity (Thornley 1972). With their transport, the chemical substrates involved in relations of whole-plant nutrition carry information. If in a split root system of fence sitter plants (Fig. 7) one chamber is kept under low or lacking provision of nutrients while roots in the other chamber are well supplied, the plants only develop a proliferate root system in the latter and reduce much of the growth of roots in the former (Gersani et al. 1998). Systemic signaling of sulfur status within the plants is mediated by the S-compound glutathione (Herschbach and Rennenberg 1994).

Photosynthetic Products

While nitrate and sulfate nutrition in the whole plant effectively is powered by source-sink relations of reduction potential (section "Nutrients"), the origin of source-sink concepts actually goes back to attempts of understanding long-distance phloem transport of photosynthetic products particularly sucrose, the major transport substance in the phloem (Münch 1930; Ziegler 1956). The mechanism is a hydraulic pressure-driven mass flow in the conductive elements, the sieve tubes of the phloem. The pressure gradient is maintained by the osmotic forces of sugar loading in the source tissues and unloading in the sink tissues, respectively. The source is photosynthetic production of sugars. The sinks are given by all kinds of metabolic consumption as well as storage of carbohydrates. The source is regulating the sinks via supply, and the sinks are regulating the sources via demand. The sink demand has feedback on source activity (Körner 2012). Not only partitioning and allocation of photosynthetic products but also of mineral nutrients are involved in the source-sink regulation. The net rate of photosynthetic CO_2 uptake by leaves is reduced at low concentrations of NO_3^-, inorganic phosphate and K^+ (Longstreth and Nobel 1980).

For whole-plant integration of source-sink states, the concept of recirculation is decisive. It is based on phloem-xylem cycling of all kinds of solutes. Parenchyma rays mediate the exchange between the two pathways of long-distance transport, i.e., phloem and xylem (van Bel 1990). The concept has been developed and much propagated by Pate and Jeschke and their collaborators in the 1970s. The important cycling of mineral nutrients has been much studied by them (Jeschke et al. 1985, 1987; Peuke et al. 1996; Jeschke and Pate 1991b) as well as by other authors who have also contributed to the demonstration of the cycling of organic nitrogen and sulfur compounds (Cooper and Clarkson 1989; da Silva and Shelp 1990; Larsson et al. 1991; Geßler et al. 2003).

The role of concentration gradients of solutes and of pressure gradients resulting from them implies that information for the regulation of the cycling in whole plants is directly contained in the substrates themselves. Arrival of a substrate in a sink is a signal from the source. Demand of a sink is a signal to the source. Evidently, however, these are rather coarse signaling functions although their effectiveness should not be neglected. Beyond that, there is also more finely tuned signaling where substrates and metabolites participate. Nitrogen, phosphorus and sulfur compounds are involved in metabolic signaling systems (Longstreth and Nobel 1980; Scheible et al. 2004). Of particular importance in systemic signaling are sugars, not only the monosaccharide glucose but also the disaccharides sucrose and trehalose (Rolland et al. 2006). Glucose affects mitotic activity and acts analogous to a morphogen controlling development. It also interacts with regulation of hormone signaling at the level of transcriptional regulation by phytohormones. Sucrose affects gene expression and activity of enzymes and regulates processes of differentiation and development of plants. Sensors involved are various enzymes of sugar metabolism and sugar transporters in membranes (see Lüttge et al. 2010).

Developmental Relations

Intimate integration of parts is characterized by building up correlations between them. With "correlation," we mean here the systemic control of development (Ziegler 1998; Kost 2014). Correlations are mutual physical and functional relations and interactions of parts within whole plants. These come about by the determination of development. If a part or module of the integrated whole of a plant is subject to such determination from outside of itself, we call it aitionomic determination. It can be either positive or negative, i.e., an induction or a blockade. Aitionomic determination from neighboring cells or tissues can be an induction of the formation of similar cells or tissues, named homoiogenetic induction, or an induction of the formation of different types of cells or tissues, named heterogenetic induction.

An illustrative example of aitionomic correlative blockade or inhibition is apical dominance (Fig. 8). Normally, the bud at the top of a plant grows faster than the lateral buds. This drives the vertical growth of the plant. It is due to a domination of the lateral buds by the terminal bud. The latter produces the phytohormone indole acetic acid (IAA), which is transported downward and carries the information for the lateral buds to retard growth or even remain dormant. If the terminal bud is injured or removed experimentally, IAA transport from the top ceases. Then the lateral buds germinate and increase their growth until one of them wins the race, seizes the lead in producing a downward stream of IAA and takes over apical dominance again. Aitionomic correlative stimulation is observed in trees at the beginning of the growing season. Developing buds of branches when starting to grow out are signaling to the cambium in the stems to resume their secondary growth.

In secondary growth of the trunks of trees, we also see correlations (Fig. 8). Within intact stems, the meristem cylinder of the cambium produces the elements of the wood toward the center and the elements of the cortex toward the periphery. Other potential developments are suppressed by aitionomic correlative blockade. Such blockade is relieved if, for example, a fallen beech tree is cut in spring time. Then one can see that the cells of the cambium are in fact totipotent (Fig. 9). When the correlative inhibition is removed, the cambium cells at the cut surface begin to divide, first forming a callus of non-differentiated cells, which, however, rapidly begin differentiation and regenerate complete small beech trees.

Another example is induction of flowering (Fig. 10). We distinguish short day or long night (SD) and long day or short night (LD) plants. They flower if the duration of the night is above (SD plants) or below (LD plants) a certain threshold length which is specific for individual plant species. A flowering signal, the phytohormone-like florigen which is presumably the transcription factor "flowering locus T," is elicited by the leaves if night length is appropriate and transported to the top of the plant inducing the terminal bud to produce flowers. Experimentally, it is enough to put one leaf under the appropriate SD or LD conditions to induce flowering in the SD or LD plants.

The orientation of plants in space is achieved by tropisms (Fig. 11). These are growth reactions directed along external signals. Positive phototropisms of shoots

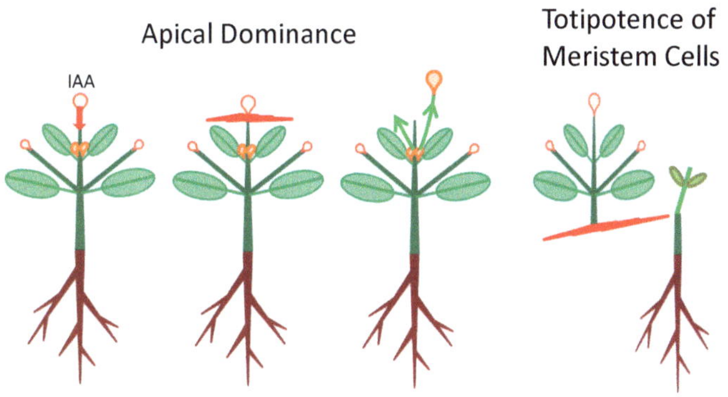

Fig. 8 Correlations

Fig. 9 Cut stem of a beech tree in spring with regenerating beech plants from the cambium

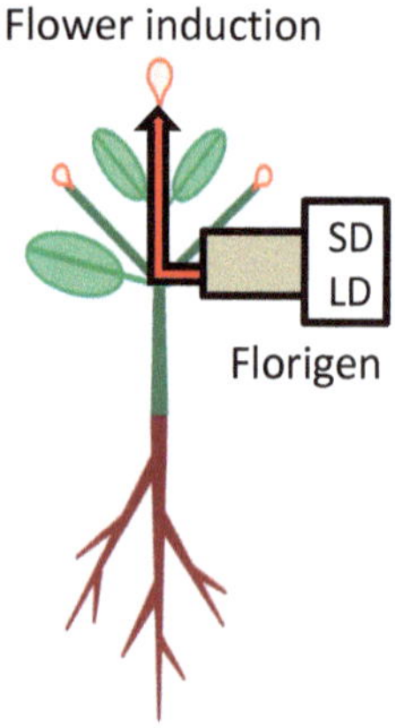

Fig. 10 Flower induction. If one leaf (dark box) is given short days (SD) or long days (LD), the plant is induced to flower if it is a short day or long day plant, respectively

are directed toward the light and negative phototropisms away from it. Gravitropisms directed to gravity are orthotropic positive of main roots and orthotropic negative of main shoots, plagiotropic (horizontal) of rhizomes and tillers, negative plagiotropic of side branches and positive plagiotropic of side roots.

The positive gravitropism of primary roots is particularly well studied at the molecular level. This is also an example for the translation of electrical and chemical phytohormone signals into each other in whole-plant performance. When the roots of the plants are in the normal vertical position, there is a symmetrical distribution of IAA coming from the shoot and back to the shoot from the roots. This is regulated by

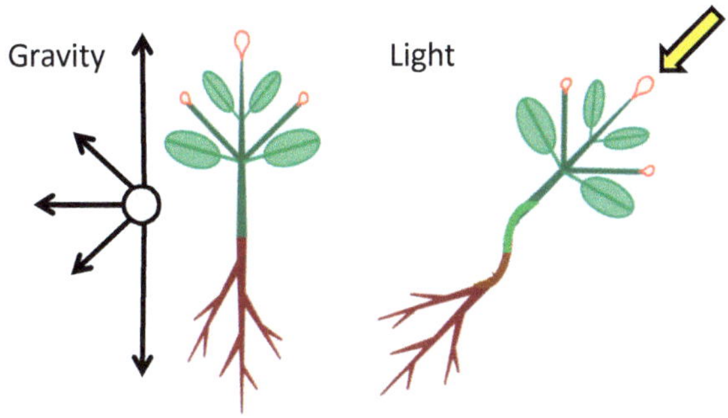

Fig. 11 Orientation in space by gravitropism (left) or phototropism (right)

the *PIN* genes. The name originates from the needle- or pin-like growth observed in *PIN*-deficient mutants, which cannot form leaves (Paponov et al. 2005; Grieneisen et al. 2007). *PIN1* and *PIN4* genes regulate IAA transport from the top down into the root tips within the central cylinder, *PIN2* genes control transport back up in the cortex, while *PIN3* genes determine distribution within the root tip (Friml 2003; Benková et al. 2003; Friml et al. 2003; Blilou et al. 2005). When the root is moved out of the vertical position, the signal of the changed direction of gravity is perceived by a system of amyloplasts functioning as statolytes and lying on cushions of endoplasmic reticulum in specific cells of the root cap, the statocysts. The products of the *PIN* genes are IAA efflux facilitators of cell membranes. Responses to changed direction of gravity lead to building up an asymmetrical distribution of IAA where IAA levels are higher at the physically lower side of cells. Electrical signals are associated with asymmetrical IAA distribution. The lower side of the roots gets less negative (Brauner and Bünning 1930; Behrens et al. 1985; Stenz and Weisenseel 1991, 1993). While signal perception is located in the root cap, the reaction of gravitropic bending, however, occurs further up in the elongation zone, where growth is inhibited at the lower side of the roots, and asymmetric cell elongations in the upper and lower sides lead to bending down of the root and acquiring the normal vertical position again. The signaling from the root tip to the elongation zone involves the asymmetric PIN3-regulated backflow of IAA which is enhanced at the lower side and flow of another phytohormone, i.e., cytokinin, from the root tip upward (Aloni et al. 2004) functioning as a growth inhibitor at the lower side. Electrical signals most likely are also involved.

These few examples mainly chosen from the textbook literature (Ziegler 1998; Kost 2014; Lüttge et al. 2010) may suffice here to illustrate that developmental biology provides principal concepts and is extraordinarily rich in examples underlining the tight interactions between the parts integrated in the emergent whole-plant systems (see also chapter by Layer 2019 in this book).

Environmental Relations

The establishment of plants in their environment with their responses to abiotic factors and their biotic interactions with other organisms is a wide field. Plants can only perform effectively as integrated unitary organisms. Since this essay restricts itself to emergence of the whole plant per se, it would go beyond our scope to consider higher ecological scaling levels (see, e.g., the holobionts and the holobiont-like systems (chapter zu Castell et al. 2019)). Instead, two examples shall be chosen to illustrate the within-whole-plant integrations during responses to environmental challenges, namely herbivory—defense and salinity (Fig. 12).

Defense

Defense under biotic attack requires structural modifications and in plants particularly chemical efforts. These are costly, and the energy and resources going into defense mechanisms are lost for growth. This dilemma is addressed by the growth-defense-balance hypothesis (GDB; Loomis 1953; Herms and Mattson 1992; Matyssek et al. 2002, 2005). The choice between defense and growth is a typical whole-plant accomplishment. The GDB is based on observations which suggest that at high affluence of resources and energy provided by vigorous photosynthesis, the choice is preferential propagation of growth as an irreversible increase in plant biomass, while at scarcity of resources and low energy supply, activation of defense is preferred to maintain and preserve the status given by resources already incorporated. Although due to the very high plasticity of the performance of plants the dilemma of growth-defense is not all that absolute and strict and the GDB hypothesis needs to be modified in this respect (Matyssek et al. 2012a, b, 2013; Souza and Lüttge 2015), responses can only be orchestrated by within-plant integrations, which in view of versatility are even more subtle than for the strict dilemma alone.

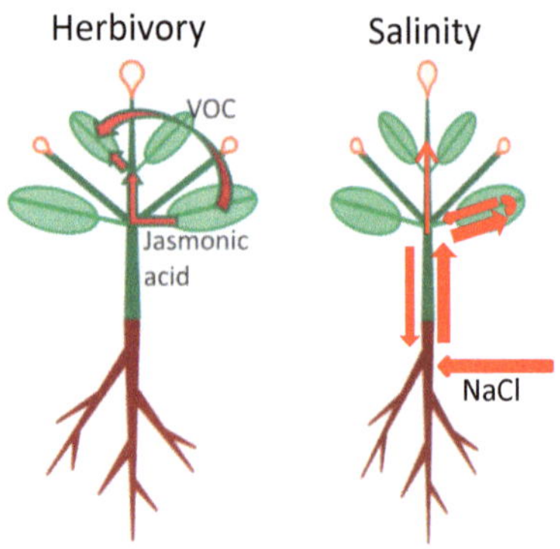

Fig. 12 Whole-plant performance under environmental stress of herbivory (left) or salinity (right)

In addition to this general view, an example of within-plant signaling is worth considering here. Plants often react to herbivore attack by producing chemical defense compounds that repel the aggressors when they start feeding. Such compounds are, for example, phenolics, phenyl-propane derivatives, terpenoids or tannins (Oßwald et al. 2012; Kolosova and Bohlmann 2012). When biting by herbivores occurs, there is systemic signaling to other parts of a plant to elicit production of defense compounds there before the herbivores arrive. The stress hormone jasmonic acid is carrying such information. However, it needs to be transported over long distances within the plant away from the site of first attack toward other parts to be conditioned and prepared for possible attack. This long-distance transport takes time which may prove to be too long for immediate protection. This is overcome by production of volatile signaling compounds, so-called pheromones, which alert the parts of a plant almost instantaneously via diffusion in the gas phase (Heil 2010). The methylated form of jasmonic acid, methyl jasmonate, is also volatile (Nincovic 2010). In ecological systems, this signaling proves to be also effective between different plants of the same and even of other species establishing preparedness intra- and inter-specifically (Ruther and Kleier 2005; von Dahl and Baldwin 2007; Baluška and Nikovic 2010; Bruce 2010; Nincovic 2010).

Salinity

Salinity, especially of sodium chloride (NaCl), is a common environmental problem in the substratum of plants. Plants can adapt (i) by excluding salt uptake at the root level (salt excluders), (ii) by taking salt up and excreting it via salt hairs and salt glands or (iii) by taking salt up and sequestering it in the vacuoles of cells at sites of salt storage (salt includers). The salt excluders run into problems of osmoregulation and need specific other adaptations. The salt includers can use the cheaply available NaCl for stabilizing their osmotic relations. They are particularly worth mentioning here in relation to whole-plant integrations because they use salt recirculation based on phloem-xylem cycling (see section "Photosynthetic Products") of NaCl. Salt is partitioned in the whole plant in a way protecting the most important centers of metabolism from salinity stress. Salt transported to the leaves in the transpiration stream via the xylem is recycled via the phloem and accumulated in the less sensitive parts. In an experiment with plants of *Ricinus communis* (Jeschke and Pate 1991a), more than two-third of the NaCl taken up was kept within the roots. During transport in the xylem, much Na^+ was accumulated in the old lower leaves and most of the Na^+ arriving in upper mature leaves was recycled in the phloem. The salt load of the growth zone of the apex and the young leaves and also leaves which were still growing and importing assimilates for growth was kept very low.

Higher Plants: Unitary Organisms Without a Neuronal System

Higher land plants cannot move around to visit optimal external conditions. Plastic responses to the dynamically variable features of and the input from the environment where they are firmly established are essential. Obviously for this, the simple stems, also called telome, of the psilophytes *Rhynia* (Fig. 1) were not a solution sophisticated enough. Specialization for the different sections of the environment, most notably the soil and the atmosphere, respectively, appeared advantageous. Thus, selection from morphogenetic patterns bore out the evolution of roots and leaves. By their integration via the stems, the modern higher land plants became emergent on the structural level of roots, stems and leaves (Figs. 1 and 2). Integration of a plethora of capacities completed the emergence of the plants as unitary organisms at the functional level (Figs. 6, 8 and 10). With the exception of neuronal systems, all kinds of mechanisms of integration are expressed in plants. There is systemic signaling (Figs. 5, 6, 8, 10 and 11). Partitioning and allocation of resources are optimizing the fulfillment of the demands of the parts in close mutual interactions (Fig. 6). The parts cannot perform independently and maintain their existence and integrity. The whole is more than the sum of its parts (Fig. 3).

References

Alberts B, Johnson A, Lewis J, Raff M, Roberts K, Walter P (2002) Molecular biology of the cell, 4th ed. Garland Science, Taylor and Francis Books

Aloni R, Langhans M, Aloni E, Ullrich CI (2004) Role of cytokinin in the regulation of root gravitropism. Planta 220:177–182

Bano A, Dörffling K, Bettin D, Hahn H (1993) Abscisic acid and cytokinins as possible root-to-shoot signals in xylem sap of rice plants in drying soil. Aust J Plant Physiol 20:109–115

Baluška F, Ninkovic V (2010) Plant communication from an ecological perspective. Springer, Berlin

Beck E (2019) Ecology: ecosystems and biodiversity. In: Wegner LH, Lüttge U (eds) Emergence and modularity in life science. Springer, Heidelberg pp 195–213

Behrens HM, Gradmann D, Sievers A (1985) Membrane potential responses following gravistimulation in roots of *Lepidium sativum* L. Planta 163:463–472

Benková E, Michniewicz M, Sauer M, Teichmann T, Seiferotová D, Jürgens G, Friml J (2003) Local, efflux-dependent auxin gradients as a common module for plant organ formation. Cell 115:591–602

BenZioni A, Vaadia Y, Lips SH (1971) Nitrate uptake by roots as regulated by nitrate reduction products of the shoot. Physiol Plant 24:288–290

Blilou I, Xu J, Wildwater M, Willemsen V, Paponov I, Friml J, Heidstra R, Aida M, Palme K, Scheres B (2005) Nature 433:39–44

Bolker JA (2000) Modularity in development and why it matters to evo-devo. Am Zool 4:770–776

Brauner L, Bünning E (1930) Geoelektrischer Effekt und Elektrotropismus. Ber Dtsch Bot Ges 48:470–476

Bruce TJA (2010) Exploiting plant signals in sustainable agriculture. In: Baluška F, Ninkovic V (eds) Plant communication from an ecological perspective. Springer, Berlin Heidelberg, pp 215–227

Buchanan BB, Gruissem W, Jones RL (2000) Biochemistry and molecular biology of plants. Amer Soc Plant Physiol, Rockville, Maryland
Cermak J, Matyssek R, Kucera J (1993) Rapid response of large, drought–stressed beech trees to irrigation. Tree Physiol 12:281–290
Clarkson DT, Smith FW, Vanden Berg PJ (1983) Regulation of sulphate transport in a tropical legume, *Macroptilium atropurpureum*, cv. Siratro. J Exp Bot 34:1463–1483
Comstock JP (2002) Hydraulic and chemical signaling in the control of stomatal conductance and transpiration. J Exp Bot 53:195–200
Cooper HD, Clarkson DT (1989) Cycling of amino-nitrogen and other nutrients between shoots and roots in cereals. A possible mechanism integrating shoot and root in the regulation of nutrient uptake. J Exp Bot 40:753–762
Davies E (2004) New functions for electrical signals in plants. New Phytol 161:607–610
Davies WJ, Mansfield TA, Hetherington AM (1990) Sensing of soil water status and the regulation of plant growth and development. Plant Cell Environ 13:709–719
Davies WJ, Tardieu F, Trejo CL (1994) How do chemical signals work in plants that grow in drying soil? Plant Physiol 104:309–314
Davies WJ, Zhang J (1991) Root signals and the regulation of growth and development of plants in drying soil. Annu Rev Plant Biol Plant Mol Biol 42:55–76
Friedman WE, Diggle PK (2011) Charles Darwin and the origins of plant evolutionary developmental biology. Plant Cell 23:1194–1207
Friml J (2003) Auxin transport—shaping the plant. Curr Opin Plant Biol 6:7–12
Friml J, Vieten A, Sauer M, Weijers D, Schwarz H, Hamann T, Offringa R, Jürgens G (2003) Efflux-dependent auxin gradients establish the apical-basal axis of *Arabidopsis*. Nature 426:147–153
Fromm J, Eschrich W (1993) Electric signals released from roots of willow (*Salix viminalis* L.) change transpiration and photosynthesis. J Plant Physiol 141:673–680
Fromm J, Fei H (1998) Electrical signaling and gas exchange in maize plants of drying soil. Plant Sci 132:203–213
Fromm J, Lautner S (2007) Electrical signals and their physiological significance in plants. Plant Cell Environ 30:249–257
Frost WB, Blevins DG, Barnett NM (1978) Cation pretreatment effects on nitrate uptake, xylem exudates, and malate levels in wheat seedlings. Plant Physiol 61:323–326
Gersani M, Abramsky Z, Falik O (1998) Density-dependent habitat selection in plants. Evol Ecol 12:223–234
Geßler A, Weber P, Schneider S, Rennenberg H (2003) Bidirectional exchange of amino compounds between phloem and xylem during long-distance transport in Norway spruce trees (*Picea abies* [L.] Karst.) J Exp Bot 54:1389–1397
Gil PM, Gurovich L, Schaffer B, Alcayaga J, Rey S, Iturriaga R (2008) Root to leaf electrical signaling in avocado in response to light and soil water content. J Plant Phys 165:1070–1078
Gould JG (2002) The structure of evolutionary theory. Harvard University Press, Cambridge, Massachusetts
Grams TEE, Koziolek C, Lautner S, Matyssek R, Fromm J (2007) Distinct roles of electric and hydraulic signals on the reaction of leaf gas exchange upon re-irrigation in *Zea mays* L. Plant Cell Environ 30:79–84
Grams TEE, Lautner S, Felle HH, Matyssek R, Fromm J (2009) Heat-induced electrical signals affect cytoplasmic and apoplastic pH as well as photosynthesis during propagation through the maize leaf. Plant Cell Environ 32:319–326
Grieneisen VA, Xu J, Marée AFM, Hogeweg P, Scheres B (2007) Auxin transport is sufficient to generate a maximum and gradient guiding root growth. Nature 449:1008–1013
Haukioja E (1991) The influence of grazing on the evolution, morphology and physiology of plants as modular organisms. Philos Trans Roy Soc London Ser B Biol Sci 333:241–247
Heil M (2010) Within-plant signalling by volatiles triggers systemic defences. In: Baluška F, Ninkovic V (eds) Plant communication from an ecological perspective. Springer, Berlin, pp 99–112

Herms DA, Mattson WJ (1992) The dilemma of plants: to grow or defend. Q Rev Biol 67:283–335
Herschbach C, Rennenberg H (1994) Influence of glutathione (GSH) on net uptake of sulphate and sulphate transport in tobacco plants. J Exp Bot 45:1069–1076
Hütt M-Th (2013) A network view on patterns of gene expression and metabolic activity. Nova Acta Leopoldina NF 114(391):183–199
Jeschke WD, Atkins CA, Pate JS (1985) Ion circulation via phloem and xylem between root and shoot of nodulated white lupin. J Plant Physiol 117:319–330
Jeschke WD, Pate JA (1991a) Cation and chloride partitioning through xylem and phloem within the whole plant of *Ricinus communis* L. under conditions of salt stress. J Exp Bot 42:1105–1116
Jeschke WD, Pate JS (1991b) Modelling of the uptake, flow and utilization of C, N and H_2O within whole plants of *Ricinus communis* L. based on empirical data. J Plant Physiol 137:488–498
Jeschke WD, Pate JS, Atkins CA (1987) Partitioning of K^+, Na^+, Mg^{++}, and Ca^{++} through xylem and phloem to component organs of white lupin under mild salinity. J Plant Physiol 128:77–93
Johnson BR, Lam SK (2010) Self-organization, natural selection, and evolution: cellular hardware and genetic software. Bioscience 60:879–885
Kadereit JW, Körner C, Kost B, Sonnewald U (2014) Strasburger. Lehrbuch der Pflanzenwissenschaften, 37th edn. Springer, Heidelberg
Kirkby EA, Knight AH (1977) Influence of the level of nitrate nutrition on ion uptake and assimilation, organic acid accumulation, and cation-anion balance in whole tomato plants. Plant Physiol 66:349–353
Kolosova N, Bohlmann J (2012) Conifer defense against insects and fungal pathogens. In: Matyssek R, Schnyder H, Oßwald W, Ernst D, Munch JC, Pretzsch (eds) Growth and defence in plants. Resource allocation at multiple scales. Ecological Studies, vol 220. Springer, Heidelberg, pp 85–109
Körner C (2012) Biological diversity—The essence of life and ecosystem functioning. Nova Acta Leopoldina NF 116(394):147–159
Kost B (2014) Systemische Kontrolle der Entwicklung. In: Kadereit JW, Körner C, Kost B, Sonnewald U (eds) Strasburger, Lehrbuch der Pflanzenwissenschaften, 37th edn. Springer, Heidelberg, pp 282–283
de Kroon H, Huber H, Stuefer JF, van Groenendael JM (2005) A modular concept of phenotypic plasticity in plants. New Phytol 166:73–82
Larsson C-M, Larsson M, Purves JV, Clarkson DT (1991) Translocation and cycling through roots of recently absorbed nitrogen and sulphur in wheat (*Triticum aestivum*) during vegetative and generative growth. Physiol Plant 82:345–352
Lautner S, Grams TEE, Matyssek R, Fromm J (2005) Characteristics of electrical signals in poplar and responses in photosynthesis. Plant Physiol 138:2200–2209
Layer PG (2019) Brains emerging: on modularity and self organization of neuronal development *in vivo* and *in vitro*. In: Wegner LH, Lüttge U (eds) Emergence and modularity in life science. Springer, Heidelberg, pp 145–169
Longstreth DJ, Nobel PS (1980) Nutrient influences on leaf photosynthesis. Effects of nitrogen, phosphorus and potassium for *Gossypium hirsutum* L. Plant Physiol 65:541–543
Loomis WE (1953) Growth and differentiation—an introduction and summary. In: Loomis WE (ed) Growth and differentiation in plants. Iowa State College Press, Ames, pp 1–17
Lorenz K (1977) Die Rückseite des Spiegels. Deutscher Taschenbuchverlag, München, Versuch einer Naturgeschichte menschlichen Erkennens
Lucas M, Laplace L, Bennett MJ (2011) Plant systems biology: network matters. Plant Cell Environ 34:535–553
Lüttge U (2012) Modularity and emergence: biology's challenge in understanding life. Plant Biol 14:865–871
Lüttge U (2013) Whole-plant physiology: synergistic emergence rather than modularity. Progr Bot 74:165–190
Lüttge U (2016) Physics and the molecular revolution in plant biology: union needed for managing the future. AIMS Biophy 3:501–521

Lüttge U, Higinbotham N (1979) Transport in plants. Springer, New York, Heidelberg, Berlin
Lüttge U, Kluge M, Thiel G (2010) Botanik. Die umfassende Biologie der Pflanzen: Wiley-VCH, Weinheim
Marr C, Geertz M, Hütt M-T, Muskhelishvili G (2008) Dissecting the logical types of network control in gene expression profiles. BMS Syst Biol 2:18
Matyssek R, Agerer R Ernst D, Munch J-C, Oßwald W, Pretzsch H, Priesack E, Schnyder H, Treutter D (2005) The plant's capacity in regulating resource demand. Plant Biol 7:560–580
Matyssek R, Lüttge U, Rennenberg H (eds) (2013) The alternatives growth and defense: resource allocation at multiple scales in plants. Nova Acta Leopoldina NF 114/No 391
Matyssek R, Maruyama S, Boyer JS (1991) Growth-induced water potentials may mobilize internal water for growth. Plant Cell Environ 14:917–923
Matyssek R, Schnyder H, Elstner E-F, Munch J-C, Pretzsch H, Sandermann H (2002) Growth and parasite defence in plants: the balance between resource sequestration and retention. Plant Biol 4:133–136
Matyssek R, Schnyder H, Oßwald W, Ernst D, Munch JC, Pretzsch (eds) (2012a) Growth and defence in plants. Resource allocation at multiple scales. Ecological studies, vol 220. Springer, Heidelberg
Matyssek R, Gayler S, zu Castell W, Oßwald W, Ernst D, Pretzsch H, Schnyder H, Munch JC (2012b) Predictability of plant resource allocation: New theory needed? In: Matyssek R, Schnyder H, Oßwald W, Ernst D, Munch JC, Pretzsch (eds) Growth and defence in plants. Resource allocation at multiple scales. Ecological studies, vol 220. Springer, Heidelberg, pp 433–449
Morris, SK (2003) Life's solution. Inevitable humans in a lonely universe. Cambridge University Press, New York
Müller GB (2007) Evo-devo: extending the evolutionary synthesis. Nat Rev Genet 8:939–949
Münch E (1930) Die Stoffbewegungen in der Pflanze. Gustav Fischer, Jena
Ninkovic V (2010) Volatile interaction between undamaged plants: A short cut to coexistence. In: Baluška F, Ninkovic V (eds) Plant communication from an ecological perspective. Springer, Berlin Heidelberg, pp 75–86
Oßwald W, Fleischmann F, Treutter D (2012) Host-parasite interactions and trade-offs between growth- and defence-related metabolism under changing environments. In: Matyssek R, Schnyder H, Oßwald W, Ernst D, Munch JC, Pretzsch (eds) Growth and defence in plants. Resource allocation at multiple scales. Ecological studies, vol 220. Springer, Heidelberg, pp 53–83
Paponov IA, Teale WD, Trebar M, Blilou I, Palme K (2005) The PIN auxin efflux facilitators: evolutionary and functional perspectives. Trends Plant Sci 10:170–177
Peuke AD, Glaab J, Kaiser WM, Jeschke WD (1996) The uptake and flow of C, N and ions between roots and shoots in *Ricinus communis* L. IV. Flow and metabolism of inorganic nitrogen and malate depending on nitrogen nutrition and salt treatment. J Exp Bot 47:377–385
Pitman MG (1975) Whole plants. In: Baker DA, Hall JL (eds) Ion transport in plant cells and tissues. North Holland Publ Comp, Amsterdam Oxford, pp 267–308
Rennenberg H, Schmitz K, Bergmann L (1979) Long-distance transport of sulfur in *Nicotiana tabacum*. Planta 147:57–62
Rolland F, Baena-Gonzalez E, Sheen J (2006) Sugar sensing and signaling in plants: conserved and novel mechanisms. Ann Rev Plant Biol 57:675–709
Ruther J, Kleier S (2005) Plant-plant signaling: ethylene synergizes volatile emission in *Zea mays* induced by exposure to (Z)-3-hexen1-ol. J Chem Ecol 21:2217–2222
Scheible W-R, Morcuende R, Czechoswski T, Fritz C, Osuna D, Palacios-Rojas N, Schindelasch D, Thimm O, Udvardi MK, Stitt M (2004) Genome-wide reprogramming of primary and secondary metabolism, protein synthesis, cellular growth processes, and the regulatory infrastructure of *Arabidopsis* in response to nitrogen. Plant Physiol 136:2483–2499
Schurr U, Schulze E-D (1996) Effects of drought on nutrient and ABA transport in *Ricinus communis*. Plant Cell Environ 19:665–674

Shabala S, Pang J, Zhou M, Shabala L, Cuin T, Nick P, Wegner LH (2009) Electrical signalling and cytokinins mediate effects of light and root cutting on ion uptake in intact plants. Plant Cell Environ 32:194–207

Shabala S, White RC, Djordjevic MA, Ruan Y-L, Mathesius U (2016) Root-to-shoot signalling: integration of diverse molecules, pathways and functions. Funct Plant Biol 43:87–104

da Silva MC, Shelp BJ (1990) Xylem-to-phloem transfer of organic nitrogen in young soybean plants. Plant Physiol 92:797–801

Souza GM, Lüttge U (2015) Stability as a phenomenon emergent from plasticity-complexity-diversity in eco-physiology. Progr Bot 76:211–239

Souza GM, Bertolli SC, Lüttge U (2016) Hierarchy and information in a system approach to plant biology: explaining the irreducibility in plant ecophysiology. Progr Bot 77:167–186

Stenz H-G, Weisenseel MH (1991) DC-electric field affects the growth direction and statocyte polarity of root tips (*Lepidium sativum*) J Plant Physiol 138:335–344

Stenz H-G, Weisenseel MH (1993) Electrotropism of maize (*Zea mays* L.) roots. Facts and artifacts. Plant Physiol 101:1107–1111

Sutcliffe JF (1976a) Regulation in the whole plant. Enc Plant Physiol 2B (Springer, Berlin) 394–417

Sutcliffe JF (1976b) Regulation of ion transport in the whole plant. In: Sunderland N (ed) Perspectives in experimental biology, vol II, Botany. Pergamon Press, Oxford, p 542

Tang A-C, Boyer JS (2003) Root pressurization affects growth-induced water potentials and growth in dehydrated maize plants. J Exp Bot 54:2479–2488

Tardieu F, Davies WJ (1993) Integration of hydraulic and chemical signalling in the control of stomatal conductance and water status of droughted plants. Plant Cell Environ 16:341–349

Tardieu F, Zhang J, Gowing DJG (1993) Stomatal control by both [ABA] in the xylem sap and leaf water status: a test of a model for droughted or ABA-fed field-grown maize. Plant Cell Environ 16:413–420

Thornley JHM (1972) A balanced quantitative model for root : shoot ratios in vegetative plants. Ann Bot 36:431–441

Turner NC, Schulze E-D, Gollan T (1985) The responses of stomata and leaf gas exchange to vapour pressure deficits and soil water content. II. In the mesophytic herbaceous species *Helianthus annuus*. Oecologia 65:348–355

van Bel AJE (1990) Xylem-phloem exchange via the rays: the undervalued route of transport. J Exp Bot 41:631–644

Volkov AG (2000) Green plants: electrochemical interfaces. J Electroanalytical Chemistry 483:150–156

von Dahl CC, Baldwin IT (2007) Deciphering the role of ethylene in plant-herbivore interactions. J Plant Growth Regul 26:201–209

Wallace W, Pate JS (1967) Nitrate assimilation in higher plants with special reference to the cocklebur (*Xanthium pennsylvanicum* Wallr.) Ann Bot 31:213–228

Wartinger A, Heilmeier H, Hartung W, Schulze E-D (1990) Daily and seasonal courses of leaf conductance and abscisic acid in the xylem sap of almond trees [*Prunus dulcis* (Miller) D.A. Webb] under desert conditions. New Phytol 116:581–587

Weston DJ, Hanson PJ, Norby RJ, Tuskan GA, Wullschleger SD (2012) From system biology to photosynthesis and whole-plant physiology. Plant Signal Behav 7:260–262

Wikipedia (2012) Emergenz. http://de.wikipedia.org/wiki/Emergenz: Aristoteles, Metaphysik, Buch 8.6. 1045a: 8–10

Wilkinson S, Davies WJ (1997) Xylem sap pH increase: a drought signal received at the apoplastic face of the guard cell which involves the suppression of saturable ABA uptake by the epidermal symplast. Plant Physiol 113:559–573

Wilkinson S, Corlett JE, Oger L, Davies WJ (1998) Effects of xylem pH on transpiration from wild-type and *flacca* tomato leaves: a vital role for abscisic acid in preventing excessive water loss even from well-watered plants. Plant Physiol 117:703–709

Zhang J, Davies WJ (1990) Changes in the concentration of ABA in xylem sap as a function of changing soil water status can account for changes in leaf conductance and growth. Plant Cell Environ 13:277–285

Ziegler H (1956) Untersuchungen über die Leitung und Sekretion der Assimilate. Planta 47:447–500

Ziegler H (1998) Physiologie des Formwechsels. In: Sitte P, Ziegler H, Ehrendorfer F, Bresinsky A (eds) Strasburger, Lehrbuch der Botanik. Gustav Fischer, Stuttgart, pp 362–419

Zimmermann W (1930) Die Phylogenie der Pflanze. Gustav Fischer, Jena

zu Castell W, Lüttge U, Matyssek R (2019) Gaia—a holobiont like system emerging from interaction. In: Wegner LH, Lüttge U (eds) Emergence and modularity in life science. Springer, Heidelberg, pp 255–279

Ecology: Ecosystems and Biodiversity

Erwin Beck

Abstract Using ecosystems as examples, this chapter engages with the emergence of understanding life by producing and assembling modules of knowledge, and finally linking them to create a holistic picture of the entire system. Ecosystems as theoretical units of arbitrary size are understood to consist of abiotic and biotic components on the one hand and of the interactions of the components on the other. The latter is extraordinarily complex but generates functionality in the system as a basis of its properties and services. Functionality can be further partitioned into processes, such as flow of energy and matter, resulting from food chains or webs. Functional diversity is considered as a composite variable that includes all significant physiological information as processes and/or traits, weighted by their abundances in a community whose composition has been filtered by environmental conditions. Two types of ecological experiments can be used to unravel the significance of the interactions of species in a functional community: The analytical approach by intentional disturbance, i.e., a change of an external condition, or the synthetic approach by using artificial species compositions in an otherwise natural environment. Both approaches allow the characterization of functional modules in an ecosystem. Due to the complexity of even simple appearing modules like biomass production, models are required for a comprehensive insight. The more so linking modules to achieve a higher level of integration is unthinkable without comprehensive synthesis models. Examples are presented for each step in the emerging knowledge about, and understanding of ecosystems.

Introduction—Another Focus

This chapter is not dedicated to the emergence of ecosystems and their progressive development by the arrival of new species, like the stepwise assembly of the vegetation on uncovered glacier moraines after the ablation of the ice (Chapin III et al.

E. Beck (✉)
Department of Plant Physiology and Bayreuth Center of Ecology and Ecosystem Research, University of Bayreuth, Bayreuth, Germany
e-mail: erwin.beck@uni-bayreuth.de

L. H. Wegner and U. Lüttge (eds.), *Emergence and Modularity in Life Sciences*,
https://doi.org/10.1007/978-3-030-06128-9_9

1994) or by the planting of a selected pioneer flora on restoration areas in former lignite mining areas (Baasch et al. 2012). It rather engages with a second aspect of the complex "understanding life", the emergence of understanding by creating, assembling and linking of modules of knowledge, a process, often termed synthesis. Ecosystems are prime examples in that respect due to their composition of abiotic features and biotic components on the one hand and the mutual interactions of these components on the other, which effect the ecosystem's functionality contributing to the system's continuance and stability. Putting both into a whole, an ecosystem is (much) more than the sum of its (abiotic and biotic) components. In most, and in particular the terrestric ecosystems, the biotic components are exceedingly numerous and a complete list is virtually beyond imagination, especially with regard to microbial life. Nevertheless, investigating an ecosystem has to start with the identification of the components, mostly the species, or if the individuals are not tangible, e.g., soil organisms, at least their DNA-sequential data, termed operational taxonomic units (OTUs). Usually, there is some additional knowledge on the species together with the name, e.g., about habitat and growth conditions, pollinators, seed dispersers and so on which is very useful information for the next emerging level of an ecosystem study.

The Next Level. Functionality in an Ecosystem, Flow of Matter and Energy

Focusing on the ecosystem, the players can be labeled by ecosystem functions which result from their metabolic and developmental activities. All organisms which by photo- or chemosynthesis produce biomass from CO_2, water and inorganic nutrients belong to the category of the producers which by their own growth and reproduction serve as foodstuff for the large group of consumers, starting with the herbivores as the basic group of a pyramid containing several so-called trophic levels (primary, secondary, tertiary consumers…). The (pseudo)steady state of the ecosystem is ultimately maintained by the heterogeneous group of the destruents or decomposers, animals, fungi, microorganisms which chop, degrade and mineralize the dead organic matter in the ecosystem, forming a parallel, top-down pyramid, the decomposers' food chain (Schuldt et al. 2017). Components of both food chains interact, effecting community and ecosystem-level processes and properties (Wardle et al. 2004). Pathogens and parasites may be considered as consumers at various levels of the consumer pyramid. Associating a species—as individual or population—with one of these functional classes is relatively easy. Identifying their position in the consumers' or destruents' pyramids is, however, often difficult, as the so-called linear food chains are in reality food webs, especially in the kingdom of the invertebrates and microbes. Likewise, omnivorous animals may not maintain a fixed position in a food chain or web.

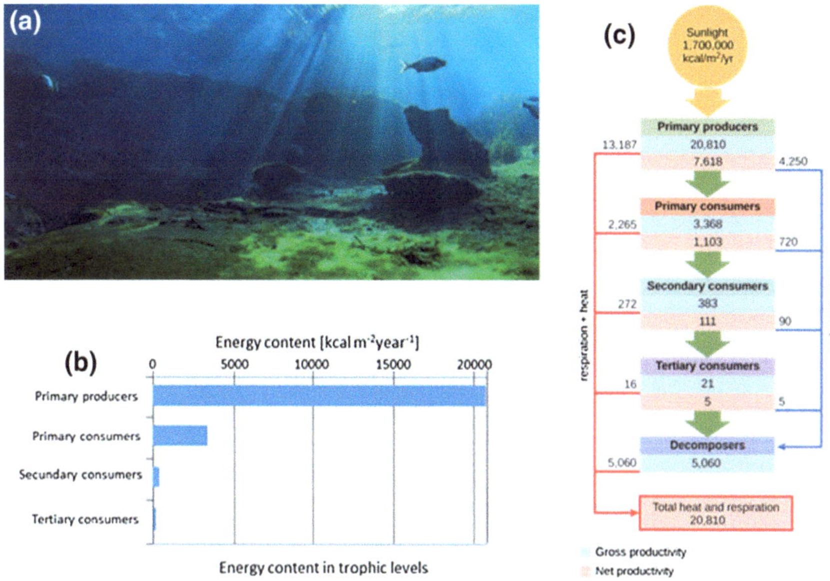

Fig. 1 The Silver Springs aquatic ecosystem: **a** view Silver Springs State Park (http://www.silversprings.com/), **b** energy content of trophic levels in a Silver Springs ecosystem, Florida, is shown. The available energy decreases from level to level, supporting fewer organisms at the subsequent level. **c** A conceptual model of the flow of energy through a spring ecosystem in Silver Springs, Florida. Notice that the energy decreases with each increase in trophic level. *Source* (**b**) and (**c**) Biology Libre texts Library, Chapter 46 Ecology of Ecosystems, Creative Commons Attribution-Noncommercial-Share Alike 3.0 United States License

The connections between the individual species (mostly as populations) might be apprehended by the flow of matter or energy associated with matter: Eating and being eaten. The classical example is that presented by H. T. Odum for the Silver Springs in Florida in 1957, an aquatic ecosystem of a manageable size, with more or less clearly defined functional groups. The energy content of each level as calculated from the amount of biomass shows that almost 90% are lost upon transition from one level to net higher level in the food chain through biological processes such as respiration, food search and mating activities. As a consequence, the pyramid of trophic levels ends when the amount of energy remaining in the food chain is insufficient for another higher level. In contrast to the energy which is finally lost as heat, nutrients are recycled by the action of the destruents.

The relatively simple Silver Spring system, structured by energy flow and trophic levels provides the design for a quantitative analysis of the interaction of processes constituting an ecosystem; in this case, feeding connections are considered (Fig. 1). However, these webs also include indirect and non-trophic relationships (Scherber et al. 2010; Kéfi et al. 2012) that necessarily remain undetected by analyses of feeding interactions, but play important roles in structuring ecosystems (Rzanny et al. 2013).

Dimensions of Ecosystems

Another important question is about the size and delimitation of the investigated ecosystem, which per se is not fixed, but depends on the purpose of research. If the structure of the food chains and the energy flow between the trophic levels are of interest, an ecosystem may be as small as a flower head of a thistle with seed predators and their parasites and parasitoids (Zwölfer 1987). On the other side, a forest, a region and even earth can be treated as ecosystem, depending on the scientific problem to be addressed. For laboratory investigations, micro- and mesocosm-systems are frequently used. Both are artificial ecosystems maintained under controlled external conditions like temperature, illumination and moisture. Microcosms are simplified miniature ecosystems, e.g., litterbags buried in the soil, which are used to simulate the performance of natural ecosystems, e.g., litter decomposition, under selected close-to nature or artificial conditions. Mesocosms are pieces of a natural ecosystem, e.g., a water column in the sea, separated for experimental purposes from the surrounding water (Fig. 2). Such mesocosms are used, e.g., for investigating the effects of the acidification of the ocean water by the increase of the atmospheric and dissolved CO_2 (Zark et al. 2017).

Fig. 2 **a** A series of **K**iel **O**ff-**S**hore **M**esocosms for **O**cean **S**imulations (KOSMOS) in the Northern Baltic Sea. The mesocosms were successfully employed in long-term experiments in different climate zones, from the high Arctic to the subtropics. Experimental perturbations include the addition of inorganic nutrients or organic compounds, CO_2 enrichment, manipulation of mixed layer depth, simulation of deep-water upwelling, species exclusion and addition of invasive species. **b** A mesocosm, consisting of a floatation frame and a flexible bag 2 m in diameter and 20 m in length. After deployment, the bag unfolds by weights pulling down its lower end to depth, thereby enclosing an undisturbed water column 55 m^3. The bag is then closed at the bottom by a full-diameter sediment trap (Riebesell et al. (2013). https://www.geomar.de/en/research/fb2/fb2-bi/infrastructure/kosmos-kiel-off-shore-mesocosms-for-oceanographic-studies). With permission by Gerd Hoffmann-Wieck

Functionality and the Use of Traits

In the following approaches in ecosystem research are addressed, which consider especially those aspects that render an ecosystem more than the sum of its components, namely the interactions, part of which has already been described above as flux of energy or matter. However, there are many more interactions, in particular the abiotic-biotic interactions and those biotic interactions which in addition to matter flow have other important functions in an ecosystem, e.g., mutualistical useful or detrimental relations. These are also subsumed under the term functionality in the ecosystem. Functional diversity is considered as a composite variable that includes all significant physiological information as processes and/or traits, weighted by their abundances in the community. The measured traits, e.g., transpiration also allow upscaling from the leaf to the plant and to the plot level. Trait-based approaches can thus be used to describe the relationship between community performance and ecosystem functioning on a higher level. The diversity of functional traits in natural communities depends on assembly processes driven by environmental filtering, such as competitive exclusion or facilitation, as well as on the size of the regional species pool (Keddy 1992; Diaz et al. 1998). Natural or anthropogenic changes of the environment can alter the strength and mesh size of the filters and thus the conditions of the assembly process. Rising temperatures for instance could cause an altitudinal upwards shift of the vegetation belts concomitant with a change in the competitive strength of the placeholder and the newcomer species. Mountain summits exhibiting characters of an island are considered as highly sensitive to climate change. Plant communities of the alpine and nival zones are adapted to lower temperatures and with respect to global warming their migration ability to colonize higher regions is geomorphologically limited. A worldwide monitoring web **Global Observation Research Initiative in Alpine Environments (GLORIA)** for the comparative study of climate change impacts on mountain biodiversity, encompassing 115 active target regions worldwide is being established since the year 2000 (http://www.gloria.ac.at:80/?a=9; Pauli et al. 2012). For the European mountains, a decline of cold-adapted and an increase of warm-adapted plant species in the summit regions have been documented between 2001 and 2008 showing an upwards shift of the thermoclines ("thermophilization") as the major environmental filters (Gottfried et al. 2012).

Due to the genetic and trait variability in the populations, many interactions can only be represented by models which must allow for the variability of the responses to the triggering abiotic or biotic factors. Many models are "only" partial models for a selected type of interaction or response, e.g., the response of the plant's photosynthetic carbon acquisition to environmental factors or of a plant—animal interaction resulting in pollination or seed dispersal. Proper modeling requires knowledge about the underlying principle or mechanism of the selected interaction, as well as of its quantitative range. To that end, analytical studies are necessary either in a mechanistical or a statistical approach like the response and effect framework (Lavorel and Garnier 2002) or via ecological experiments. Finally, partial models are the modules which must be put together in the synthesis step to arrive at a higher level of ecosystem understanding.

Ecological Experiments: Analytical Approach

Ecological experiments are intentionally planned disturbances of the original ecosystem, thus differing from the mesocosms (Fig. 2), which are by some provision separated from the natural ecosystem. In the emergence of ecosystem understanding, ecological experiments are used to elucidate mechanisms operating in an ecosystem and to produce the basis for models and predictions. A vast multitude of scientific problems can be tackled by such field experiments, finances commonly limiting their spatial and temporal dimensions. The most well-known ecological experiments are the FACE-experiments (**Free Air Carbon dioxide Enrichment** experiments) where the CO_2 concentration in the free air is artificially enhanced (Fig. 3) (Ainsworth and Long 2005).

This is not the place to discuss the results of the various FACE-experiments in forests, grass- and cropland; however, one message is clear: If CO_2 is no longer limiting productivity, another environmental factor becomes limiting, in many cases nitrogen or phosphate. However, on the forest ecosystem level the so-called progressive nitrogen limitation hypothesis modified the simple explanation by Liebig's law of the decreasing effect of CO_2-enriched air on tree growth. It assumes that plant growth in elevated CO_2 sequesters nitrogen in wood or soil organic matter, resulting in reduced N availability and the observed negative feedback on growth (Norby and Zak 2011). For ecosystem research, other results than productivity may be interesting as well, e.g. the reduction of transpiration (Ainsworth and Rogers 2007) or the stimulating effect on ectotrophic but not on endotrophic mycorrhiza (Garcia et al. 2008).

Other frequently installed ecological experiments are the rain-out shelter experiments simulating drought accompanying global warming by climate change (Beier et al. 2012), which are, however neither easy to perform in a natural habitat (e.g., a forest) nor are the results easy to explain (Vogel et al. 2013), as the microclimate conditions under the shelter differ in more than the moisture aspect from the open situation. Provided a sufficiently long experimental time-span of several years, interesting effects, not only on ecosystem processes like litter degradation have been found, but also, e.g., on the different genetic adaptability of plant species (populations via selection, Ravenscroft et al. 2015) or soil animals (Bataillon et al. 2016). Such response effects of ecosystem components to a climate change in turn may have an effect on their interactions with other components, e.g., herbivores.

Fig. 3 Aerial view of the Aspen FACE experiment (after 2010 Northern Forest Ecosystem Experiment) showing the control facilities (middle left row), and the 12 atmospheric treatment rings (air, CO_2, ozone) for four treatments with three replicates each. Each 30 m-ring consists of vertical vent-pipes which disperse carbon dioxide, ozone or normal air alone or in combination into the center of the ring. *Photo* www.aspenface.mtu.edu, by courtesy of Janet Pikkarainen

Other diligently planned treatments of natural ecosystems, in particular tropical forests, investigate the effects of global change, e.g., the rise of the temperature in the "TRACE"-experiment in Costa Rica with infrared radiators and soil heaters. It should have started in 2016, but due to drought was postponed. Another interesting ecological experiment, the so-called NUMEX experiment (**Nutrient Manipulation Experiment**) in the tropical mountain rainforest of the South Ecuadorian Andes, simulates the atmospheric input of plant nutrients from fires in the Amazon or dust from the Sahara by the almost permanent easterly winds. Moderate fertilization of the mountain forest with nitrogen, phosphate and calcium resulted in a differentiated response of trees, some of which showed growth inhibition, others growth stimulation (Homeier et al. 2012, 2013; Werner et al. 2013). N and P additions resulted also in an altered soil microbial biomass and a reduced fine root biomass, but no generalizable effect on the rate of fine root colonization by arbuscular mycorrhizal fungi was detected. Partly contrasting responses of tree species (stem growth and litter production) suggest future changes in species composition of these forests by atmospheric nutrient inputs in the long term.

Such ecological experiments investigate responses to artificially set environmental changes of real ecosystems and their components and thus follow a top-down approach, addressing also the question of ecosystem stability and resistance.

Ecological Experiments: Synthetic Approach

More fundamental questions about the role of biological diversity for the functioning of ecosystems require another approach, namely by synthesis of biodiversity, in other words a bottom-up strategy. Two big and long-term experiments follow this approach, the Jena Experiment in Germany (16 years) and the Cedar Creek experiment in the USA (35 years), both using grassland ecosystems. Preset parameters are the numbers (1, 2, 4, 8, 16, 32, 60) of perennial grasses or herbs or other functional non-woody species (e.g., leguminous herbs) cultivated on a unit plot. The Jena Experiment, e.g., comprises more than 80 unit plots (Fig. 4).

No animals or microorganisms are added to the plots as these organisms are expected to enter the developing vegetation in a natural way. On the other hand, the soil is not sterilized before planting or sowing. Care is not taken to avoid extreme events such as drought or flooding; however, weeding is the major and costly task, as the natural seed-rain is permanently challenging the protocol of the experiments. Since not all components of the different quasi-steady state ecosystems arrive already during the first year, the full assembly of the ecosystem components even in a grassland takes several years. Moreover special treatments, e.g., fertilization or mowing can help to unveil basic relationships. Without going into details, a few typical results shall demonstrate the kind of questions addressed by such bottom-up approach:

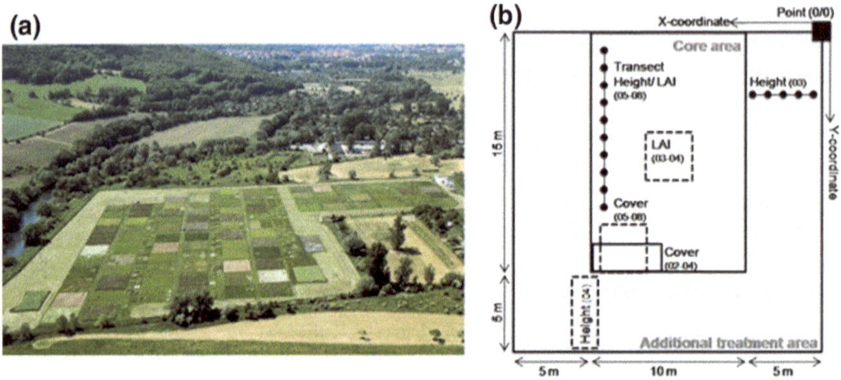

Fig. 4 The Jena experiment **a** Panoramic view over the entire facility in the river plain of the Saale near Jena (background). **b** Structure of one of the 20 × 20 m plots with a core area surrounded on three sides by lateral areas for experiments in the same vegetation as on the core area. LAI, leaf area index. *Source* (**a**) The Jena Experiment, by courtesy of N. Eisenhauer; (**b**) Weisser et al. (2017), by courtesy of the author

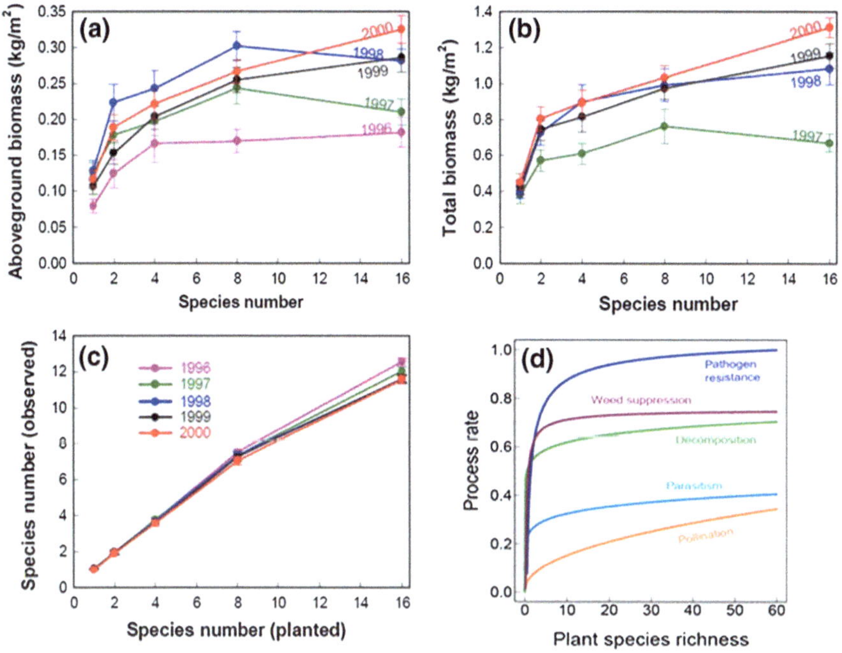

Fig. 5 The Jena experiment. **a**, **b** Effects of the number of plant species on the aboveground and total plant biomass in the sequence of the years. **c** Preservation of the original species composition over the years by annual weeding campaigns. **d** Effects of the plant diversity on various ecosystem processes based on biotic interactions. From Weisser et al. (2017), by courtesy of the author

- Plant diversity shows a persistent positive impact on ecosystem functioning, stability, and productivity (Fig. 5; Roscher et al. 2005; Proulx et al. 2010; Weisser et al. 2017). Increasing plant diversity from one to 16 species per plot had a similar positive effect on biomass production than standard agronomic fertilizing (Weigelt et al. 2009).
- Species diversity and carbon storage rates in the soil are positively correlated (Eisenhauer et al. 2010).
- Multitrophic interactions in highly diverse plant communities concur with more complex above- and below ground animal networks: Diversity creates diversity, the effect of species diversity being stronger than of species abundance (Scherber et al. 2010; Ebeling et al. 2011).
- Due to a high microbial activity, highly diverse plant communities show a closer nitrogen cycling (Oelmann et al. 2011), attenuating the leaching of N from agricultural soils into the groundwater and thus reducing the need of fertilization (Oelmann et al. 2007).
- About 45% of the examined ecosystem processes were found to correlate with the plant diversity value (Allan et al. 2013).

- Plasticity of most plant traits, both above- and below ground, depends on plant diversity in a complex way. Therefore, a caveat turns out for using plain database traits for associating plant traits with particular functions in an ecosystem.

Grassland and crop fields are ecosystems that are relatively open to experimental management and whose responses to interferences usually show up in short to moderate time spans.

A similar systematically structured experiment, however with subtropical woody plant species, termed biodiversity–ecosystem functioning (**BEF**) experiment, was established in 2009/2010 in the mountain area of Jiangxi Province, China. The goal is to assess the "role of tree and shrub diversity for production, erosion control, element cycling, interactions with other trophic levels and strata (herb layer, soil macrofauna, herbivores, decomposers, mycorrhiza, soil microorganisms), and species conservation in Chinese subtropical Forest ecosystems". About 200,000 saplings (50 species each of trees and shrubs, 50 species of each evergreen and deciduous plant life-forms were planted on 100 ha of previously cleared forest area. In a parallel observational approach, 27 comparative study plots were set up in existing forests in an adjacent National Nature Reserve. It is expected that the species and functional richness of shrubs and trees will lead to increased productivity, which may in turn lead to increases in other ecosystem variables.

The Highest Level: Ecological Models

Processes taking place in an ecosystem can be analyzed by suitable measurements and experiments, however, upscaling in space and time requires models which capture the processes under varying external conditions. Nevertheless, a model is in principle a reductionistic approach, as usually not all variables can be included that are effective in nature. Nevertheless, a model can show the real effect of a process, e.g., on biomass production, whereas understanding the mechanisms and reactions driving the process is another dimension or category. To illustrate it by analogy: Understanding the control of stomatal conductance by the interplay of several biophysical and biochemical partial reactions constitutes the process; in contrast, quantifying the rate or the amount of transpired water at any moment or over a longer time period can only be accomplished by a model. To stay with the analogy: Plant ecophysiologists are interested in stomatal conductance, ecologists in transpiration. Because any kind of process taking place in an ecosystem can be modeled, given sufficient analytical data, countless models (conceptual, analytical and simulation models) have been developed and published. In this essay, only numerical simulation models will be considered. The so-called land surface models (LSM) linking soil-vegetation-atmosphere processes and the vegetation (plant communities) are usually considered as an assembly of only a few (global) Plant Functional Types (PFT; Bonan et al. 2002; Sakschewski et al. 2016).

In order to create and parameterize the model, it is not necessary to know any detail of the involved processes. To simplify the models, proxies can be used which have proven to portray particular processes and their interdependences in a realistic way. For instance in a model of photosynthetic carbon acquisition, instead of the maximal carboxylation activity the nitrogen content of RubisCO as share of the leaf's total nitrogen content can be used (Oleson et al. 2013), allowing to transfer the measurement from a tree to the lab. Models, however, have to be forced by the most effective drivers, in many cases by climate or weather, data of which can be collected continuously.

A relatively simple example shall illustrate the architecture and use of a Land Surface Model (LSM): In the tropical Andes, as in many other tropical high mountains, anthropogenic pastures are often invaded and finally ruined by a highly competitive weed, the bracken fern (*Pteridium* spp.). Competition of the widely used tropical pasture grass *Setaria sphacelata* (a C4 plant) with the C3 plant bracken fern was modeled using photosynthetic biomass production as proxy for growth (Silva et al. 2012). The architecture of the model followed the two-big leaf approach (Dai et al. 2004), which is a derivative of the Common or Community Land Models (CLM, Lawrence and Fisher 2013), however, differentiating between the portions of sunlit and shaded leaves on a plot. The model involves five modules, as shown in Fig. 6 and is parameterized by 20 physiological and ecological parameters, e.g., the light- and CO_2-responses of photosynthetic net carbon uptake and stomatal conductance. Recording microclimate and radiation continuously over a year at an elevation of 2100 m a.s.l, growth of both plant species could be realistically modeled for that year 2008, showing better growth of the grass at dry and warm weather whereas the fern grew better under wet and cool conditions. Over the entire year growth of both species was nearly balanced with a slight advantage of the grass (Fig. 7). The advantage of that model is that it provides also data on the belowground biomass. Thus for the year 2008, the total biomass production (dry matter) of *Setaria* was calculated as 58.8 Mg ha^{-1} a^{-1} and for bracken 55.5 Mg ha^{-1} a^{-1} (Beck et al. 2013).

The architecture of the model shows biophysical and biochemical parameters as the heart of the modules, but biological parameters such as the plants' capacity of (macro)nutrient extraction from the soil or the sensitivity for pathogen attack is completely missing. Especially, bracken is highly susceptible to infection by a fungus (*Cryptomycella pteridis* (Kalchbr.) Hohn) which kills the leaves within a few months (Roos et al. 2011). More modules are necessary to adjust the model better to the natural situation (see below).

While the development of a mature "forest" requires more than a few decades, partial modules can be addressed in shorter time, such as the influence of the plant species composition and age of the forest on the soil fungal community composition (Wu et al. 2013). Such modules can later on be integrated in a comprehensive land surface model.

Models need truthing by ground observations for assessing their correctness and accuracy. For models estimating the state of the vegetation, truthing by remote sensing has proven as a useful approach, if the absorbance, transmittance and reflection characteristics of the vegetation are known from flied spectroscopy (Bendix et al. 2010).

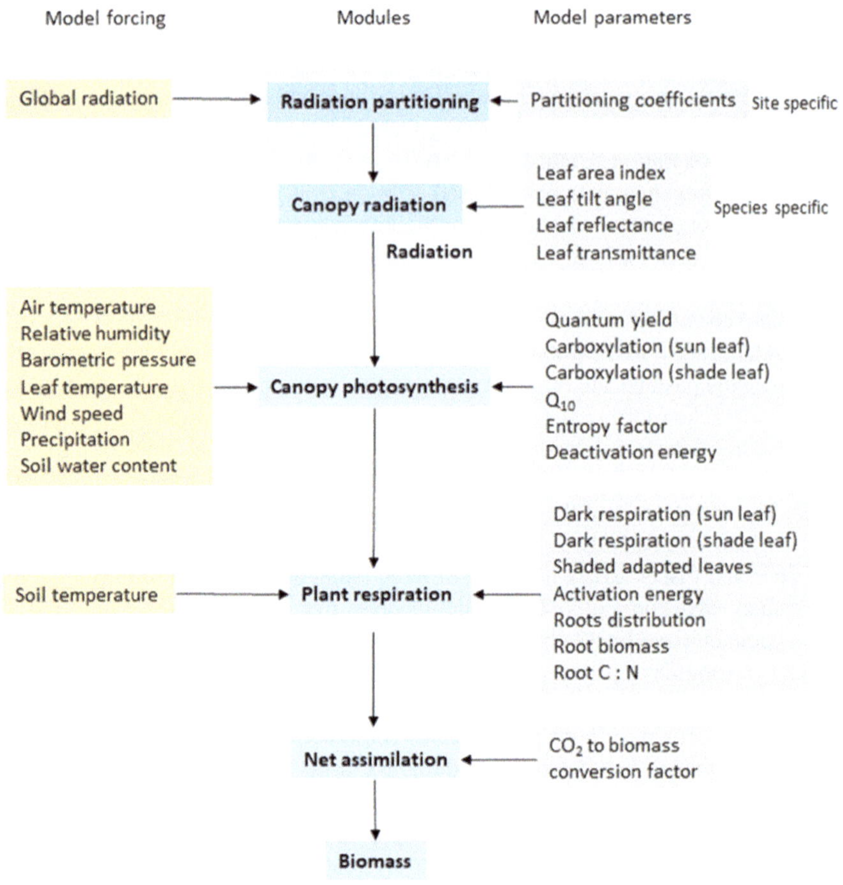

Fig. 6 Architecture of the southern bracken competition model (Silva et al. 2012, modified)

With those data of *Setaria* and bracken the increase of the dominance of bracken with increasing elevation could be demonstrated which agrees with the results of the model when forced with the altitudinal decreasing temperatures (Curatola et al. 2013; Göttlicher et al. 2009). Therefore, at least the most critical parameters have been used in the model.

Community Land Models (CLMs), as the basis for Land Surface Models (LSMs) undergo a continuous evolution by refining the involved modules and adding new modules. The latest CLM version 4.5 by Oleson and Lawrence et al. (2013) simulates in 24 Chapters (420 pages) landscape elements and processes, conveying an impression of the inexhaustibility of the approach. A few examples are shown:

- Surface characterization including land type heterogeneity and ecosystem structure,
- Absorption, reflection, and transmittance of solar radiation,

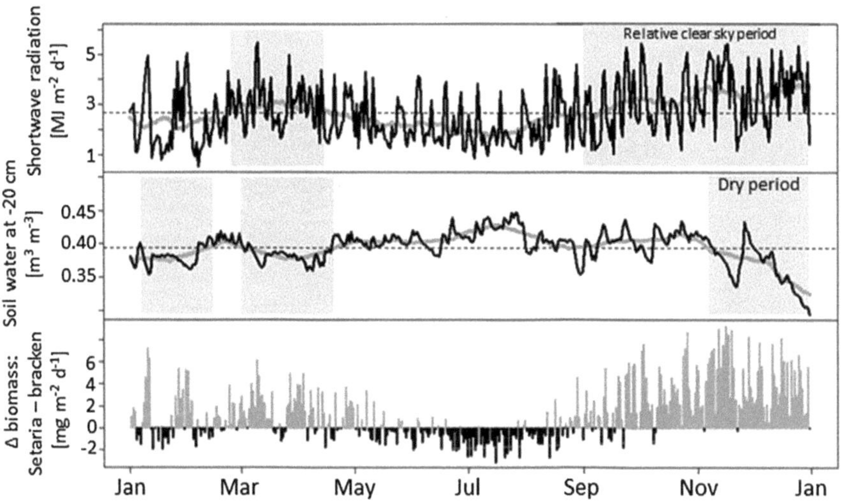

Fig. 7 Daily accumulated solar radiation, soil water content at 20 cm depth and difference between the biomass production (dry matter) of the two competing species *Setaria sphacelata* and bracken fern. The bold gray lines in the upper two panels represent means of 30 days determined with a moving window which were used for weather characterization. The dashed lines show the annual means. The gray bars in the lower panel show when and to which extent *Setaria* produces more biomass than bracken, the black bars indicate better growth of bracken. From Beck et al. (2013), slightly modified

- Momentum, sensible heat (ground and canopy), and latent heat (ground evaporation, canopy evaporation, transpiration) fluxes,
- Canopy hydrology (interception, throughfall and drip),
- Soil hydrology (surface runoff, infiltration, redistribution of water within the column, sub-surface drainage, groundwater),
- Stomatal physiology and photosynthesis,
- Urban energy balance and climate,
- Vegetation carbon and nitrogen allocation and respiration,
- Vegetation phenology,
- Soil and litter carbon decomposition,
- Plant mortality,
- Dynamic global vegetation distribution,
- Carbon isotope fractionation.

The modules and partial models work with fixed parameters, e.g., only 24 types of vegetation units are differentiated; however, many of them share the same parameters or coefficients and the spectrum of the units is rather coarse, e.g., there is only one type of tropical broadleaf evergreen forests and one type of tropical deciduous forests. As a consequence, the global PFTs fail in specific environments so that a regionalization with more specific PFTs is required (Wullschleger et al. 2014). Moreover, the traits used for PFT parameterization frequently show a high variability or

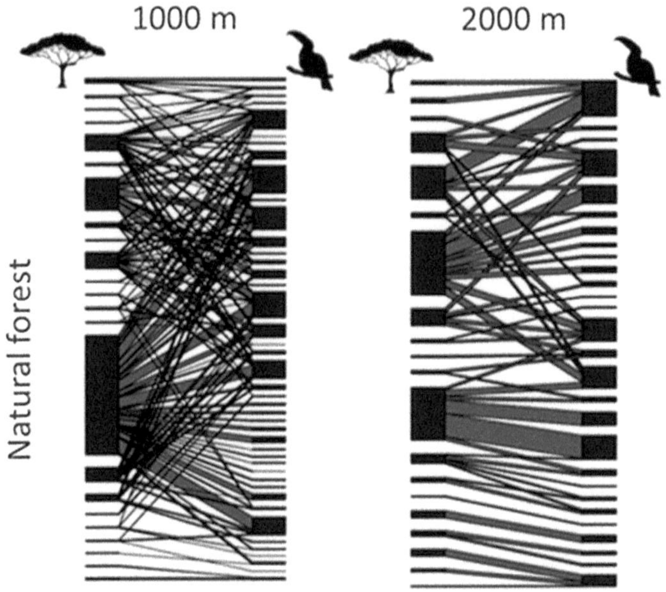

Fig. 8 Plant-frugivore interaction networks at two elevations (1000, 2000 m above sea level) in the natural mountain rain forest of the Podocarpus National Park (South Ecuador). Plant species are shown on the left side, frugivorous bird species on the right side of each plot. Each line between a plant and a bird indicates a visit during which the bird consumes fruits of the plant. The thicker the line, the more feeding events have been recorded of the two mutualistic partners. From Neuschulz (2016. Figure by M. Quitián). By courtesy of Eike L. Neuschulz

dynamics, thus, the inclusion of trait variability in LSMs is indispensable (Kattge et al. 2011; Sakschewski et al. 2016). Finally, biotic interactions, such as seed dispersal and plant regeneration, are usually oversimplified in the models and relevant biotic interactions responsible for the indirect internal feedback loops such as herbivory or pathogen attack are hardly implemented (Bonan 2008). Thus, even these very detailed current Land Surface Models have several flaws and need refinement.

To provide an idea, how manifold and complex even principally easy interactions can be, such as seed dispersal by birds, quantitative observations in an evergreen tropical mountain rain forest are shown in Fig. 8.

Those interactions, bearing per se many biologically interesting aspects, can be used as proxies for forest regeneration by the pattern of the seeds dispersed by the birds.

The challenge is, how to include this function in the modules listed above, e.g., vegetation phenology, vegetation carbon and nitrogen allocation and respiration, and plant mortality. This becomes even more complicated, if the dynamics of ecosystem traits have to be overlaid by biological interaction modules, like the change of the mycorrhizal interactions during the growth of a forest (Haug et al. 2010).

Conclusion

Learning to understand nature in its relation to space is by considering an ecosystem, its components and their interactions. Addressing the "nature of vegetation", Tansley (1935) for the first time suggested to comprehend nature as a living "Quasi-Organism" which he termed "Ecosystem". Nevertheless, due to the enormous complexity of ecosystems serious investigations must first concentrate on individual elements or modules which finally must be properly combined by a procedure termed synthesis. Whereas the investigation of individual modules in most cases already requires modeling the integration of the modules into a whole is unthinkable without a comprehensive model.

Emergent understanding of ecosystems by investigating its abiotic and biotic components, their functionality and their interactions as modules has been presented here by a few examples. Considering the case of the Land Surface Models, understanding has arrived at a level where simulations produce near-to-reality results, as revealed by truthing examinations. At that state, cautious predictions about the reactions of the ecosystem to changes in the environment, e.g., by climate change appear possible. However, it should be kept in mind that extrapolation is always risky as services and stability of an ecosystem are different traits. Services might be properly described using only a few modules of a system, while stability may depend on many more components and their interactions; and that is the reason why more complex ecosystems are considered more stable but not necessarily more effective. The question is whether focusing only on the principles, e.g., the dependence of C3- and C4-photosynthesis on the temperature would come up with similar results as the model explained above, in other words whether a similar result could be obtained with a simpler approach. The example of the competition between bracken and the pasture grass *Setaria* shows that under "normal" conditions components of the microclimate and of the soil might be sufficient for realistic modeling results, however, that more modules will be required if unusual events take place.

Proven models can not only be used for forecasts, but can also serve to propose an improvement of human land use (Knoke et al. 2014), in particular also with respect to global change.

References

Ainsworth EA, Long SP (2005) What have we learned from 15 years of free-air CO_2 enrichment (FACE)? A meta-analytic review of the responses of photosynthesis, canopy properties and plant production to rising CO_2. New Phytol 165(2):351–371

Ainsworth EA, Rogers A (2007) The response of photosynthesis and stomatal conductance to rising CO_2: mechanisms and environmental interactions. Plant Cell Environ 30(3):258–270

Allan E, Jenkins T, Fergus AJF, Roscher C, Fischer M, Petermann J, Weisser WW, Schmid B (2013) Experimental plant communities develop phylogenetically overdispersed abundance distributions during assembly. Ecology 94:465–477

Baasch A, Kirmer A, Tischew S (2012) Nine years of vegetation development in a postmining site: effects of spontaneous and assisted site recovery. J Appl Ecol 49:251–260

Bataillon T, Galtier N, Bernard A, Cryer N, Faivre N, Santoni S, Severac D, Mikkelsen TN, Larsen KS, Beier C, Sørensen JG, Holmstrup M, Ehlers BK (2016) A replicated climate change field experiment reveals rapid evolutionary response in an ecologically important soil invertebrate. Glob Change Biol 22(7):2370–2379. https://doi.org/10.1111/gcb.13293 Epub 2016 Apr 24

Beck E, Bendix J, Silva B, Rollenbeck R, Lehnert L, Hamer U, Potthast K, Tischer A, Roos K (2013) Future provisioning services: repasturisation of abandoned pastures, problems, and pasture management. In: Bendix J, Beck E, Bräuning A, Makeschin F, Mosandl R, Scheu S, Wilcke W (eds) Ecosystem services, biodiversity and environmental change in a tropical mountain ecosystem of South Ecuador. Springer, Berlin, Heidelberg. Ecol Stud 221:355–370

Beier C, Beierkuhnlein C, Wohlgemuth T, Penuelas J, Emmett B, Körner C, de Boeck H, Christensen JH, Leuzinger S, Janssens IA, Hansen K (2012) Precipitation manipulation experiments—challenges and recommendations for the future. Ecol Lett 15(8):899–911. https://doi.org/10.1111/j.1461-0248.2012.01793.x Epub 2012 May 4

Bendix J, Silva B, Roos K, Göttlicher DO, Rollenbeck R, Nauß T, Beck E (2010) Model parameterization to simulate and compare the PAR absorption potential of two competing plant species. Int J Biometeorol 54:283–295

Bonan GB (2008) Forests and climate change: forcings, feedbacks, and the climate benefits of forests. Science 320:1444–1449

Bonan GB, Levis S, Kergoat L, Oleson KW (2002) Landscapes as patches of plant functional types: an integrating concept for climate and ecosystem models. Glob Biogeochem Cycles 16(2)

Chapin FS III, Walker LR, Fastie C, Sharman LC (1994) Mechanisms of primary succession following deglaciation at Glacier Bay, Alaska. Ecol Monogr 64:149–175

Curatola Fernández G, Silva B, Gawlik J, Thies B, Bendix J (2013) Bracken fern frond status classification in the Andes of Southern Ecuador: combining multispectral satellite data and field spectroscopy. Int J Remote Sens 34:7020–7037

Dai Y, Dickinson RE, Wang Y-P (2004) A two-big-leaf model for canopy temperature, photosynthesis, and stomatal conductance. J Clim 17:2281–2299

Diaz S, Marcelo C, Fernando C (1998) Plant functional traits and environmental filters at a regional scale. J Veget Sci 9:113–122

Ebeling A, Klein AM, Tscharntke T (2011) Plant-flower visitor interaction webs: temporal stability and pollinator specialization increases along an experimental plant diversity gradient. Basic Appl Ecol 12:300–309

Eisenhauer N, Bessler H, Engels C, Gleixner G, Habekost M, Milcu A, Partsch S, Sabais ACW, Scherber C, Steinbeiss S, Weigelt A, Weisser WW, Scheu S (2010) Plant diversity effects on soil microorganisms support the singular hypothesis. Ecology 91:485–496

Garcia MO, Ovasapyan T, Greas M, Treseder KK (2008) Mycorrhizal dynamics under elevated CO_2 and nitrogen fertilization in a warm temperate forest. Plant Soil 303:301–310. https://doi.org/10.1007/s11104-007-9509-9

Goettlicher D, Obregón A, Homeier J, Rollenbeck R, Nauß T, Bendix J (2009) Land cover classification in the Andes of southern Ecuador using landsat ETM+ data as a basis for SVAT modeling. Int J Remote Sens 30:1867–1886

Gottfried M, Pauli H, Futschik A, Akhalkatsi M, Barancok P, Alonso JLB, Coldea G, Dick J, Erschbamer B, Fernandez Calzado MRF, Kazakis G, Krajci J, Larsson P, Mallaun M, Michelsen O, Moiseev D, Moiseev P, Molau U, Merzouki A, Nagy L, Nakhutsrishvili G, Pedersen B, Pelino G, Puscas M, Rossi G, Stanisci A, Theurillat J-P, Tomaselli M, Villar L, Vittoz P, Vogiatzakis I, Grabherr G (2012) Continent-wide response of mountain vegetation to climate change. Nat Clim Change. https://doi.org/10.1038/nclimate1329

Haug I, Wubet T, Weiß M, Aguirre N, Weber M, Günter S, Kottke I (2010) Species-rich but distinct arbuscular mycorrhizal communities in reforestation plots on degraded pastures and in neighboring pristine tropical mountain rain forest. Trop Ecol 51(2):1–25

Homeier J, Hertel D, Camenzind T, Cumbicus NL, Maraun M, Martinson GO et al (2012) Tropical Andean Forests are highly susceptible to nutrient inputs—rapid effects of experimental N and P addition to an Ecuadorian Montane Forest. PLoS ONE 7(10):e47128. https://doi.org/10.1371/journal.pone.0047128

Homeier J, Leuschner C, Bräuning A, Cumbicus NL, Hertel D, Martinson GO, Spannl S, Veldkamp E (2013) Effects of nutrient addition on the productivity of mountain forests and implications for the carbon cycle. In: Bendix J et al (eds) Ecosystem services, biodiversity and environmental change in a tropical mountain ecosystem of South Ecuador. Ecol Stud 221:315–329

Kattge J, Diaz S, Lavoral S, Prentice IC, Leadley P, Bönisch G, Garnier E, Westoby M, Reich PB et al (2011) TRY—a global database of plant traits. Glob Change Biol 17:2905–2935

Keddy P (1992) Assembly and response rules: two goals for predictive community ecology. J Veg Sci 3:157–164

Kéfi S, Berlow EL, Wieters EA, Navarrete SA, Petchey OL, Wood SA, Boit A, Joppa LN, Lafferty KD, Williams RJ, Martinez ND, Menge BA, Blanchette CA, Iles AC, Brose U (2012) More than a meal… integrating non-feeding interactions into food webs. Ecol Lett 15:291–300

Knoke T, Bendix J, Pohle P, Hamer U, Hildebrandt P, Roos K, Gerique A, Lopez Sandoval M, Breuer L, Tischer A, Silva B, Baltazar C, Aguirre N, Castro LM, Windhorst D, Weber M, Stimm B, Gunter S, Palomeque X, Mora J, Mosandl R, Beck E (2014) Afforestation or intense pasturing improve the ecological and economic value of abandoned tropical farmlands. Nat Comm 5. https://doi.org/10.1038/ncomms6612

Lavorel S, Garnier E (2001) Predicting changes in community composition and ecosystem functioning from plant traits: revisiting the Holy Grail. Funct Ecol 16:545–556

Lawrence DM, Fisher R (2013) The community land model philosophy: model development and science applications. iLEAPS Newslett 13:16–19

Neuschulz EL (2016) Plant-animal interactions in tropical mountain forests. In: Bogner F, Bendix J, Beck E (eds) Biodiversity hotspot tropical mountain rainforest. NCI Foundation, Loja, Ecuador, pp 98–103. ISBN 978-9942-14-538-3

Norby RJ, Zak DR (2011) Ecological lessons from free-air CO_2 enrichment (FACE) experiments. Annu Rev Ecol Evol Syst 42:181–203

Odum HT (1957) Trophic structures and productivity of Silver Springs, Florida. Ecol Monogr 27:55–112

Oelmann Y, Buchmann N, Gleixner G, Habekost M, Roscher C, Rosenkranz S, Schulze E-D, Steinbeiss S, Temperton VM, Weigelt A, Weisser WW, Wilcke W (2011) Plant diversity effects on above- and belowground N pools in temperate grassland ecosystems: development in the first five years after establishment. Glob Biogeochem Cycles 25(2). https://doi.org/10.1029/2010gb003869

Oelmann Y, Wilcke W, Temperton VM, Buchmann N, Roscher C, Schumacher J, Schulze E-D, Weisser WW (2007) Soil and plant nitrogen pools as related to plant diversity in an experimental grassland. Soil Sci Soc Am J 71:720–729

Oleson KW, Lawrence DM et 24 al (2013) Technical description of version 4.5 of the community land model (CLM). NCAR Earth Systems Laboratory, Climate and Global Dynamics Division. ISSN Print Edition 2153-2397, ISSN Electronic Edition 2153-2400

Pauli H, Gottfried M, Dullinger S, Abdaladze O, Akhalkatsi M, Benito Alonso JL, Coldea G, Dick J, Erschbamer B, Fernández Calzado R, Ghosn D, Holten JI, Kanka R, Kazakis G, Kollár J, Larsson P, Moiseev P, Moiseev D, Molau U, Molero Mesa J, Nagy L, Pelino G, Puscas M, Rossi G, Stanisci A, Syverhuset AO, Theurillat J-P, Tomaselli M, Unterluggauer P, Villar L, Vittoz P, Grabherr G (2012) Recent plant diversity changes on Europe's mountain summits. Science 336(6079):353–355. https://doi.org/10.1126/science.1219033

Proulx R, Wirth C, Voigt W, Weigelt A, Roscher C, Attinger S, Baade J, Barnard RL, Buchmann N, Buscot F, Eisenhauer N, Fischer M, Gleixner G, Halle S, Hildebrandt A, Kowalski E, Kuu A, Lange M, Milcu A, Niklaus PA, Oelmann Y, Rosenkranz S, Sabais A, Scherber C, Scherer-Lorenzen M, Scheu S, Schulze E-D, Schumacher J, Schwichtenberg G, Soussana J-F, Temperton

VM, Weisser WW, Wilcke W, Schmid B (2010) Diversity promotes temporal stability across levels of ecosystem organization in experimental grassland. PLoS ONE 5:e13382

Ravenscroft CH, Whitlock R, Fridley JD (2015) Rapid genetic divergence in response to 15 years of simulated climate change. Glob Change Biol 21(11):4165–4176. https://doi.org/10.1111/gcb.12966PMCID:PMC4975715

Riebesell U, Czerny J, von Bröckel K, Boxhammer T, Büdenbender J, Deckelnick M, Fischer M, Hoffmann D, Krug S, Lentz U, Ludwig A, Muche R, Schulz KG (2013) Technical note: a mobile sea-going mesocosm system—new opportunities for ocean change research. Biogeosciences (BG) 10:1835–1847. https://doi.org/10.5194/bg-10-1835-2013

Roos K, Rödel HG, Beck E (2011) Short- and long-term effects of weed control on pastures infested with Pteridium arachnoideum and an attempt to regenerate abandoned pastures in South Ecuador. Weed Res 51:165–176. https://doi.org/10.1111/j.1365-3189,2010.00833.x

Roscher C, Temperton VM, Scherer-Lorenzen M, Schmitz M, Schumacher J, Schmid B, Buchmann N, Weisser WW, Schulze E-D (2005) Overyielding in experimental grassland communities—irrespective of species pool or spatial scale. Ecol Lett 8:419–429

Rzanny M, Kuu A, Voigt W (2013) Bottom-up and top-down forces structuring consumer communities in an experimental grassland. Oikos 122:967–976

Sakschewski B, von Bloh W, Boit A, Poorter L, Pena-Claros M, Heinke J, Joshi J, Thonicke K (2016) Resilience of amazon forests emerges from plant trait diversity. Nat Clim Change 6:1032–1036

Scherber C, Eisenhauer N, Weisser WW, Schmid B, Voigt W, Fischer M, Schulze E-D, Roscher C, Weigelt A, Allan E, Beßler H, Bonkowski M, Buchmann N, Buscot F, Clement LW, Ebeling A, Engels C, Halle S, Kertscher I, Klein AM, Koller R, König S, Kowalski E, Kummer V, Kuu A, Lange M, Lauterbach D, Middelhoff C, Migunova VD, Milcu A, Müller R, Partsch S, Petermann JS, Renker C, Rottstock T, Sabais A, Scheu S, Schumacher J, Temperton VM, Tscharntke T (2010) Bottom-up effects of plant diversity on multitrophic interactions in a biodiversity experiment. Nature 468:553–556

Schuldt A, Bruelheide H, Buscot F, Assmann T, Erfmeier A, Klein AM, Ma K, Scholten T, Staab M, Wirth C, Zhang J, Wubet T (2017) Belowground top-down and aboveground bottom-up effects structure multitrophic community relationships in a biodiverse forest. Sci Rep 7, Article number 4222. https://doi.org/10.1038/s41598-017-04619-3

Silva B, Roos K, Voss I, König N, Rollenbeck R, Scheibe R, Beck E, Bendix J (2012) Simulating canopy photosynthesis for two competing species of an anthorpogenic grassland community in the Andes of Southern Ecuador. Ecol Model 239:14–26

Tansley AG (1935) The use and abuse of vegetational concepts and terms. Ecology 16:284–307

Vogel A, Fester T, Eisenhauer N, Scherer-Lorenzen M, Schmid B, Weisser WW, Weigelt A (2013) Separating drought effects from roof artifacts on ecosystem processes in a grassland drought experiment. PLoS ONE 8(8):e70997

Wardle DA, Bardgett RD, Klironomos JN, Setälä H, van der Putten WH, Wall DH (2004) Ecological linkages between aboveground and belowground biota. Science 304(5677):1629–1633. https://doi.org/10.1126/science.1094875

Weigelt A, Weisser WW, Buchmann N, Scherer-Lorenzen M (2009) Biodiversity for multifunctional grasslands: equal productivity in high-diversity low-input and low-diversity high-input systems. Biogeosciences 6:1695–1706

Weisser WW et al (2017) Biodiversity effects on ecosystem functioning in a 15-year grassland experiment: patterns, mechanisms, and open questions. Basic Appl Ecol 23(Supplement C):1–73

Werner FA, Jantz N, Krashevska V, Peters T, Behling H, Maraun M, Scheu S, Brehm G (2013) Climate change: effects on biodiversity and ecosystem functioning. In: Bendix J et al (eds) Ecosystem services, biodiversity and environmental change in a tropical mountain ecosystem of South Ecuador. Ecol Stud 221:247–263

Wu YT et al (2013) Forest age and plant species composition determine the soil fungal community composition in a Chinese subtropical forest. PLoS ONE 8:e66829. https://doi.org/10.1371/journal.pone.0066829

Wullschleger SD, Epstein HE, Box EG, Euskirchen ES, Goswami S, Iversen CM, Kattge J, Norby RJ, van Bodegom PM, Xu H (2014) Plant functional types in Earth system models: past experiences and future directions for application of dynamic vegetation models in high-latitude ecosystems. Ann Bot 114:1–16

Zark M, Broda NK, Hornick T, Grossart HP, Riebesell U, Dittmar T (2017) Ocean acidification experiments in large-scale mesocosms reveal similar dynamics of dissolved organic matter production and biotransformation. Front Mar Sci 4, Art. Nr. 271. https://doi.org/10.3389/fmars.2017.00271

Zwölfer H (1987) Species richness, species packing and evolution in insect-plant systems. In: Schulze E-D, Zwölfer H (eds) Potentials and limitations of ecosystem analysis. Springer, Berlin, Heidelberg. Ecol Stud 61:301–319

Part III
Ultimate Integration

Emergent Properties and Stability in Hierarchical Biosystems: There Is no Privileged Level of Causation

Gustavo M. Souza and Marcelo N. do Amaral

Abstract The main question herein is about the extent that reductionist approach can reach higher-level explanation as well as predict behavior from lower level of systems organization when concerning stability. Starting from a theoretical discussion on the complex hierarchical organization of biological systems, we offer a variety of examples showing the interchanging between bottom-up and top-down events underlying the regulation of the system. Moreover, we discuss and illustrate with different examples, how system stability under externally changing conditions are reached from many strategies without a specific level of causation, highlighting the role of downward causation with upward causation. Finally, we would like to demonstrate that there should be no dogmas when it comes to causation in biological systems.

Introduction

The basis of the classical paradigm of biological sciences, which supposes that the understanding of the parts building up an organism could allow to comprehend the whole system, has been challenged by an emergent antireductionist perspective (Baetu 2012). The classical reductionist view is founded on the assumption that the parts integrating a system are mechanically connected (in a linear straightforward way) allowing reliable and predictable explanation towards a higher-level organization (e.g., from set of molecules to whole metabolism). One of the fundamental aspects of biological systems is their evident nested hierarchical organization, in which lower levels are contained in higher levels (e.g., from a set to whole metabolism), such as Russian dolls (Souza et al. 2016a). Noble (2011) has argued that this approach can reinforce a metaphor on the scientific thinking with a biased view to represent the lower levels as more concrete, supporting a reductionist perspective that the micro-scale components (e.g., molecules) are the cause of the higher-level properties (a bottom-up causation).

G. M. Souza (✉) · M. N. do Amaral
Department of Botany, Institute of Biology, Federal University of Pelotas, Pelotas, RS, Brazil
e-mail: gmsouza@ufpel.edu.br

L. H. Wegner and U. Lüttge (eds.), *Emergence and Modularity in Life Sciences*,
https://doi.org/10.1007/978-3-030-06128-9_10

On the other hand, the systemic approach (Alcocer-Cuarón et al. 2014; Souza and Lüttge 2015; Souza et al. 2016b) starts from the assumption that biological systems constitute of a set of self-organized components interacting with each other in a non-linear way far from thermodynamical equilibrium, engendering emergent properties at higher levels of organization. The concept of emergence, as a set of phenomena that cannot be deduced from underlying laws, in an epistemic sense, presupposes the existence of levels of organization. Ontologically, emergence presupposes new higher levels of entities arising and having casual powers not possessed by the parts (Kauffman and Clayton 2006). Therefore, the real meaning of the possible effects from lower-scale organizations (e.g., molecular level) should be considered at the proper higher-scale context (e.g., metabolic) (Kohl et al. 2010). In other words, the context, in which a given biological process takes place, matters for its own regulation as well as for its role in the system organization as a whole as well (Amaral and Souza 2017).

The main question herein is about the extent that reductionist approach can reach higher-level explanation as well as predict behavior from below. In the middle of the twentieth century, Ludwig von Bertalanffy published the seminal book General System Theory (von Bertallanffy 1968), in which some principles of the systemic thinking were formalized, specially taking into account the biological systems. Recently, system biolog is gaining renewed interest due to the progress in molecular biology, particularly in genome sequencing and high-throughput measurements, inaugurating the "OMICS" age (Amaral and Souza 2017). The focus is on the understanding of the system's structure and dynamics, because a system is not just an assembly of genes and proteins, its properties cannot be fully understood merely by drawing diagrams of their interconnections, such as static "road maps." The knowledge of how changes to one part of a system may affect other parts, or how its control is performed, is not satisfactorily provided by road maps (Kitano 2002).

In the following sections, we explore the systemic hierarchical view in biological thinking and some consequences, specially, toward the concepts of downward causation and stability.

Hierarchy as the Fundamental Shape of Biological Systems Organization: Upward X Downward Causation

The concept of hierarchical levels and its implications for the biological organization have been early recognized (Novikoff 1945), and different hierarchical models of biological organization have been proposed (MacMahon et al. 1978; Korn 1999; Rojdestvenski et al. 1999; Ravasz et al. 2002; Souza et al. 2016a). However, the trend, starting at the end of twentieth century, to examine biological systems mostly from the molecular perspective has overshadowed the integrated view of living beings. On the other hand, recent discoveries have led to the identification and characterization of the systemic aspects of biological organization, identifying network hierarchical modularity at cellular level (Ravasz et al. 2002; Dietz et al. 2010; Lucas et al. 2011).

The bases for complexity are modular hierarchical structures, leading to emerging levels of structures and functions based on lower-level networks. These three aspects, modularity, hierarchy, and structure, are crucial in the emergence of complexity through interactions between simpler units (Booch 1994; Ellis 2008). The basic principle is that a complex system is divided into ever simpler subsystems, and assigning these systems to specific modules. Each module is again divided into submodules until it reaches a base level, where the required tasks are simple operations that can be performed by simple mechanisms. This is the level where the actual work is done, each of these components feeding their results into the next higher-level components and interacting with each other until the desired result emerges at the appropriate top level. The result is a highly structured hierarchy of interacting entities (Ellis 2012).

Beyond a pure epistemological meaning, the importance of studying integration of different levels of biological organization relies on an ontological base. Schneider (1998) has summarized the main arguments: 1—spatial and temporal patterns are dependent on the scale of analysis, 2—there is more than one characteristic scale for the research, 3—experimental results cannot be directly transferred to larger scales, 4—biological interactions with the environment occur in multiple scales, and 5—environmental problems are created by the propagation of effects on different scales in the biosystem. In this context, for instance, experimental evidence provided by Vítolo et al. (2012) and Bertolli et al. (2014) showed that, in a particular experiment, in which plants are exposed to environmental changes, different variables measured at different scales of plant organization exhibited different levels of homeostatic capacity, impairing a unique diagnosis by a single variable. In other words, there was no specific representative scale of organization accounting for the whole system.

Each level of the hierarchy consists of modules that interact with other modules through weaker connections and lower-frequency interaction dynamics, so each module can be composed of an interactive network of lower-level modules. The emergence of a top-level system from these lower-level modules occurs when a reliable top-level behavior arises from the lower-level actions that occur in the context of that structure. Each of these different levels of hierarchical function can only be described and behave appropriately only at that level (for example, the basic concepts of molecular biology cannot be described in the language of quantum physics) (Ellis 2012). In other words, while the links among components of a specific subsystem are mostly strong (supporting a local pattern of organization), the links among different subsystems cross scale are mostly weaker. Ultimately, all types of links can be taken as informational insofar as the links act as "signals" among different parts of the system, enabling integration, i.e., hierarchical organized networks. Inasmuch as natural networks are not ideal, because there are often reversible feedback relationships, the downstream flow of information is ambiguous, bringing uncertainty to the system, making the accesses to the higher-scale traits of the system from its basic components unsettled. This is a strong limitation for the reductionist approach (Corominas-Murtra et al. 2013; Souza et al. 2016a). Therefore, ideas concerning lower-level causality

Fig. 1 Scheme of the traditional view of a bottom-up hierarchical system exhibiting upward causation

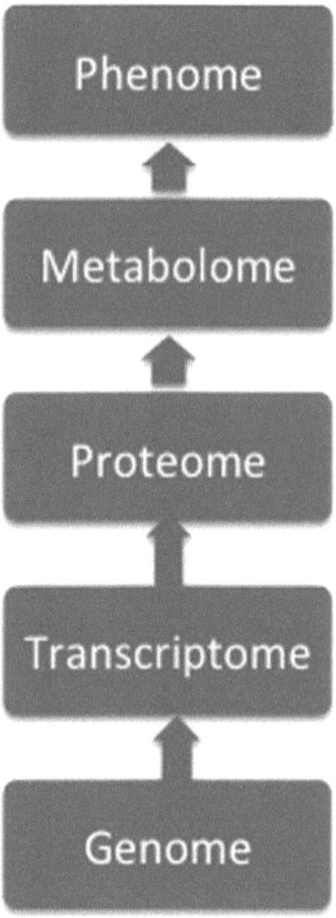

alone cannot explain higher-level behavior, since the concepts employed are simply not appropriate to the above-mentioned types of causality. Higher-level entities have causal power in their own right, which partially determine what happens at lower levels in the hierarchy (Ellis 2012). Thus, the upper levels of the system organization work as a framework for entities of the lower levels (Ahl and Allen 1996).

As discussed above, regardless the epistemological perspective, it is very common in biological sciences to think in terms of hierarchy of levels, with genes occupying the lowest level, and the networks of proteins and metabolites, organelles, cells, tissues, organs representing different intermediate levels, and the organism as a whole occupying the highest level of an individual organization. The reductionist causal chain is, then, represented in a single bottom-up direction (Fig. 1). However, there is no literal sense in which genes are below cells, for example. In fact, the only reason for DNA sequences to be at the bottom of the hierarchy is that they exist on a smaller (i.e., molecular) scale in biological systems. The formation of networks, cells, tissues, and organs should be seen as the creation of processes at different scales (Noble 2012).

The central dogma of molecular biology itself (Crick 1970) leads us to a bottom-up view of systems biology by stating that information flows from DNA to RNA, RNA to proteins, which can then form protein networks, and so on through biological levels until it reaches the whole organism. That is, from the knowledge about genes and proteins, we could reconstruct all other levels (Noble 2008a). Actually, the simple idea of a "dogma" is a contradiction, since in science there is no "truths" and the knowledge is in continuous movement of construction.

Contrary, many evidences seem to lead to the idea that the development of very complex systems, such as those in biology, requires downward causation in order to construct the necessary biological information (Kuppers 1994; Roederer 2005). This information cannot be originated in an ascending way because it implicitly incorporates information about environmental niches. It would be different in a different environment. Thus, upper-level conditions influence what happens on the lower scales, even if those lower scales do the work. This is what Ellis (2012) characterizes as top-down causation, in a similar way as downward causation (Noble 2008a).

For instance, DNA sequences form a list of proteins and RNAs that can be made in a given organism, as well as regulatory elements incorporated within it. These regulatory elements allow the regulation of gene group expression to be coordinated (Jacob 1960). Therefore, the genome behaves as a database that can be organized in response to cellular and environmental signals (Shapiro 2011). However, what proteins and RNAs will be produced at what time and place is not fully specified. DNA sequences do absolutely nothing until they are triggered to do so by a variety of transcription factors, which activate and deactivate genes, and various other forms of epigenetic control. All of these, and the processes of cells, tissues, and organs that determine when they are produced and used, "control" the genome. That is why it is possible for several different cell types to originate using exactly the same genome. A heart cell is made using precisely the same genome at its core as the various cells in the human body (Noble 2008a).

This does not imply that this bottom-up approach is mostly wrong. It is conceivable that it may function in some cases in simpler organisms, such as prokaryotes. This simply means that this approach and the vast amount of data coming from genomic, transcriptomic, and proteomic techniques need to be complemented by other approaches. In the case of eukaryotes, many of the interactions among the components are restricted by the complex cellular structure, including a system of membranes and organelles. The direct approach would necessarily include such structures, in which case it is no longer purely from the bottom up because, as we have already seen, many of these structural features are inherited independently of the DNA sequences. Higher levels of DNA and proteins would be required for successful modeling (Kohl et al. 2010).

One of the most classic examples to report the importance of non-genomic elements in biological processes was demonstrated and discussed by Noble (2012) using heart rate. Heart rate is generated by the electrical potential in nerve cells and is constituted by the so-called Hodgkin cycle of the oscillating ion current through a membrane. The oscillations occur through channels of proteins that cross the cell membrane. However, the control of these channels is, in turn, controlled by cell potential (a cell-level parameter) that is influenced by intercellular coupling and the dynamics of other processes across the membrane. Consequently, the behavior of the system cannot be understood from an analysis of the constituents alone or on the lower scale. This is a multi-level loop. Potential is a cell-level parameter; the openings and closures of the ion channel are protein level parameters. With rupture of feedback (downward causality) the cellular potential, ion channel gating, and cellular rhythm are eliminated (Hodgkin and Huxley 1952; Gross and Green 2017). The sequence of events, including feedback between cell potential and protein activity, is simply heart rate. It is a property of the interactions between all components of the system. It does not make sense to talk about heart rate at the protein and DNA levels, and it makes no sense to assume there is a separate program that "runs" the rhythm. Without the restrictions imposed by the cellular structure itself, mainly of the membranes and their compartments, besides other processes that maintain the ionic concentrations, the rhythm would not occur. There is, therefore, a complex and continuous feedback between processes at different levels (Noble 2012).

There are several other examples of biological processes regulated by feedbacks between two or more levels. In *Arabidopsis thaliana*, the circadian rhythm is regulated by three interlocking feedback loops, involving several transcription factors that, according to the time of day, repress or activate other genes, functioning as links between environmental factors and the regulation of gene expression (Alabadí et al. 2001; McClung 2006; Nakamichi et al. 2010; Nusinow et al. 2011; Sanchez and Kay 2016). In addition to the transcription–translation feedback cycles, the circadian rhythm is regulated by many other mechanisms, including hormones, epigenetics, and post-transcriptional and post-translational modifications (Seo and Mas 2014; Nolte and Staiger 2015; Romanowski et al. 2015), in addition to the communication between organelles (nucleus–chloroplast) (Hassidim et al. 2007). Both phytochromes and cryptochromes provide light input to the clock, although the signal transduction pathways are incompletely defined. Interestingly, photoreceptor expression is itself rhythmic, indicating that the clock gates its sensitivity to light (e.g., Tóth et al. 2001), although bulk phytochrome protein levels do not oscillate (Sharrock and Clack 2002). However, the interaction and hierarchical structures between levels of regulation still need to be clarified (Sanchez and Kay 2016).

In addition, there are several other mechanisms that demonstrate the importance of downward causation (top-down regulation). These include triggers of cell signaling (via hormones and transmitters), control of gene expression (via transcription factors), and the protein machinery that reads genes and continually repairs copying errors and makes the genome reliable and epigenetic control (Noble 2012).

Genes store hereditary information, but their actual expression usually depends on other factors, both physical and chemical, external to DNA. The term epigenetic was coined to denote this higher level, extra-DNA, activity level, and control. Epigenetic changes can occur due to chemical signals received from other genes, or from chemical and physical signals originating from other cells, organs or the external environment. These signals can serve to activate or deactivate genes using molecular markers or the physical rearrangement of DNA (Davies 2012).

Additionally, genes can be switched off, not only by proteins that bind to DNA, but also by several small molecules attached to specific points in the DNA. For example, the methyl group is often used to effectively mute a gene by being attached to the GC sites. The methylation pattern in a genome can be copied and inherited by the daughter cell, ensuring that the same genes are active in stem cells and daughters of the same type. This is an example of epigenetic heritage. Methylation of DNA may itself physically prevent binding of transcriptional regulators to the gene, and more importantly, methylated DNA participates in the formation of chromatin through interactions with various other epigenetic modifications such as the histone code, polycomb complexes, nucleosome positioning, non-coding RNA, and ATP-dependent chromatin remodeling proteins (Wade 2001; Suzuki and Bird 2008; Jin et al. 2011). Other epigenetic mechanisms involve several post-translational modifications of proteins. For example, histones, which will form nucleosomes, have small attached tails, which can be modified in this way, leading to thousands of different variations in the tail pattern of each nucleosome. Modification of specific histone tails is associated (though not necessarily) with altered chromatin structure and behavior (Henikoff and Shilatifard 2011).

For genetic instructions to be read from DNA, the reading machine must have physical access to the gene in question. This is difficult if the gene is physically inaccessible within a highly compacted structure. Chromatin is therefore incessantly compacted and uncompressed to expose specific genes that need to be transcribed at a given time. The genetic signals originating from DNA and their biological consequences depend, therefore, on the three-dimensional organization of the chromatin (Davies 2012). An important example of the physical structure of chromatin is given by cancer cells, in which chromatin can become drastically and visibly reorganized in the nucleus, a gross abnormality that correlates with the oncogenes being expressed and the tumor suppressor genes being switched off (Misteli 2010). Some chromatin architecture can also be inherited, providing yet another form of epigenetic cell memory. The silencing of genes due to this physical factor seems to be a crucial feature of the regulation and control of eukaryotic genes. Chromatin marks the intersection of causality up and down, because its structure and behavior are influenced by both the genes it contains and by the macroscopic forces acting on it from the rest of the cell and the cell environment (Davies 2012). Recently, evidence has also emerged that the dynamics of nucleosomes along the chromatin chains have functional importance for epigenome maintenance, gene regulation, and control of DNA replication (Deal et al. 2010). Davies (2012) defines this genome–epigenome relationship brilliantly: “In a nutshell, genetics deals with what genomes are and epigenetics deals with what they do.”

However, unlike the genome, which has a “physical headquarters” in DNA molecules, in the case of epigenetics, there is no physical “seat,” no physical place from which epigenomic instructions emerge. The epigenome should not be found in one location and the final source of epigenetic information cannot be localized specifically; instead, it is distributed throughout the cell. To be sure, the epigenome is manifested in particular structures (histone tails, nucleosome patterns, methylation patterns, and chromatin packing), but it does not originate there. The epigenome is

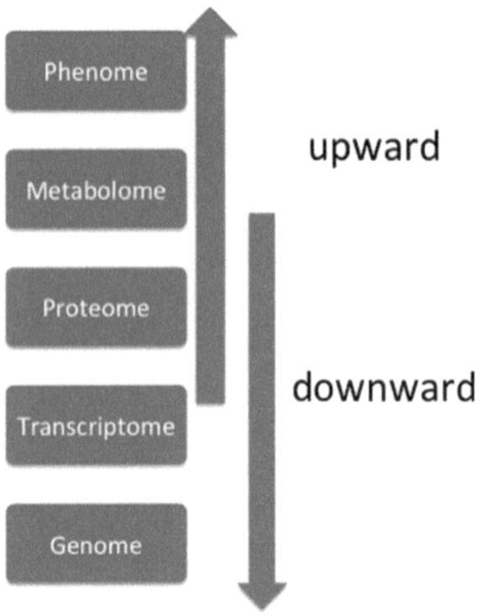

Fig. 2 Scheme showing the perspective of a hierarchical system exhibiting no privilege level of causation. Both upward and downward causation events regulate the whole systems organization altogether

everywhere and nowhere; it is a global, systemic entity. Expressed more starkly, the epigenome is a "virtual object" (Davies 2012).

As we have seen so far, top-down causality occurs because of the crucial role of context in influencing the lowest level organization. Higher levels of organization restrict and regulate lower-level interactions. A key feature here is the multi-level capability of performing higher-level functions and the consequent existence of lower-level variable equivalence classes with respect to higher-level actions. An equivalence class identifies all lower-level states that correspond to the same top-level state. A European team (Wegscheid et al. 2006) performed in vivo complementation experiments on the bacteria *Escherichia coli* and *Bacillus subtilis*, which have significantly different conformational structures that are associated with important biophysical and biochemical differences in vitro. However, in spite of these structural and functional differences in vitro, it has been shown that they can be substituted in vivo without loss of functionality, at least under standard growth conditions, demonstrating functional equivalence. This shows that the traditional reductionist view is insufficient to explain complex biological phenomena that involve different scales of organization, and downward causation (Fig. 2) plays an essential role on biological organization.

Summarizing, Noble (2008b) proposed some principles, among others, of the system biology approach: 1—biological functionality is multi-level; 2—transmission of information in not one way; and 3—theory of biological relativity: there is no privileged level of causality.

Therefore, according to Noble (2008b) "*it is simply a prejudice that inclines us to give some causal priority to lower-level, molecular events. The concept of level in biology is itself metaphorical. There is not literal sense in which genes and proteins lie underneath cells, tissues and organs. It is a convenient form of biological classification to refer to different levels, and we would find it very hard to do without the concept. But we should not be fooled by the metaphor into thinking that 'high' and 'low' here have their normal meanings. From the metaphor itself, we can derive no justification for referring to one level of causality as privileged over others. That would be a misuse of the metaphor of level.*"

Moreover, there is a relationship between decreasing restrictions among different organization scales in a biological system insofar the highest levels, somehow, restrict the organization of the lowest levels (Korn 1999). This relationship can be correlated to system regulation, which creates greater global robustness (general stability system) and a buffer effect between the superior levels and the inferior ones in relation to the potential destructive effects of environmental disturbances (Ahl and Allen 1996). For instance, the study of Vítolo et al. (2012) showed that the deleterious effects of high temperature on plants were expressed most on growth level, indicating that, somehow, the effects on plant growth acted as a buffer allowing photosynthetic maintenance. The understanding of the precise processes that allow such buffer effects needs future attention, but Souza et al. (2009) suggest that such buffer control could be related to changes in the links among plant subsystems. Robustness of the system is related to a spatial–temporal disconnection between the different levels of organization (Rojdestvenski et al. 1999), which confines the environmental disturbances into certain modules (subsystems) of the system that are directly influenced by the external environment. This could be reached by increasing the number of weak links among network motifs (Csermely 2006).

Stability as a Trade-off Between Plasticity and Robustness: There Is no Privileged Level of Causation

Biological stability has been defined as a global emergent property from complexity, diversity, and plasticity according to the "quadruped" model of Souza and Lüttge (2015). In this sense, stability is a dynamical property allowing biological systems to face different environmental conditions by interchanging strategies to acclimate. There is no unique solution to deal with external disturbances, regardless of the specific situation; plasticity (flexibility) of metabolic networks or resistance to change by homeostatic mechanisms (robustness) could work properly well for different types of organisms. For instance, different plant species can show different strategies to face water shortage, which is one of the most limiting factors for plants life. In the same environment, while some species maintain their water status relatively constant controlling water loss by an efficient stomatal dynamics of closing and opening (*isohydric* species), other species support significant water content fluctuations without constrains on their physiological functioning (*anisohydric* species) (Tardieu and Simonneau 1998; Braga et al. 2016). However, on the other hand, different species can show a convergence of strategies to maintain stability, as observed by Souza et al. (2009). In a study with different tropical tree species (early and late successional), growing in environments significantly different (understory and gap conditions), the same pattern of physiological network connectance was observed, indicating that there is a conservative pattern of photosynthesis regulation based on network modulation and environmental coupling, regardless of plant species. This suggests that changes in network connectance may not be specific of a functional group but rather a more general response to environmental fluctuations, strongly related to system stability (Souza et al. 2009).

The ability of a system to maintain its organization stable when subjected to external changes relies on the capacity to control networks properties. The system simply "absorbs" the effects of the perturbation, as a network, by compensating it through internal reciprocal adjustments between tightly coupled constraints, together with the metabolites and reaction processes involved, while the whole dynamics is maintained in the initial attractor or shifts to a new available one (metastability) (Souza and Lüttge 2015; Bich et al. 2016). In the literature, this kind of response is usually referred to as dynamic stability (Rosen 1970), the capability to counterbalance the displacement of the system from a certain initial state, provoked by a perturbation, and end up in the same final state. On the other hand, regulatory control requires a subsystem that is sufficiently independent of the dynamics of the controlled processes, and which can be altered without disrupting these processes (Bechtel 2007; Bich et al. 2016). A regulatory control is that the former, in order to reach a compensatory response, the changes from many local processes may need to cross the entire system, thus the time taken to implement a response can be proportional to the size of the system. On the other hand, regulatory control allows a decoupled subsystem to induce the appropriate collective pattern of behavior in a more rapid and efficient way. Its compensatory action is more efficient and robust because, instead of involving each time the progressive accommodation of the constitutive organization to the change, it can rapidly switch between available regimes through dedicated mechanisms, acting in relation to specific perturbations. The efficiency of this kind of response is not negatively affected by the size of the system (Bich et al. 2016).

The stability of the constitutive regime against disturbances can also be improved by the contribution of additional interactions or mechanisms that are still part of the basic metabolic network and self-maintenance through feedback events, which are therefore considered as the essence of regulation. Thus, feedbacks are especially relevant to the stability of a system because they perform circular causal relationships between the output of the system and one or more of its inputs, such that the functioning of the system becomes dependent on the effects of its own actions. This type of response depends only on the interaction between the components, reactions, and control subsystems that already participate in the constitutive regime, without resorting to additional and dedicated mechanisms. When biological systems are found in dynamically stable regimes, they present a basic form of robustness as an inherent ability to respond to perturbations through highly distributed endogenous compensation patterns to remain within their viability region (Bich et al. 2016).

Because organisms constantly face changing environmental conditions, they have developed systems that allow phenotypic changes or prevent them. At one extreme, as quoted above, there are some traits that exhibit little or no phenotypic change despite an environmental challenge. These traits are robust (Waddington 1961; Lempe et al. 2012). At the other extreme, there are traits that exhibit a high degree of plasticity—programmed changes in the phenotype in response to different environments (Pigliucci 2001), which are important for physical fitness.

Within the biological networks, there are at least two components that influence the position of a trait in the robust-to-plastic spectrum: redundancy and topology of the interactions. Redundancy influences phenotypic robustness and plasticity. For example, duplication of a gene may provide robustness through a redundant function or may promote plasticity through its subfunctionalization. In the same way, the network topology also influences phenotypic robustness and plasticity. In most of the existing literature, the modulation of robustness and plasticity has been attributed to specific mechanisms. Ultimately, traits are shaped by a blend of robustness and plasticity acting at different times (Lachowiec et al. 2015).

In many species, the robustness in several traits has been attributed to the function of genes that play crucial cellular functions and are involved in cellular homeostasis (Queitsch et al. 2002; Levy and Siegal 2008). Examples include chaperone proteins (Rutherford and Lindquist 1998; Queitsch et al. 2002; Yeyati et al. 2007), genes that function in proper maintenance of the genome, and genes that modify the expression of many others (Lehner et al. 2006; Levy and Siegal 2008). Importantly, these genes may affect several networks and specific regulatory motifs without being an integral component of these networks (Whitacre 2012; Lachowiec et al. 2015).

Hsp90 (Heat shock protein 90) is a highly connected and evolutionarily conserved protein chaperone that facilitates the maturation of certain proteins (clients) and keeps them poised for activation (Zhao et al. 2005; McClellan et al. 2007; Jarosz and Lindquist 2010; Gopinath et al. 2014). Hsp90 clients, among them many kinases, TFs and receptors, are conformationally flexible, which is the property that the chaperone is thought to recognize (Taipale et al. 2010). The role of Hsp90 in maintaining phenotypic robustness (Rutherford and Lindquist 1998; Queitsch et al. 2002; Sangster et al. 2007; Yeyati et al. 2007; Jarosz and Lindquist 2010) is fundamentally an epistasis phenomenon on a network scale. The interaction of Hsp90 with its many clients maintains robustness, yet the chaperone's property of enabling signal transduction proteins to perceive and transduce signals is also of fundamental importance for phenotypic plasticity, in particular programmed plasticity. Environmental stresses such as increased temperature, drought, oxidative stress, and altered conductivity compromise protein folding and increase demands for Hsp90. Under these conditions, Hsp90 clients will be less functional and the genetic network will be perturbed at many nodes. Hence, it is not surprising that Hsp90 function is environmentally responsive. Hsp90 levels are strongly upregulated in conditions that compromise protein folding to bolster protein homeostasis (Gerspacher et al. 2009; Cid et al. 2010; Park et al. 2010; Pratt et al. 2010; Sekimoto et al. 2010; Stensløkken et al. 2010).

The classical response to thermal shock (stress) (HSR) was attributed to denaturation of proteins, including membrane proteins. However, induction of HSPs occurs in many circumstances where protein denaturation is not observed (de Marco et al. 2005). Recently, considerable evidence has been accumulated in favor of the "Membrane Sensor Hypothesis," which predicts that the level of HSPs can be altered as a result of changes in the plasma membrane. Small temperature stresses can modulate significantly transient receptor potential (TRP), an important group of cationic channels that play an important role in sensory physiology and some members are very sensitive to temperature, and are also influenced by lipids. In addition, stress hormones often modify plasma membrane structure and function and thus initiate a cascade of events, which may affect HSR. The major transactivator heat shock factor-1 (HSF) integrates the signals originating from the plasma membrane and orchestrates the expression of individual heat shock genes (Török et al. 2014).

Members of the HSF family control the expression of HSPs. Under optimal growth conditions, HSF1 shuttles as a latent monomer between the cytoplasm and the nucleus in complex with Hsp90, p23 and immunophilin. Upon elevated temperatures, this complex disintegrates; HSF1 accumulates in the nucleus, quickly undergoes several post translational modifications, trimerizes as a homo- or heterotrimer together with HSF2 and binds to its recognition site (heat shock element, HSE) in the HSP-promoter region and ultimately drives HSP expression (Sandqvist et al. 2009; Anckar and Sistonen 2011). Understanding of how the signal from heat perception at the plasma membrane is transduced toward HSF1 activation, ultimately driving the expression of HSPs, is still incomplete and fragmentary. However, over the last decades a complex interplay of multiple pathways emerged in which derivatives of plasma membrane phospholipids, sphingolipids, and cholesterol all play important roles (Török et al. 2014).

The biochemical process underlying cells organization, in particular the role of the enzymes in biosynthesis, results in a complex non-linear dynamics, since a single enzyme may function in many interconnected enzymatic pathways. Moreover, the diffusion and the number of signaling molecules (e.g., plant hormones) among developing cells follows a complex self-organized non-homogenous dynamics, inducing minimal differences among cells (Choi et al. 2016). Nonlinear complex systems, such as cells whole metabolism, are sensitive to small perturbation, mainly under suboptimal conditions. Thereupon, random perturbations tend to become magnified, increasing the differentiation among groups of cells and creating independence from each other (Møller and Swaddle 1997). Moreover, the complex electrical network, named "electrome," underpinning cellular activities, shall display a major role on the whole system organization (De Loof 2016). For instance, recent analyses of plant electrome (Saraiva et al. 2017; Souza et al. 2017) have uncovered the high complexity involved in plant electrophysiological dynamics. The electrical signaling flowing through plant's body exhibits a range of color noises, with short time correlation, varying from reddened to black noise, showing that the temporal dynamics of the micro-volted electrical signals are not random, with some traits similar to human brain EEG. Furthermore, such complexity can be affected by environmental stimuli (Saraiva et al. 2017), for instance, decreasing complexity under stressful conditions. Moreover, under some specific constraining conditions (for instance, under drought), plant electrome can be pushed to a self-organized critical state, showing bursts of electrical "spikes" without a characteristic size (Souza et al. 2017).

Overall, as aforementioned, such "epigenetic noise" (including epi-regulation of gene expression, short and long-distance signaling regulating different scales within the system) allows a variation of responses, from the cell to the whole plant level, to a plethora of environmental cues. Such plasticity and complexity, as well as the diversity of individual components, are the bases of system stability, conferring multifunctional regulatory capacity for plants, without a unique direction of causation (Noble 2008b; Souza and Lüttge 2015).

Concluding Remarks

As exemplified above, notwithstanding the environment triggering responses in the organism, the pattern of such responses is determined by the internal dynamics of the system itself. This internal dynamics is integrated in a complex metabolic network that operates out of some rules of interactions. The rules that specify the interactions among the system components are performed using local information, without reference to a pre-existing global pattern (Camazine et al. 2001), which can be spreading throughout the whole system by different processes of signaling (Choi et al. 2016). In complex systems, such interactions are typically non-linear processes based on negative and positive feedback loops. The negative feedback plays a crucial role in maintaining homeostasis of the system, whereas the positive feedback operates propagating and amplifying signals throughout the system. Both processes work together in the formation and stabilization of new patterns of organization, which makes the prediction of their global behavior difficult (Camazine et al. 2001; Lüttge 2012). There are multiple feedbacks from higher levels to lower levels in addition to those from lower to higher levels. Through the existence of multiple back-up mechanisms, many DNA changes, such as knockouts, do not have a phenotypic effect on their own (Noble 2011). For instance, about 80 per cent of the knockouts in yeast are normally "silent" by alternative metabolic pathways (Hillenmeyer et al. 2008). Such a dynamical process of organization, operating through the different scales of the organization of the system, produces stable emergent properties and, hence, makes the system as whole non-reducible to its components at smaller scales of organization (Souza et al. 2016b).

The limitations of the reductionist molecular approach have become increasingly evident. It became clear that biological systems cannot be explained only at the genetic level. Instead, they should be understood as complex systems resulting from dynamic interactions of different components at different levels, each individually functioning as wholes, which eventually control the phenotype. Complex systems exist at different levels of biological organization ranging from the subatomic realm to individual organisms to whole populations and beyond. Hence, a need arose for an integrative framework, which provides a holistic understanding of the biological systems (Sheth and Thaker 2014).

Finally, according to Noble (2011) a downward form of causation is not a simple reverse form of upward causation; it is better seen as completing a feedback circuit. "*There should be no dogmas when it comes to causation in biological systems.*"

References

Ahl V, Allen TFH (1996) Hierarchy theory. A vision, vocabulary, and epistemology. Columbia University Press, New York

Alabadí D, Oyama T, Yanovsky MJ, Harmon FG, Mas P, Kay SA (2001) Reciprocal regulation between TOC1 and LHY/CCA1 within the Arabidopsis circadian clock. Science 293:880–883

Alcocer-Cuarón C, Rivera AL, Castano VM (2014) Hierarchical structure of biological systems: a bioengineering approach. Bioengineered 5:73–79. https://doi.org/10.4161/bioe.26570

Amaral MN, Souza GM (2017) The challenge to translate OMICS data to whole plant physiology: the context matters. Front Plant Sci. https://doi.org/10.3389/fpls.2017.02146

Anckar J, Sistonen L (2011) Regulation of HSF1 function in the heat stress response: implications in aging and disease. Annu Rev Biochem 80:1089–1115. https://doi.org/10.1146/annurev-biochem-060809-095203

Baetu TM (2012) Emergence, therefore antireductionism? a critique of emergent antireductionism. Biol Philos 27:433–448. https://doi.org/10.1007/s10539-011-9290-2

Bechtel W (2007) Biological mechanisms: organized to maintain autonomy. In: Boogerd F, Bruggerman F, Hofmeyr JH, Westerhoff HV (eds) Systems biology: philosophical foundations, 1st edn. Elsevier, Amsterdam, pp 269–302

Bertolli SC, Mazzafera P, Souza GM (2014) Why is so difficult to identify a single indicator of water stress in plants? a proposal for a multivariate analysis to access emergent properties. Plant Biol 16:578–585

Bich L, Mossio M, Ruiz-Mirazo K, Moreno A (2016) Biological regulation: controlling the system from within. Biol Philos 31:237–265

Booch G (1994) Object oriented analysis and design with applications. Addison Wesley, New York

Braga NS, Vitória AP, Souza GM, Barros CF, Freitas L (2016) Weak relationships between leaf phenology and isohydric and anisohydric behavior in lowland wet tropical forest trees. Biotropica 48(4):453–464

Camazine S, Deneubourg J-L, Franks NR, Sneyd J, Theraulaz G, Bonabeau E (2001) Self-organization in biological systems. Princeton University Press, Princeton

Choi W, Hilleary R, Swanson SJ, Kim SH, Gilroy S (2016) Rapid, long-distance electrical and calcium signalling in plants. Annu Rev Plant Biol 67:287–307. https://doi.org/10.1146/annurev-arplant-043015-112130

Cid C, Garcia-Descalzo L, Casado-Lafuente V, Amils R, Aguilera A (2010) Proteomic analysis of the response of an acidophilic strain of Chlamydomonas sp. (Chlorophyta) to natural metal-rich water. Proteomics 10:2026–2036

Corominas-Murtra B, Goñi J, Solé RV, Rodríguez-Caso C (2013) On the origins of hierarchy in complex networks. Proc Natl Acad Sci 110:13316–13321

Csermely P (2006) Weak links. In: Stabilizers of complex systems from proteins to social networks. Springer, Berlin

Crick F (1970) Central dogma of molecular biology. Nature 227:561–563

Davies PCW (2012) The epigenome and top-down causation. Interface Focus 2:42–48. https://doi.org/10.1098/rsfs.2011.0070

De Loof A (2016) The cell's self-generated "electrome": the biophysical essence of the immaterial dimension of life? Commun Integr Biol 9(5):e1197446. https://doi.org/10.1080/19420889.2016.1197446

de Marco A, Vigh L, Diamant S, Goloubinoff P (2005) Native folding of aggregation-prone recombinant proteins in Escherichia coli by osmolytes, plasmid- or benzyl alcohol-overexpressed molecular chaperones. Cell Stress Chaperones Winter 10(4):329–339

Deal RB, Henikoff JG, Henikoff S (2010) Genome-wide kinetics of nucleosome turnover determined by metabolic labeling of histones. Science 328(5982):1161–1164

Dietz KJ, Jacquot JP, Harris G (2010) Hubs and bottlenecks in plant molecular signaling networks. New Phytol 188:919–938

Ellis GFR (2008) On the nature of causation in complex systems. Trans R Soc S Afr 63:69–84

Ellis GFR (2012) Top-down causation and emergence: some comments on mechanisms. Interface Focus 2(1):126–140. https://doi.org/10.1098/rsfs.2011.0062
Gerspacher C, Scheuber U, Schiera G, Proia P, Gygax D, Di Liegro I (2009) The effect of cadmium on brain cells in culture. Int J Mol Med 24:311–318
Gopinath RK, You S-T, Chien K-Y et al (2014) The Hsp90-dependent proteome is conserved and enriched for hub proteins with high levels of protein–protein connectivity. Genome Biol Evol 6:2851–2865
Gross F, Green S (2017) The Sum of the parts: large-scale modeling in systems biology. Philos Theor Pract Biol 9:10. https://doi.org/10.3998/ptb.6959004.0009.010
Hassidim M, Yakir E, FradkinD H, Kron I, Keren N, Harir Y, Yerushalmi S, Green RM (2007) Mutations in CHLOROPLAST RNA BINDING provide evidence for the involvement of the chloroplast in the regulation of the circadian clock in Arabidopsis. Plant J 51:551–562
Henikoff S, Shilatifard A (2011) Histone modification cause or cog? Trends Genet 27:295–342. https://doi.org/10.1016/j.tig.2011.06.006
Hillenmeyer ME, Fung E, Wildenhain J et al (2008) The chemical genomic portrait of yeast: uncovering a phenotype for all genes. Science 320:362–365. https://doi.org/10.1126/science.1150021
Hodgkin AL, Huxley AF (1952) A quantitative description of membrane current and its application to conduction and excitation in nerve. J Physiol 117:500–544
Jacob F, Perrin D, Sanchez C, Monod J, Edelstein S (1960) The operon: a group of genes with expression coordinated by an operator. C R Hebd Seances Acad Sci 250:1727–1729
Jarosz D, Lindquist S (2010) Hsp90 and environmental stress transform the adaptive value of natural genetic variation. Science 330:1820–1824
Jin B, Li Y, Robertson KD (2011) DNA Methylation superior or subordinate in the epigenetic hierarchy? Genes Cancer 2(6):607–617
Kauffman SA, Clayton P (2006) On emergence, agency, and organization. Biol Philos 21:501–521. https://doi.org/10.1007/s10539-005-9003-9
Kitano H (2002) System biology: a brief overview. Science 295:1662–1664
Kohl P, Crampin EJ, Quinn TA, Noble D (2010) Systems biology: an approach. Clin Pharmacol Therl 88(1):25–33. https://doi.org/10.1038/clpt.2010.92
Korn R (1999) Biological organization—a new look at an old problem. Bioscience 49:51–57
Kuppers BO (1994) Information and the origin of life. MIT Press, Cambridge
Lachowiec J, Queitsch C, Kliebenstein DJ (2015) Molecular mechanisms governing differential robustness of development and environmental responses in plants. Ann Bot 117(5):795–809. https://doi.org/10.1093/aob/mcv151
Lehner B, Crombie C, Tischler J, Fortunato A, Fraser AG (2006) Systematic mapping of genetic interactions in Caenorhabditis elegans identifies common modifiers of diverse signaling pathways. Nat Genet 38:896–903
Lempe J, Lachowiec J, Sullivan AM, Queitsch C (2012) Molecular mechanisms of robustness in plants. Curr Opin Plant Biol 16:62–69
Levy SF, Siegal ML (2008) Network hubs buffer environmental variation in Saccharomyces cerevisiae. PLoS Biol 6:e264
Lu SX, Knowles SM, Andronis C, Ong MS, Tobin EM (2009) CIRCADIAN CLOCK ASSOCIATED1 and LATE ELONGATED HYPOCOTYL function synergistically in the circadian clock of Arabidopsis. Plant Physiol 150:834–843
Lucas M, Laplaze L, Bennett M (2011) Plant systems biology: network matters. Plant, Cell Environ 34:535–553
Lüttge U (2012) Modularity and emergence: biology's challenge in understanding life. Plant Biol 14:865–871
McClung CR (2006) Plant circadian rhythms. Plant Cell 18:792–803
MacMahon JA, Phillips DL, Robinson JV, Schimpf DJ (1978) Levels of biological organization: an organism-centered approach. Bioscience 28:700–704

McClellan AJ, Xia Y, Deutschbauer AM, Davis RW, Gerstein M, Frydman J (2007) Diverse cellular functions of the Hsp90 molecular chaperone uncovered using systems approaches. Cell 131:121–135
Misteli T (2010) Higher-order genome organization in human disease. Cold Spring Harb Perspect Biol 2(8):a000794. https://doi.org/10.1101/cshperspect.a000794
Møller AP, Swaddle JP (1997) Asymmetry, developmental stability and evolution. Oxford University Press, Oxford.
Nakamichi N, Kiba T, Henriques R, Mizuno T, Chua N-H, Sakakibara H (2010) PSEUDO-RESPONSE REGULATORS 9, 7, and 5 Are transcriptional repressors in the Arabidopsis circadian clock. Plant Cell 22(3):594–605
Noble D (2008a) Genes and causation. Philos Trans Royal Soc A 366:3001–3015
Noble D (2008b) The first systems biologist, and the future of physiology. Exp Physiol 93:16–26
Noble D (2011) A theory of biological relativity: no privileged level of causation. Interface Focus 2(1):55–64. https://doi.org/10.1098/rsfs.2011.0067
Noble D (2012) A theory of biological relativity: no privileged level of causation. Interface Focus 2:55–64. https://doi.org/10.1098/rsfs.2011.0067
Nolte C, Staiger D (2015) RNA around the clock. Front Plant Sci 6:311. https://doi.org/10.3389/fpls.2015.00311
Novikoff AB (1945) The concept of integrative levels and biology. Science 101:209–215
Nusinow DA, Helfer A, Hamilton EE, King JJ, Imaizumi T, Schultz TF, Farre EM, Kay SA (2011) The ELF4-ELF3-LUX complex links the circadian clock to diurnal control of hypocotyl growth. Nature 475:398–402
Pigliucci M (2001) Phenotypic plasticity: beyond nature and nurture. JHU Press, Baltimore
Pratt WB, Morishima Y, Peng H-M, Osawa Y (2010) Proposal for a role of the Hsp90/Hsp70-based chaperone machinery in making triage decisions when proteins undergo oxidative and toxic damage. Exp Biol Med 235:278–289
Queitsch C, Sangster T, Lindquist S (2002) Hsp90 as a capacitor of phenotypic variation. Nature 417:618–624
Ravasz E, Somera AL, Mongru DA, Oltvai ZN, Barabási AL (2002) Hierarchical organization of modularity in metabolic networks. Science 297:1551–1555
Roederer JG (2005) Information and its role in nature. Springer, Berlin
Rojdestvenski I, Cottam M, Youn-II P, Oquist G (1999) Robustness and time-scale hierarchy in biological systems. BioSystems 50:71–82
Romanowski A, Yanovsky MJ (2015) Circadian rhythms and post-transcriptional regulation in higher plants. Front Plant Sci 6:437. https://doi.org/10.3389/fpls.2015.00437
Rosen R (1970) Dynamical system theory in biology. Stability theory and its applications. Wiley, New York
Rutherford SL, Lindquist SL (1998) Hsp90 as a capacitor for morphological evolution. Nature 396:336–342
Sanchez SE, Kay SA (2016) The plant circadian clock: from a simple timekeeper to a complex developmental manager. Cold Spring Harb Perspect Biol 8(12):a027748. https://doi.org/10.1101/cshperspect.a027748
Sandqvist A, Bjork JK, Akerfelt M, Chitikova Z, Grichine A, Vourc'h C, Jolly C, Salminen TA, Nymalm Y, Sistonen L (2009) Heterotrimerization of heat-shock factors 1 and 2 provides a transcriptional switch in response to distinct stimuli. Mol Biol Cell 20(5):1340–1347. https://doi.org/10.1091/mbc.e08-08-0864
Sangster TA, Bahrami A, Wilczek A et al (2007) Phenotypic diversity and altered environmental plasticity in Arabidopsis thaliana with reduced Hsp90 levels. PLoS ONE 2:e648. https://doi.org/10.1371/journal.pone.0000648
Saraiva GFR, Ferreira AS, Souza GM (2017) Osmotic stress decreases complexity underlying the electrophysiological dynamic in soybean. Plant Biol 19(5):702–708
Sekimoto T, Oda T, Pozo FM et al (2010) The molecular chaperone Hsp90 regulates accumulation of DNA polymerase g at replication stalling sites in UV-irradiated cells. Mol Cell 37:79–89

Seo PJ, Mas P (2014) Multiple layers of posttranslational regulation refine circadian clock activity in Arabidopsis. Plant Cell 26:79–87

Schneider DC (1998) Applied scaling theory. In: Peterson DL, Parker VT (eds) Ecological scale: theory and applications, 1st edn. Columbia University Press, New York, pp 253–269

Shapiro JA (2011) Evolution: a view from the 21st century. Pearson Education Inc, Upper Saddle River

Sharrock RA, Clack T (2002) Patterns of expression and normalized levels of the five Arabidopsis phytochromes. Plant Physiol 130:442–456

Sheth BP, Thaker VS (2014) Plant systems biology: insights, advances and challenges. Planta 240:33–54. https://doi.org/10.1007/s00425-014-2059-5

Souza GM, Lüttge U (2015) Stability as a phenomenon emergent from plasticity–complexity–diversity in eco-physiology. In: Lüttge U, Beyschlag W (eds) Progress in botany, 1st edn. Springer Science + Business Media, Berlim, pp 211–239

Souza GM, Bertolli SC, Lüttge U (2016a) Hierarchy and information in a system approach to plant biology: explaining the irreducibility in plant ecophysiology. In: Lüttge U (ed) Progress in botany, 1st edn. Springer Science + Business Media, Berlim, pp 167–186

Souza GM, Ferreira AS, Saraiva GFR, Toledo GRA (2017) Plant "electrome" can be pushed toward a self-organized critical state by external cues: evidences from a study with soybean seedlings subject to diferent environmental conditions. Plant Signal Behav 12(3):e1290040

Souza GM, Prado CHBA, Ribeiro RV, Barbosa JPRAD, Gonçalves NA, Habermann G (2016b) Toward a systemic plant physiology. Theor Exp Plant Physiol 28:341–346. https://doi.org/10.1007/s40626-016-0071-9

Souza GM, Ribeiro RV, Prado CHBA, Damineli DSC, Sato AM, Oliveira MS (2009) Using network connectance and autonomy analyses to uncover patterns of photosynthetic responses in tropical woody species. Ecol Complex 6:15–26

Stensløkken K-O, Ellefsen S, Larsen HK, Vaage J, Nilsson GE (2010) Expression of heat shock proteins in anoxic crucian carp (Carassius carassius): support for cold as a preparatory cue for anoxia. Am J Physiol Regul Integr Comp Physiol 298:1499–1508

Suzuki MM, Bird A (2008) DNA methylation landscapes: provocative insights from epigenomics. Nat Rev Genet 9(6):465–476

Taipale M, Jarosz D, Lindquist S (2010) HSP90 at the hub of protein homeostasis: emerging mechanistic insights. Nat Rev Mol Cell Biol 11:515–528

Tardieu F, Simonneau T (1998) Variability among species of stomatal control under fluctuating soil water status and evaporative demand: Modeling isohydric and anisohydric behaviours. J Exp Bot 49:419–432

Török Z, Crul T, Maresca B et al (2014) Plasma membranes as heat stress sensors: from lipid-controlled molecular switches to therapeutic applications. Biochim Biophys Acta 1838:1594–1618. https://doi.org/10.1016/j.bbamem.2013.12.015

Tóth R, Kevei É, Hall A, Millar AJ, Nagy F, Kozma-Bognár L (2001) Circadian clock-regulated expression of phytochrome and cryptochrome genes in Arabidopsis. Plant Physiol 127:1607–1616

Vítolo HF, Souza GM, Silveira J (2012) Cross-scale multivariate analysis of physiological responses to high temperature in two tropical crops with C3 and C4 metabolism. Environ Exp Bot 80:54–62

Von Bertalanffy L (1968) General system theory. George Braziller Publ, New York

Waddington CH (1961) Genetic assimilation. Adv Gen 10:257–293

Wade PA (2001) Methyl CpG-binding proteins and transcriptional repression. BioEssays 23(12):1131–1137

Wegscheid B, Condon C, Hartmann RK (2006) Type A and B RNase P RNAs are interchangeable in vivo despite substantial biophysical differences. EMBO Rep 7:411–417

Whitacre JM (2012) Biological robustness: paradigms, mechanisms, and systems principles. Front Genet 3:67. https://doi.org/10.3389/fgene.2012.00067

Yeyati PLL, Bancewicz RMM, Maule J, Van Heyningen V (2007) Hsp90 selectively modulates phenotype in vertebrate development. PLoS Genet 3:e43

Zhao R, Davey M, Hsu YC et al (2005) Navigating the chaperone network: an integrative map of physical and genetic interactions mediated by the hsp90 chaperone. Cell 120:715–727

Emergence and Sustainment of Humankind on Earth: The Categorical Imperative

Ulrich Lüttge and Fabio Rubio Scarano

Abstract This essay argues that the "sustainment" of humankind on Earth will require short-term ruptural transformation of rooted individual and societal attitudes and ethics into a new categorical imperative. We establish a difference between "sustainment" of humankind and "sustainability". Sustainment is the process of dynamic state of maintenance of conditions without destruction of prerequisite requirements for well and healthy performance of organisms and systems of life emergent on Earth (i.e. ecological stability). Sustainability is as a moral value related to intergenerational justice, and therefore a societal goal or utopia, backed up by specific science and policy tools. To address sustainment of humankind on Earth, we first provide a brief history on origins and extinction processes, then we discuss existing problems and challenges for sustainment, and we finally provide a reflection on plausible futures with humankind sustained. We aimed to balance out futures perspectives such as those related to the Gaia hypothesis, to the evolution of consciousness, and to trans-humanism, as constrained by the short time left for humankind to halt or reverse the trajectory of planetary degradation it imposed on the planet.

Introduction: Confrontation of Tribal Attitudes by Global Ethics

How optimistic can one be regarding the sustainment of humankind on Earth? The answer to this question depends on understanding the extent to which humankind can replace evolved group and tribal attitudes by global ethics. On the one hand, prospects of self-organization, self-regulation, self-repair and regeneration of nature give reason for some optimism (Lüttge et al. 2013; Scarano and Garbin 2013). On the other hand, the way humankind is currently exploiting the biosphere and Gaia, and

U. Lüttge (✉)
Darmstadt University of Technology, Darmstadt, Germany
e-mail: luettge@bio.tu-darmstadt.de

F. R. Scarano
Universidade Federal do Rio de Janeiro, Rio de Janeiro, Brazil
e-mail: fscarano@biologia.ufrj.br

L. H. Wegner and U. Lüttge (eds.), *Emergence and Modularity in Life Sciences*,
https://doi.org/10.1007/978-3-030-06128-9_11

the short time span available for behaviour change before an even more acute climate crisis (see also Scarano 2019, in this book), leads to an increasingly pessimistic tone of a majority of scientific authors in environmentalism, ecology, philosophy, social sciences and other fields.

The glaciologist Jean Jouzel believes that in view of the climate problem humankind shall make a giant jump changing attitude and behaviour in its relation to Gaia (Robert and Jouzel 2015). With his book "The Future of Life", Edward O. Wilson presents a fascinating journey through the originally so perfectly functioning ecosystems of the Earth. He describes many examples of the disastrous exploitations performed by man all over the globe and views all the grave problems. He winds up with a sad testament that we leave to the next generations with the statement of how the world will look like at the end of the twenty-first century if the current developments continue. It is a solitary world with artificial landscapes and deserted bush-land instead of rainforests, and it is recurring to genetically engineered organisms for constructing new ecosystems if that is feasible at all (Wilson 2002, 2004, p. 104). Nevertheless, E. O. Wilson remains semi-optimistic in his book. He thinks that there are still many resources on Earth and that the human spirit and inventiveness with the potency of science and technology can manage to sustain our planet undamaged and maintain the wonderful richness of life. If humankind only wishes and is forced to do so, it could make the appropriate decisions (Wilson 2002, 2004, pp. 195, 222).

Others are outright pessimistic about humankind on Earth. Lovelock (2009) assumes that humankind may survive the sixth extinction wave with a greatly reduced population size of a few hundred million people. Many scientists think that life will persist but without man. Alan Weisman is very pessimistic. He develops a series of scenarios of how the planet may develop and look like after an assumed sudden disappearance of man. Even after a few days without pumping, the entire metro of New York shall be flooded. After one year, plants will grow on roads. Within 50–100 years, buildings will break down. In 300 years, cities along coastlines will be washed away, and in 500 years, woods will overgrow everything with remnants of aluminium and stainless steel. Humankind will fade away, while there remains optimism for life as such. Although humankind will take many species away with it, life will continue on Earth with new evolution possibly beginning from bacteria. Life as such in general has a great potential of regeneration power (Weisman 2007a, b). Perhaps elsewhere in the universe, consciousness and empathy may be sustained, but we shall not know.

This chapter aims to balance out optimistic and pessimistic perspectives into a realistic analysis of what it would take to sustain humankind on Earth, while faced by ecological, socio-economic and sociopolitical menaces and practices. Such an analysis required a broad review of literature, across time, geographical space, and disciplines and fields of knowledge. In view of the ongoing interference of humankind with unscrupulous exploitation of the environment, our central argument is that the sustainment of humankind on Earth will require short-term breaking transformation of rooted individual and societal attitude and ethics into a new categorical imperative. We address specifically two types of questions regarding sustainment. First, at which levels must sustainment be achieved? The most critical levels appear to be social levels, as the political and the economic levels, and in a strict biological sense the ecological level (in alignment with the tripod of sustainability—social, economic and environmental). Second, what can be sustained? Will this be specific forms and specific integrated systems of life? Or will this be just life as such with new forms evolving from survivors of extinction? We also speculate on plausible futures depending on whether or not humankind is successful in producing such deep transformations.

Sustainment and Sustainability: Definitions

We establish a difference between "sustainability" and "sustainment". Sustainability is a concept and a moral value, which arguably emerged out of the need for an intergenerational ethics, both human and non-human, in many ways similar to Hans Jonas' (1979) "ethics of responsibility" (Jonas 2003). However, sustainability has become an inflated fashionable term frequently used and often not much reflected for marking demands and challenges in very many respects. The emergence of sustainability in multiple branches—science, politics, technology, moral value—is explored in detail in other chapters of this book, i.e. in relation to the emergence of biomimetic products (Speck and Speck 2019, this book), and by Scarano (2019, this book) who actually argues that sustainability is currently a new utopia. While sustainability is a concept, a value, and possibly a new utopia, we propose "sustainment" as the process whereby humankind might come to realize its own permanence in the planet and that of life as we know it.

Sustainability fits the equation social justice + environmental balance + economic prosperity. Applying it to life of organisms, we must be aware of the fact that all living organisms are open systems through which a steady flow of matter and energy is occurring. Organisms are not in thermodynamic steady-state equilibrium. They are in dynamic pseudo-steady state. Similarly, all systems of life built up from organisms are dynamic open systems. Thus, we can talk about stability where the dynamics of pseudo-steady states are maintained in space and time (Souza and Lüttge 2015; for the debate, see also chapter of Schmidt 2019, this volume). With this thought, here we apply the term "sustainment" meaning the process of dynamic state of maintenance of conditions without destruction of prerequisite requirements for well and healthy performance of organisms and systems of life on Earth.

Origins and Extinctions

The question of sustainment of humankind is closely related to the question of life itself, because humankind is part of life on our planet. We have no clear knowledge about the origin of our universe. We have no access to physical laws beyond the

Planck time of 5.39×10^{-44} s. We can observe the dynamics of the universe, but regarding its origin, the big bang remains a hypothesis and other alternatives are also discussed. Similarly, we have no knowledge about the origin of life. In his philosophy of biology, Karl Popper notes that life in the universe is a cosmological emergence, and he suspects that even on Earth it must have originated multiple times. The major problem with that is if it is adapted to the environment where it arises; the adaptation of the new organism to its environment is one of the major problems with the question of the origin of life (Niemann 2013; Popper 2013). The theory of evolution explains the uncovering and development of life over the last 4×10^9 years but not its origin. Therefore, we also have no distinct definition of life. Life remains a philosophical and not a scientific category (Nachtigall 2010). We can only scientifically describe life by its relevant properties and listing its most outstanding features that have evolved gradually. Life emerged from the integration of building blocks or modules (Bolker 2000; Lüttge 2012).

The oldest fossils of organisms that we know are calcareous crusts of biofilms of prokaryotic cyanobacteria so-called stromatolites from pre-Cambrian times. Stromatolites found in Australia date back 3.5×10^9 years and in South-Africa 3.4×10^9 years (Allwood and Brown 2004). Fossils of multi-cellular organisms discovered in Africa are 2.1×10^9 years old (El Albani et al. 2010; Donoghue and Antcliffe 2010). However, then later followed an almost instantaneous outburst of complex multi-cellular life as documented by the fossils of the Ediacara fauna $580–540 \times 10^6$ years ago (Grazhdankin 2014), and subsequently metazoan life showed a sudden proliferation in the so-called Cambrian explosion $535–525 \times 10^6$ years ago (Gould 1989). We do not understand why evolution occurred in such outbursts. Perhaps it is an artefact due to the absence of fossilization of earlier forms. On the other hand, possibly sudden self-organization of modules may have really led to the brisk emergence of higher developed metazoan life.

For a reflection on the dynamics of sustainment and extinction of life or of particular forms of life, the Ediacara fauna gives us an early lesson. From the fossils found in the Ediacara hills of the Flinders ranges in Southern Australia which gave the name to this fauna, but also worldwide, e.g. in Newfoundland, Canada, in Russia, in Namibia, Brazil and China, we know about 100–120 species (Grazhdankin 2014). They had tubular structures frequently with a fern-frond like organization and were several centimetres up to almost 2 m large. A lot of them were non-mobile and anchored in the ocean floor; others had creeping mobility. They were feeding by osmotrophically absorbing organic molecules from the ocean water or by foraging mats of micro-organisms like the stromatolites. Most of the investigators of this fauna regard these organisms as the first multi-cellular metazoans. A minority view is that they might have been fungi, lichens or even just giant protists (Retallack 1994; Peterson et al. 2003; Seilacher et al. 2003). On the botanical side, we can assume that certainly also multi-cellular algae were living at the time. After only 40 million years, there was a mass extinction of this fauna. Its species were not surviving into the Cambrian. As one reason, it is assumed that they were outcompeted by the new Cambrian fauna the species of which possibly used the environment to the effect that living conditions of the Ediacaran organisms were destroyed.

Fossils of the Cambrian explosion were specifically found in the Burgess Shale of the Rocky Mountains in British Columbia, Canada. They comprise an enormous diversity of anatomical structures and building principles. We know more than 20 different forms. Among them were the first chorda animals, i.e. chordates, and thus, precursors of the vertebrates. *Pikaia gracilens* was 5 cm large and had a backbone cord. It was the only one of the organisms of the Cambrian explosion which survived their extinction. Without it, vertebrates would not have evolved including humankind which would not have appeared in this form on the planet (Gould 1989).

Thus, we know from the fate of these early faunas that life of individual species is a coming and going. Species are appearing and dying out again. During the entire history of life on Earth of more than 3.5×10^9 years, 4×10^9 species have lived and the survivals of those to date are 1%. Extinction comes in smaller and larger waves. Five major extinction waves (and there are claims for a sixth wave, which we discuss later) of higher organized species occurred at the transitions between geological ages (Courtillot 1995; Butterfield 2007; Alroy 2008; Melott and Bambach 2011; Matyssek and Lüttge 2013):

- 444 million years ago at the transition from the Ordovician to the Silurian not so long after the Cambrian explosion, i.e. less than 100 million years later.
- 364 million years ago at the transition from the Devonian to the Carbonic.
- 251 million years ago at the transition from the Permian to the Triassic 90 when 95% of all existing species were extinct. This included the well-known trilobites.
- 206 million years ago at the transition from the Triassic to the Jurassic. This extinction was associated with the appearance of the dinosaurs.
- 65 million years ago at the transition from the Cretaceous to the Tertiary when 75% of all existing species were extinct. This extinction included the ammonites, but remarkably also the dinosaurs after their geologically short time of 140 million years which were so dominant at their times. This led to the tremendous proliferation of the mammals on Earth.

Extinction waves were due to changes of environmental conditions caused by volcanism and meteorites. They were always followed by very rapid evolution. They created room for the selection of new forms of life such as in the above enumeration the dinosaurs and after extinction of the dinosaurs the mammals, which then included man. The extinction waves were global events.

As a consequence of the exploitation of the biosphere by man (with all the problems discussed next), some argue that we are presently experiencing a sixth extinction wave which is now anthropogenic, self-made by humankind, and therefore substantially different from the previous five waves. Habitats are altered or destroyed so that they become uninhabitable for many species which are dying out at an accelerated rate. Will humankind vanish with it?

As we have seen above, 99% of all species that have occurred on our planet have been subject again to extinction. All species are threatened by dying out. This also applies to man. The evolutionary history of hominides is more than 7 million years long. However, our own genus *Homo* with *Homo rudolfensis* did not appear earlier than 2×10^6 years ago and species of *Homo* lived on Earth at times as follows (see Matyssek and Lüttge 2013):

- *Homo rudolfensis* 1.9 million years back,
- *Homo habilis* 1.9–1.44 million years back,
- *Homo erectus* 1.9–0.5 million years back,
- *Homo heidelbergensis* 0.8–0.3 million years back,
- *Homo neanderthalensis* 0.2–0.03 million years ago,
- *Homo sapiens* 0.2 million years back to present.

All of them except us, *Homo sapiens*, are extinct. Species of *Homo* clearly were not exempt from the destiny of extinction. Our own species is the only survivor to date, more than 60 million years after the disappearance of the dinosaurs on Earth. The question is if humankind is not taking itself along with the current anthropogenic sixth wave of major extinctions while only once again life as such and with other forms is sustained?

Problems

The pressures behind what is possibly a sixth wave of extinction largely derive from a process known as "The Great Acceleration", which takes place since the 1950s and means that Earth System indicators show shifts beyond the range of variability of the Holocene, all driven by human activities. Thus, humanity has transgressed planetary boundaries especially as regards climate change and biosphere integrity (Steffen et al. 2015; see also Scarano 2019, in this book). In this section, we examine some key problems, such as overpopulation, pollution, land-use change, invasive species, sea-level rise and food insecurity.

Overpopulation

Overpopulation of the planet by the dominating mammal *Homo sapiens* is a primary problem. All the other problems are secondary consequences of the activities of life of humankind that has been and continues growing too much. Human population sizes can be listed as follows:

- in the year 1700 6×10^8,
- in the year 2014 7×10^9, of which 1×10^9 are suffering from hunger, in Sub-Saharan Africa 30% of the population,
- extrapolation for the year 2050:
  - at current fecundity and propagation 12.8×10^9,
  - prognosis by the UN in 2014 9.6×10^9, provided that family planning continues and is successful, then stagnation or slight reduction.

How can the phase of stagnation be reached? In a humane way by family planning? Or with dreadful horrifying humane tragedies some of which the globe is already experiencing to a certain extent, i.e. epidemics, hunger, wars and natural catastrophes, where in critical regions natural evil is multiplied by local overpopulation. If stagnation is not reached, would it still be possible to reduce socio-economic inequalities and their consequences such as hunger?

Pollution

Pollutants are deteriorating all major parts of the biosphere's environments: air and the atmosphere, soils and the lithosphere, water and the hydrosphere. Major pollutants are gaseous ones (Matyssek et al. 2013a) because they all are the so-called greenhouse gas molecules contributing to global warming which even a wide public is now recognizing to be one of the foremost menaces:

- Carbon dioxide (CO_2) predominantly comes from fossil sources of energy.
- Methane (CH_4) comes from cattle farming but is also liberated from wet fields of rice; it develops with new energy sources, such as fracking for gas and harvesting marine methane clathrates.
- Nitrogen oxide (N_2O) develops in nitrogen-fertilized crops and is part of one of the vicious cycles of intensive agriculture needed to feed a growing human population of the globe.
- Ozone (O_3) in the troposphere is a secondary pollutant formed in atmospheric chemistry from emissions of traffic (Matyssek et al. 2013a, b), industry and agriculture, besides acting as a greenhouse gas it exerts oxidative stress.

Pollutants of water and soil are heavy metals and all kinds of drugs and chemicals in human waste. Among the major ones increasingly polluting the oceans is plastic not only in its macroscopic form but broken down to the small millimetre-sized building blocks of plastic and found as pellets and also further to microscopic particles (Weisman 2007a, b) which via food chains also accumulate in fish, one of the prominent sources of food of humankind. Agrochemicals include pesticides, fertilizers, plant-growth regulators, repellents and attractants, which are added to the environment to boost agricultural productivity. They are the most common pollutants in 20% of the Earth's land surface (Sanchez-Bayo and Tennekes 2015).

Land-Use Change

The recent special report on Land Degradation of the Intergovernmental Panel on Biodiversity and Ecosystem Services (IPBES 2018) informs that by 2014, more than 1.5 billion hectares of natural ecosystems had been converted to croplands. Although agricultural expansion is the main driver of conversion of natural habitats, cities cover

2% of the planet surface use 75% of its natural resources (Puppim de Oliveira et al. 2011). Less than 25% of the Earth's land surface has escaped substantial impacts of human activity—and by 2050, the IPBES experts estimate this will have fallen to less than 10% (IPBES 2018). This has created 35 biodiversity hot spots in the planet, i.e. areas with high species diversity, high levels of endemisms, but highly threatened (Myers et al. 2000; Zachos and Habel 2011). Such degradation impacts the well-being of at least 3.2 billion people and has immediate links with climate change (IPBES 2018).

Invasive Species

Invasive species develop into destructive pests. Before the times of man's interference migration of organisms across oceans was difficult and very limited. Rare arrivals could evolve, and a characteristic endemic flora and fauna could develop, for example, on young marine volcanic islands such as Galapagos and Hawaii. However, under anthropogenic influence, wittingly for agriculture and husbandry and unwittingly as carried on passengers, plants and animals were and are increasingly rapidly distributed traverse the oceans. Where they arrive, the foreign species can propagate rapidly with numerous adverse effects, for example:

- destruction of native habitats,
- destruction of native flora and fauna by replacement of native species,
- interference by a growing human population as consequence of agriculture and husbandry with pollution, land use and overexploitation (Wilson 2002, 2004, p. 74).

In Tahiti, the tree species *Miconia calvescens* DC. (Melastomataceae), primarily introduced for its ornamental appeal and originally native to tropical areas of South America, is now called "green cancer". It forms dense woods up to 15 m high and covers two-thirds of the entire vegetation of the island. It also threatens the tropical forests of Hawaii (Wilson 2002, 2004, p. 100). Thus, similarly to having talked about "stem species" for renewal and repair (Scarano and Garbin 2013; see also section on *Gaia*, next), we can here also speak of "cancer-species" for destruction.

Sea-Level Rise

By heating up and by dilution of the ocean waters but particularly by melting of glaciers of mountains, of Greenland and of the West Antarctic, the level of the oceans is continuously rising, i.e. currently about 3 mm per year. However, with different global warming scenarios, more dramatic increases are predicted, i.e. 3 m up to 2300 or even 7 m if Greenland and the West Antarctic are melting completely (Robert and

Jouzel 2015; Hansen et al. 2016). This will take away extended areas of land now inhabited by large human populations.

Food Insecurity

Food insecurity results from the primary problem of overpopulation, particularly considering the persistence of inequality and inefficiency in resource distribution. For instance, 30% or more of the food produced every year is lost or wasted (FAO 2011).

Regarding the size of the human population predicted by 2050, we would need to double the productivity of highly developed agriculture. This is one of the major current challenges of humankind (Lüttge 2013). Anthropogenic management of nature in agriculture and forestry also needs to become sustainable to stop loss of arable land (Lüttge 2016). On the one side, there are promising advances of research; on the other side, limitations are building up. One of them is increasing pollution. This causes tremendous reductions of productivity. Considering ozone alone, for example, which in India reduces productivity of wheat by 2–59% and of rice by 7–55% (Pandey et al. 2013), we can see that stopping just this one pollution we could well advance towards reaching the aim of necessary increased food supply. Another limitation is vanishing resources exemplified as follows.

Arable land and soil: At present 1.4×10^9 ha or 17% is irrigated and provides 30–40% of global food production. Losses are advancing due to salinization by 200,000 ha/year, erosion, degradation, sealing and urbanization by 1,000,000 ha/year.

Water: Water is one of the most critical resources for agriculture. It is already an issue of international conflicts, and it is predicted that wars may arise over it. According to the global education project (www.theglobaleducationproject.org/earth/fresh-water.php), the total global water amounts to $1\,400 \times 10^6$ km^3. Most of it is saline, largely in the oceans. Freshwater is 35×10^6 km^3. Much of it is in glaciers, snow and ice. The major reserve for human use including agriculture is fresh ground water with 10.5×10^6 km^3. It is threatened by overexploitation locally. Other sources are comparatively minimal: rain 0.1, lakes 0.09, soil 0.02 and rivers 0.002×10^6 km^3.

Nutrients: The major mineral nutrients for agriculture are nitrogen and phosphorus. Nitrogen is abundant as di-nitrogen, N_2, which makes up 80% of the planet's atmosphere. It can be reduced technically to ammonia to become available as fertilizer in the Haber–Bosch process

$$N_2 + 3H_2 \rightarrow 2NH_3.$$

It has a very high demand of energy, i.e. currently 1.4% of the world's energy consumption, as it needs high pressure (250–350 bar) and temperature (500–600 °C) and uses hydrogen, H_2, which is an important energy carrier. Currently, 95% of the phosphorus mined is for agriculture. It is estimated that the more readily exploitable reserves are depleted in 100–130 years. Exploration of new reserves will have a high energy demand. Efficient techniques of phosphorus recycling from life stock and human faces need to be developed.

Vanishing agricultural yield potential: During agricultural history yield increased considerably. For one grain of wheat, for example in antiquity 3 grains could be harvested, in the middle ages 6 grains and currently 50 grains. At the end of the twentieth century, in the decade 1987–1997, the development of yield still advanced. The global per-area production of wheat and rice increased by 20% and 17%, respectively. This has come to an end. In the following decade 1997–2007 wheat declined by 1% and rice only grew by 2%. The reasons are manifold, but pollution is a major cause as seen by the example of ozone above.

Gaia, Biosphere and the Holobiont

Regarding the problems listed and the pressures on the biosphere, it is relevant to take a brief but closer look at the Gaia hypothesis as originally posed by Lovelock (1979). Actually, viewing the entirety of life on the planet, we can look at three conceptions: Gaia, biosphere and holobiont. Named after Gaia, the goddess of Earth of the ancient Greek mythology, Lovelock (1979) has defined his scientific concept of Gaia as "complex entity involving the Earth's biosphere, atmosphere, oceans and soil". We would rather comprise with biosphere all the life on our planet, i.e. all life harboured by lithosphere, hydrosphere and atmosphere at the surface of the Earth. In this way, Gaia and biosphere can be considered synonyms (Matyssek and Lüttge 2013).

A holobiont is *a host organism (plant or animal) in interaction with all associated micro-organisms as an entity for selection in evolution*. These interactions may be symbiotic or parasitic of organisms closely living together. In this way, man is also a holobiont with carrying a good kilogram of innumerable bacteria with him with 100–150 folds the number of genes than in his own genome. When we extend this definition to organisms in general and if we stress interaction, we arrive at a definition of *organism in interaction with all associated organisms as an entity for selection in evolution*. This opens the application of the concept of holobiont to an enormous range of scaling levels. Habitats, biotopes, ecosystems and biomes are all highly integrated holobionts (Lüttge 2012; Matyssek and Lüttge 2013). For a higher scaling level, this may be illustrated by looking at forests which are entirely integrated by interconnecting the trees with the fungal hyphae of their mycorrhizae in the soil (Mathesius and Watt 2010). Man is not only biologically part of and dependent on Gaia/biosphere, but from his cultural position, he also influences it. With the extension to man's cultural means of interaction and globalization, the entire humankind as an abstraction may be considered as a holobiont.

The entire biosphere is a holobiont, and Gaia is a supra-organism, the ultimate case of a holobiont (Matyssek and Lüttge 2013; zu Castell et al. 2019, in this book). The global extinction waves described above reflect instabilities (Schmidt 2019, in this book) during geological times. Considering life as such, however, these instabilities mean dynamical stability with pseudo-steady states. Life as such remained stable after extensive disturbances of the biosphere. It was maintained, with new evolution leading to emergence of new forms. When the pressure of selection in a wider sense of evolution is not only the fight for resources but the struggle for fitness under varying ambient conditions in general, we can see that with this a system like Gaia as a holobiont is also subject to evolution even when there is no direct akin competitor.

The sustainability of Gaia is of primary relevance for the sustainment of humankind. In his initial approach to the concept of Gaia, Lovelock (1979) was very optimistic considering Gaia as "a short-hand for the hypothesis [……] that the biosphere is a self-regulating entity with the capacity to keep our planet healthy by controlling the chemical and physical environment". Knowing about the massive extinctions, this has been early criticized. However, we need to understand that as keeping the planet healthy for sustaining life as such and not particular specific forms of life. Then, Lovelock's optimistic view of 1979 is appreciated even on the background of the five waves of massive extinction.

Indeed up to date we observe pioneer species of plants establishing themselves on extreme sites, starting successions and creating new habitats for living holobionts. We have called such pioneer species stem species in analogy to the stem cells of the self-assembly processes of development of individual organisms (Lüttge et al. 2013; Scarano and Garbin 2013). They are involved in self-organization and self-management of Gaia. They play the central role in regeneration of devastated land (Scarano et al. 1998; Lüttge et al. 2013; Scarano and Garbin 2013; Lüttge 2016). Is this sufficient, however, for sustaining humankind as part of the Earth, of Gaia?

Tribal and national egoisms appear to prevent the appropriate actions of humankind. Moreover, there are also many ideologies about things that appear desirable, but for sociopolitical action, it needs to be questioned if we can afford them under the threat of the apocalypse of not being able to feed 10 billion of people. Questionable among others are (1) sexual-moral ethics where they limit family planning, (2) resource demanding meat consumption although evolved as an omnivore man cannot completely abandon meat, (3) banning particular forms of energy, (4) growing so-called energy plants, (5) bio-agriculture although ecologically sustained practices of agroforestry management need to be developed, (6) hostility against genetically modified organisms (GMOs), which provide the most promising products of rapid enough breeding for meeting the world food challenge.

Thus, in his later book (Lovelock 2009), 30 years after the first proposal of the concept of self-stabilizing Gaia (Lovelock 1979), James Lovelock abandons his original optimistic view and speaks of "a final warning". His pessimistic prognosis is that if under the influence of humankind at the current pace the self-regulating potency of Gaia declines and is extinguished, the planet may be able to support only a few hundred million people, i.e. a population of the size as around the year 1700 in the culturally so productive and creative renaissance. The selection of these survivors from a much more than one order of magnitude larger population will be an apocalypse.

Idea of Humankind on Earth

Hans Jonas calls the existence of humankind on Earth the categorical imperative (Jonas 2003). He meant this particularly in relation to responsibility and Kantian ethics, which has enormous practical demands and consequences we will refer to later. The idea of humankind on Earth may be a categorical command: a matter of life and death. Therefore, it seems appropriate to discuss briefly at least the self-conceptions of man on Earth. Self-reflections of man are imbedded in mythology, in religions, in philosophy and in the arts with painting, sculpture, literature and music. Sustaining humankind, what is it that could be preserved?

Questioning the prospects of a sustainment of humankind for an extended future, it may be pertinent when for a start we just pick up two philosophical views on the higher advance of humankind and its progress. Both are views from two fundamentally different horizons and philosophical positions arriving at different outlooks and beliefs how humankind could advance. However, both share and exemplify the profound thinking of about how humankind could evolve freeing itself from the bonds of its current existence: (1) "Übermensch" (superman) of Friedrich Nietzsche (philosopher and writer; 1844–1900), and (2) the point omega or "noosphere", jointly elaborated in 1926 by Marie-Joseph Pierre Teilhard de Chardin (anthropological palaeontologist, philosopher and Jesuit theologian; 1881–1955), Vladimir Vernadsky (Ucranian geochemist; 1863–1945) and Edouard Le Roy (French philosopher and mathematician; 1870–1954).

In his philosophy, Nietzsche focuses on the dictum "god is dead" and man has to search himself for the sense of life. In his novel "Also sprach Zarathustra" (Nietzsche 1883–1885, 1891; "Thus spoke Zarathustra") and right at the beginning of it, he introduces the "Übermensch" striving for higher advance of humankind. Superman with exaltation of the will to dominate, the will for power, is unscrupulous and pitiless, and the present man can thus be overcome and renewed.

The biologist Konrad Lorenz thinks that with the evolution of culture, humankind has the potency of unforeseen higher development (Lorenz 1977, p. 304). By great contrast to Nietzsche, Teilhard de Chardin aims at a synthesis of Christian creation theology and science of evolution. He proposed an evolution of humankind to supreme consciousness (Teilhard de Chardin 1956, 1966). From the point of the "primordial atom" of George Henri Lemaître (1894–1966), which led to the big-bang hypothesis of the origin of our universe, Teilhard de Chardin imagines us to evolve in a spiritual unification from the individual egos towards an overego. This is the "point omega" of Teilhard de Chardin, the maximum state of complexity and consciousness. In other words of Teilhard de Chardin, this is an evolution of the biosphere to a noosphere, the sphere of thinking and the domain of consciousness.

It is not possible here to go at any depth, but the two extreme positions on a putative development and longevity of humankind may exemplify the wide scope of mythology, religion and philosophy evaluating the idea of man. Humankind is destructively exploiting the environment of the planet and covers areas of the globe with gruesome devilish wars of self-destruction. Yet man is also wonderfully creative generating divine productions in the arts, literature and music. In his book "A history of the world in 100 objects" Neil MacGregor has illustrated this. He has chosen 100 objects from the British Museum and used them to narrate the entire history of humankind (MacGregor 2010).

Plausible Futures?

Harari (2017) distinguishes three central programmes on the agenda of future humankind: immortality, bliss and divinity. Divinity, upgrading man to god, is somehow related to Nietzsche's superman, discussed above. Next, we will examine the evolution of consciousness and trans-humanism in the realms of bliss and immortality.

Consciousness

There are many drafts of definitions trying to help grasping and understanding consciousness. Consciousness means sense of individuality, personal identity and reflection of the ego on interior and exterior subjects. Consciousness may not have a material basis. It can be considered as an emergent phenomenon of the human brain with its 86×10^6 neurons and their ten to one hundred thousand times more numerous connections (Jaeger 2017, pp. 328–329). Thus, consciousness is a property of highly developed brains. Therefore, we often consider consciousness as a specific attribute of man. Exploring the sustainment of humankind, it appears interesting to ask for the existence of conscious life beyond humankind.

Evolution of consciousness was gradual (Heschl 2009). There are different degrees of consciousness developed by animals with complex brains, such as primates, dolphins and birds. Dolphins and elephants have been observed to show empathy, an attribute of consciousness, in the face of death of congeners (Heschl 2009, p. 225). The presence of man on Earth may have inhibited further evolution of consciousness in such lineages, but prospects of evolution may arise if man perishes (Morris 2003, 2008; Heschl 2009).

Simon Conway Morris relies very heavily on the phenomenon of convergence when he argues that evolution of man-like conscious beings is inevitable and will occur again if humankind disappears (Morris 2003, 2008). Biology is full of examples for convergence where under the pressure of environmental conditions selection has frequently led to independent evolution of particular structures and functions in phylogenetically remote lineages of life. A prominent example among animals is the eye with lens and retina that has evolved leading to an almost identical structure among worms (Alcipodia), mollusks and mammals including man (Morris 2003, 2008). In the plant kingdom convergence of stem succulence is often depicted, the evolution of which occurred independently in several plant families including the Cactaceae and the Euphorbiaceae. It is hard to distinguish stem succulents of the two families if it were not for the flowers and the presence of latex in the Euphorbiaceae, which is lacking in the Cactaceae. Convergence not only occurred in the two phylogenetically far separated families but also on remote continents with the Euphorbiaceae in the old world and the Cactaceae in the Americas. Even cognitive abilities related to culture show convergence. The songs of birds and humpback whales have remarkable similarities to human music, with respect to both the output of songs and the underlying neurology (Noad et al. 2000; Gray et al. 2001; Morris 2003, 2008). These few examples may be sufficient to illustrate the principle of convergence. There are numerous examples, and convergence is rather the rule than exception.

When the sixth extinction takes away higher forms of life, evolution may start again on Earth beginning with bacteria and produce convergence towards formation of complex brains. In the early stages of evolution on our planet 4×10^9 years ago, bacteria have overcome extreme conditions. They were under pressures of selection of a strength and severity far beyond the environmental problems of our days. This included UV radiation and radioactivity. There are still many extant so-called extremophilic bacteria and archaea which can endure extreme conditions of stress particularly of temperature, desiccation, ionizing and UV radiation. Polyextremophilic bacteria species are resistant to several of these stresses. Among them is *Deinococcus radiodurans* Brooks and Murray which can endure extremely high rates of radioactive radiation. It possesses very effective mechanisms of DNA repair and molecular tools for protecting proteins (Anderson et al. 1956; Mattimore and Battista 1996; Battista 1997; Cox and Battista 2005).

Extant extremophilic micro-organisms, bacteria and fungi, also live deep under the surface of the Earth's crust in the so-called SLiMEs, the acronym meaning **s**ubsurface **li**tho-autotrophic **m**icrobial **e**cosystems. The thermophilic organisms can reach depths as low as 4.2 km where temperatures of 110 °C prevail. In the deep biosphere far from the Earth's surface, they have an extremely hard existence in isolation and with scarcity of energy and nutrients. They obtain their energy from inorganic chemical reactions weathering the rocks and survive with very slow rates of reproduction (Frederikson and Onstott 1996; Fyfe 1996; Kerr 1997; Wilson 2002, 2004, pp. 28–29). Where ever there is some fluid and room, such life appears possible. With these organisms a new evolution could possibly start. The observations of SLiMEs on Earth also nourish expectations that life may even exist extraterrestrially on Mars or on Jupiter's moon Europa.

However, the remaining time on Earth is short for a second evolution of conscious life. The lifetime of our solar system is limited. Cosmic processes will lead to a development of the solar system where the solar radiation in about $500–1\,000 \times 10^6$ years will heat up the Earth so much that life becomes impossible. In 5×10^9 years, the sun will then develop to a red giant before it dies and it will definitely destroy Gaia. Our universe is 13.7×10^9 years old. The age of life on Earth is 4×10^9 years. That means if the remaining time for life is 1×10^9 years, 80% of its time have elapsed already and man is with it for 0.05% of the time.

There are billions of planets elsewhere in the universe which could carry life, and the probability that life evolved in space appears to be very large (Pajot and Mayor 2015). In the view of Morris' convergence (Morris 2003, 2008), this must include conscious life. Views like those of Morris that evolution towards highest degrees of complexity with consciousness is inevitable are already expressed by Teilhard de Chardin (1956, 1966, p. 35) with the idea of encephalization laying ground to evolution towards the fullness of the noosphere. With the huge distances and travel times of information involved, it appears just as improbable, however, that we ever learn about the life in remote space as it is probable that it may exist. With remarkable fantasy Gerhard Börner develops visions of the future of humankind in the universe. He argues that the cosmic times of 5×10^9 or at least $0.5–1 \times 10^9$ years available before the Earth is burnt out will be sufficient for an inventive and creative humankind to acquire approaches for escaping and settling elsewhere (Börner 2012). Despite apparent political advances in the dialogue between nations to foster sustainability (see historical description in Scarano 2019, this book), such far-reaching projections along cosmic dimensions of space and time contrast with the fact that the time available for actions is rather short for assuring sustainment of humankind. Allowance of delay is much shorter than the cosmic times of Börner by six to seven orders of magnitude, but this very short time is decisive.

Trans-humanism or Voluntary Human Extinction?

Immortality is a fundamental desire of man, a dream, a promise of religions. Scientifically, medicine is working on man's longevity. The number of centenarians is increasing. Trans-humanism goes far beyond that. It is a doctrine that originated in the 1980s in Los Angeles, USA, and currently develops to a new movement. Longevity of man could be 140 years up to 500 years and even infinitely (Weisman 2007a, b, p. 330; Thivent 2015; Alexandre 2015). For its realization medicine techniques are envisaged to be used. This would include artificial intelligence and implanting nano-robots onto the neurons in the human brain creating technically modified people at the interface of brain/machine (Alexandre 2015). Enormous ethical problems are related to this. Trans-humanism develops ideas for sustainment of the longevity of individuals. Will this, however, not destroy the integrity of humankind (Berger et al. 2015)? Will the dominance of robots really support the quality of individual life and associated social relations of communities? On the contrary, it is even well possible that learning machines and a reign of machines with trans-human intelligence will finish the era of humankind even more effectively than the sixth extinction wave (Jaeger 2017).

The death of death will not sustain humankind as we know it. Death is part of humanism. Life and death are intimately interwoven. The death of individuals and even of entire species is necessary for development and innovation. It is the basis of evolution. Selection is correlated with death. Decease and birth are dialectical contrasts where the synthesis is fresh advance. In death is life—life is in death. Inevitable death, is it sacrifice or reconciliation?

The movement of Voluntary Human Extinction (VHEMT) founded in 1991 by Les U. Knight goes in the opposite direction, i.e. rather for an immediate termination of humankind after the currently living generations than for an infinite extension (Weisman 2007a, b, p. 328; Omrod 2011). This could be achieved by renouncing on reproduction, e.g. by developing a virus which renders all human sperm cells ineffective. The idea of VHEMT is that such Voluntary Human Extinction would avoid terrifying painful mass dying of people and would make the remaining time for the present generations rather pleasant until their end. We find it contra-positive to all evolved innate driving forces of humankind, it is certainly not on agendas such as that noted by Harari (2017), and also it would violate the categorical imperative.

Conclusions

Sustainment of Gaia and with it of the existence of humankind on Earth demands that we develop ethics of responsibility for the future (Jonas 2003). Remarkably the book on the philosophy of the principle of responsibility with an ethics for the technological civilization by Hans Jonas and the book on the scientific concept of Gaia by Jim Lovelock first appeared in the same year 1979 (Jonas 2003; Lovelock 1979). It does not appear that they knew of each other. In 1979, Lovelock still looked at Gaia in a rather optimistic way. Jonas with remarkable clarity as early as 1979 already realized all the problems that have now become grave existential concerns for humankind.

Much of the traditional ethics are built on the categorical imperative expressed by Immanuel Kant in his practical philosophy ("Die Kritik der praktischen Vernunft", 1788) according to which one should always perform in a way that the maxim of one's will could equally serve as the principle for a general jurisdiction.[1] As explained by Jonas (2003), these ethics are addressed with directness towards individual persons at present time and specific location. This chapter describes our actual situation. Due to man's interactions with Gaia, completely new ethical challenges arise. Lovelock (1979) writes that "… if Gaia does exist, then we may find ourselves and all other living things to be parts and partners of a vast being …" (p. 1), but that man assumes he "… is the possessor of this planet; if not the owner …" while in contrast "… The Gaia hypothesis implies that the stable steady state of our planet includes man as a part of, or partner in, a very democratic entity" (p. 145). Jonas (2003) speaks of the imperative of the existence of humankind on Earth and that this is the only imperative which in his view really justifies the Kantian contention of "categorical" (pp. 91–92). The new ethics must not remain restricted to persons, time and location. We need global ethics directed towards groups and populations of people, to nations and the entire humankind including present and future generations, which are our descendants. We are responsible for people who are not living yet, whom we do not know yet and of whom we are not aware of how they shall live. Edward O. Wilson picks up the distinction between short-term and long-term values and that ethically

[1] "Handle so, dass die Maxime deines Willens zugleich als Prinzip einer allgemeinen Gesetzgebung gelten könnte".

we have the difficulty of not only having to look at the presence and the immediate short-term future but that with universal environmental ethics we need to incorporate the long-term future (Wilson 2002, 2004, pp. 65, 160). Some writers argue in a very Gaian vein that such responsibility must be extended for also embracing morally relevant interests of animals and even organisms like plants and fungi (Varner 1998; James 2015). When the existence of man on Earth is the very categorical imperative, this creates categorical responsibility.

The German theologian Hans Küng around 1990 initiated the project "world ethos" (Küng 1990; Küng and Kuschel 2001). It aims for peace by coordinating the common ethical commands expressed in the great world religions with various formulations all eventually coming close to Kant's categorical imperative. A huge meeting, the "world parliament of religions", took place in 1993 in Chicago. Foundations were established in Tübingen, Germany, and in other countries. Sustainment and protection of ecosystems were explicitly articulated with the request of an "ecological world ethos" (von Weizsäcker 2001). The recent re-emergence of millenary beliefs of indigenous peoples in the global south and the incorporation of some of their principles of living well with other people and with nature into national policies is another symptom of this ongoing trend (see Scarano 2019, in this book). Like Kantian ethics determining personal performance, world ethics is asking for political action. There needs to be intergenerational justice, and increasingly intragenerational justice seems a prerequisite for that. Can humankind abandon tribal egoisms inherited in evolution and still so much practiced in nationalisms (Diamond 1991)? Can it be "… that the destiny of humankind is to become tamed, so that fierce, destructive, and greedy forces of tribalism and nationalism are fused into a compulsive urge to belong to the commonwealth of all creatures which constitute Gaia" (Lovelock 1979, p. 148)? Could humankind, under a demand of ethical integrity, give up the exclusiveness of self-interest or will it remain intellectually and ethically overwhelmed (Jaeger 2017)?

References

Alexandre L (2015) Les transhumanistes ont gagné sans livrer bataille. La Recherche 504:89–90
Allwood A, Brown A (2004) Seeking the oldest evidence of life on Earth. Microbiol Aust 25:26–27
Alroy J (2008) Dynamics of organization and extinction in the marine fossil record. Proc Natl Acad Sci USA 105(Suppl. 1):11536–11542
Anderson AW, Nordan HC, Cain RF, Parrish G, Duggan D (1956) Studies on a radio-resistant micrococcus. I. Isolation, morphology, cultural characteristics, and resistance to gamma irradiation. Food Technol 10:575–577
Battista JR (1997) Against all odds: the survival strategies of *Deinococcus radiodurans*. Ann Rev Microbiol 51:203–224
Berger F, Lethimonnier F, Sigaux F (2015) Tuer la mort est un crime contre l'humanité. La Recherche 504:84–86
Bolker JA (2000) Modularity in development and why it matters to evo-devo. Am Zool 4:770–776
Börner G (2012) Die Entwicklung des Kosmos: Vom Urknall zum komplexen Universum. In: Hacker J, Hecker M (eds) Was ist Leben? Nova Acta Leopoldina 116/394:41–68. Halle

Butterfield NJ (2007) Macroevolution and macroecology through deep time. Paleontology 50:41–55
Courtillot V (1995) La vie en catastrophes – Du hasard dans l'évolution des espèces. Arthème Fayard, Pairs
Cox MM, Battista JR (2005) *Deinococcus radiodurans*—the consummate survivor. Nat Rev Microbiol 3:882–892
Diamond J (1991) The rise and fall of the third chimpanzee. Vintage, London
Donoghue PCJ, Antcliffe JB (2010) Early life: origins of multi-cellularity. Nature 466:41–42
El Albani A, Bengtson S, Canfield DE, Bekker A, Macchiarelli R, Mazurier A, Hammarlund E, Boulvais P, Dupuy J-J, Fontaine C, Fuersich FT, Gauthier-Lafaye F, Janvier P, Javaux E, Ossa F, Pierson-Wickmann A-C, Riboulleau A, Sardini P, Vachard D, Whitehouse M, Meunier A (2010) Large colonial organisms with coordinated growth in oxygenated environments 2.1 Gyr ago. Nature 466:100–104
FAO (2011) Global food losses and food waste—extent, causes and prevention. Food and Agriculture Organization, Rome
Frederikson JK, Onstott TC (1996) Leben im Tiefengestein. Spektrum der Wissenschaft 12(1996):66–71
Fyfe WS (1996) The biosphere is going deep. Science 273:448
Gould SJ (1989) Wonderful life: the Burgess shale and the nature of history. W. W. Norton, New York
Gray PM, Krause B, Atema J, Payne R, Krumhansl C, Baptista L (2001) The music of nature and the nature of music. Science 291:52–54
Grazhdankin D (2014) Patterns of evolution of the Ediacaran soft-bodied biota. J Palaeontol 88:269–283
Hansen J, Sato M, Hearty P, Ruedy R, Kelley M, Masson-Delmotte V, Russell G, Tselioudis G, Cao J, Rignot E, Velicogna I, Tormey B, Donovan B, Kandiano E, von Schuckmann K, Kharecha P, Legrande AN, Bauer M, Lo K-W (2016) Ice melt, sea level rise and superstorms: evidence from paleoclimate data, climate modeling, and modern observations that 2° global warming could be dangerous. Atmos Chem Phys 16:3761–3812
Harari YN (2017) Homo Deus. A brief history of tomorrow. Vintage, London
Heschl A (2009) Darwins Traum. Die Entstehung des menschlichen Bewusstseins. Wiley-VCH, Weinheim
IPBES (2018) Summary for policymakers of the thematic assessment report on land degradation and restoration of the intergovernmental science-policy platform on biodiversity and ecosystem services. In: Scholes R, Montanarella L, Brainich A, Barger N, ten Brink B, Cantele M, Erasmus B, Fisher J, Gardner T, Holland TG, Kohler F, Kotiaho JS, von Maltitz G, Nangendo G, Pandit R, Parrotta J, Potts MD, Prince S, Sankaran M, Willemen L (eds) IPBES Secretariat, Bonn, Germany
Jaeger L (2017) Wissenschaft und Spiritualität. Universum, Leben, Geist –Zwei Wege zu den großen Geheimnissen. Springer, Berlin, Heidelberg
James SP (2015) Environmental philosophy. An introduction. Polity Press, Cambridge
Jonas H (2003) Das Prinzip Verantwortung. Versuch einer Ethik für die technologische Zivilisation. Suhrkamp Taschenbuch, Frankfurt a. M.
Kerr RA (1997) Geomicrobiology: life goes to extremes in the deep Earth—and elsewhere? Science 276:703–704
Küng H (1990) Projekt Weltethos. Piper Verlag, München
Küng H, Kuschel K-J (2001) (eds) Wissenschaft und Weltethos. Piper-Verlag, München
Lorenz K (1977) Die Rückseite des Spiegels. Versuch einer Naturgeschichte menschlichen Erkennens. Deutscher Taschenbuch Verlag, München
Lovelock JE (1979) Gaia. A new look at life on Earth. Oxford University Press, Oxford
Lovelock JE (2009) The vanishing face of Gaia—a final warning. Basic Books, New York
Lüttge U (2012) Modularity and emergence: biology's challenge in understanding life. Plant Biol 14:865–871

Lüttge U (2013) The planet Earth: can it feed nine billion people? In: Matyssek R, Lüttge U, Rennenberg H (eds) The alternatives growth and defense: resource allocation at multiple scales in plants. Nova Acta Leopoldina 114/391:345–364
Lüttge U (2016) Plants shape the terrestrial environment on Earth: challenges of management for sustainability. Prog Botany 77:187–217
Lüttge U, Garbin ML, Scarano FR (2013) Evo-Devo-Eco and ecological stem species: potential repair systems in the planetary biosphere crisis. Prog Botany 74:191–212
MacGregor N (2010) A history of the world in 100 objects. Penguin Books
Mathesius U, Watt M (2010) Rhizosphere signals for plant-microbe interactions: implications for field-grown plants. Prog Botany 72:125–161
Mattimore V, Battista JR (1996) Radioresistance of *Deinococcus radiodurans*: functions necessary to survive ionizing radiation are also necessary to survive prolonged desiccation. J Bacteriol 178:633–637
Matyssek R, Lüttge U (2013) Gaia: the planet holobiont. Nova Acta Leopoldina NF 114(391):325–344
Matyssek R, Clarke N, Cudlin P, Mikkelsen TN, Tuovinen J-P, Wieser G, Paoletti E (eds) (2013a) Climate change, air pollution and global challenges. Developments in environmental sciences, vol 13. Elsevier, Amsterdam
Matyssek R, Wieser G, Fleischmann F, Grünhage L (2013b) Ozone research, *quo vadis*? Lessons from the free-air canopy fumigation experiment at Kranzberg Forest, pp 103–129. In: Matyssek R, Clarke N, Cudlin P, Mikkelsen TN, Tuovinen J-P, Wieser G, Paoletti E (eds) Climate change, air pollution and global challenges. Developments in environmental sciences, vol 13. Elsevier, Amsterdam
Melott AL, Bambach RK (2011) A ubiquitous ≈62-Myr periodic fluctuation superimposed on general trends in fossil biodiversity I. Documentation. Palaeobiology 7:92–112
Morris SC (2003) Life's solution. Inevitable humans in a lonely universe. Cambridge University Press, New York
Morris SC (2008) Jenseits des Zufalls. Wir Menschen im einsamen Universum. Berlin University Press, Berlin
Myers N, Mittermeier RA, Mittermeier CG, da Fonseca GAB, Kent J (2000) Biodiversity hotspots for conservation priorities. Nature 403:853–858
Nachtigall W (2010) Bionik als Wissenschaft. Erkennen – Abstrahieren – Umsetzen. Springer, Heidelberg
Niemann H-J (2013) Karl Popper, die Mühle bei Hunstanton und die Anfänge der Molekularbiolgie. Bemerkungen zu Karl Poppers Medawar-Vorlesung, Teil II. Aufklärung und Kritik 2:7–34
Nietzsche F (1883–1885, 1891) Also sprach Zarathustra. Chemnitz & Leipzig
Noad MJ, Cato DH, Bryden MM, Jenner M-N, Jenner KCS (2000) Cultural revolution in whale songs. Nature 408:537. https://doi.org/10.1038/35046199
Omrod JS (2011) Making room for the tigers and the polar bears: biography, phantasy and ideology in the voluntary human extinction movement. Psychoanal Cult Soc 16:142–161
Pajot P, Mayor M (2015) La quête d'exoplanètes est devenue un pan majeur de l'astrophysique. La Recherche 504:34–37
Pandey V, Oksanen E, Singh N, Sharma C (2013) Impacts of air pollution and climate change on plants: implications for India. In: Matyssek R, Clarke N, Cudlin P, Mikkelsen TN, Tuovinen J-P, Wieser G, Paoletti E (eds) Climate change, air pollution and global challenges. Developments in environmental sciences, vol 13. Elsevier, Amsterdam, pp 391–409
Peterson KJ, Waggoner B, Hagadorn JW (2003) A fungal analog for newfoundland ediacaran fossils? Integr Comp Biol 43:127–136
Popper KR (2013) Eine Neuinterpretation des Darwinismus. Die erste Medawar-Vorlesung 1986. Aufklärung und Kritrik 1:7–20
Puppim de Oliveira JÁ, Balaban O, Doll CNH, Moreno-Peñaranda R, Gasparatos A, Iossifova D, Suwa A (2011) Cities and biodiversity: perspectives and governance challenges for implementing the convention on biological diversity (CBD) at the city level. Biol Cons 144:1302–1313

Retallack GJ (1994) Where the Ediacaran fossils lichens? Paleobiology 20:523–544
Robert B, Jouzel J (2015) Je crois en un sursaut de ce monde face au problème climatique. La Recherche 506:5–8
Sánchez-Bayo F, Tennekes HA (2015) Environmental risk assessment of agrochemicals—a critical appraisal of current approaches. In: Larramendy ML, Soloneski S (eds) Toxicity and hazard of agrochemicals, Intech open science. https://doi.org/10.5772/59450
Scarano FR (2019) The emergence of sustainability. In: Wegner LH, Lüttge U (eds) Emergence and modularity in life science. Springer, Cham, pp 51–74
Scarano FR, Garbin ML (2013) Stem species: plant species that function as regenerating cells of Gaia. Nova Acta Leopoldina NF 114(391):317–324
Scarano FR, Rios RI, Esteves FA (1998) Tree species richness, diversity and flooding regime: case studies of recuperation after anthropic impact in Brazilian flood-prone forests. Int J Ecol Environ Sci 24:223–225
Schmidt JC (2019) Is there anything new under the sun? Instability as the core of emergence. In: Wegner LH, Lüttge U (eds) Emergence and modularity in life science. Springer, Cham, pp 3–36
Seilacher A, Grazhdankin D, Legouta A (2003) Ediacaran biota: the dawn of animal life in the shadow of giant protists. Paleontol Res 27:43–54
Souza GM, Lüttge U (2015) Stability as a phenomenon emergent from plasticity—complexity—diversity in eco-physiology. Prog Botany 76:211–239
Speck T, Speck O (2019) Emergence in biomimetic materials systems. In: Wegner LH, Lüttge U (eds) Emergence and modularity in life science. Springer, Cham, pp 97–115
Steffen W, Richardson K, Röckstrom J et al (2015) Planetary boundaries: guiding human development on a changing planet. Science 347 https://doi.org/10.1126/science.1259855
Teilhard de Chardin P (1956) Le groupe zoologique humain. Éditions Albin Michel, Paris
Teilhard de Chardin P (1966) Die Entstehung des Menschen, 4th ed. C. H. Beck, München
Thivent V (2015) La nouvelle condition humaine. La Recherche 504:80–82
Varner GE (1998) In nature's interests? Interests, animal rights, and environmental ethics. Oxford University Press, New York
Weisman A (2007a) The world without us. Thomas Dunne Books, St. Martin's Press, New York
Weisman A (2007b) Die Welt ohne uns. Piper, München–Zürich
von Weizsäcker EU (2001) Ökologisches Weltethos. In: Küng H, Kuschel K-J (eds) Wissenschaft und Weltethos. Piper-Verlag, München, pp 337–355
Wilson EO (2002) The future of life. Alfred A. Knopf, New York
Wilson EO (2004) Die Zukunft des Lebens. Wilhelm Goldmann Verlag, München
Zachos FE, Habel JC (eds) (2011) Biodiversity hotspots. Springer, Heidelberg
zu Castell W, Lüttge U, Matyssek R (2019) Gaia—a holobiont like system emerging from interaction. In: Wegner LH, Lüttge U (eds) Emergence and modularity in life science. Springer, Cham, pp 255–279

Gaia—A Holobiont-like System Emerging From Interaction

Wolfgang zu Castell, Ulrich Lüttge and Rainer Matyssek

> *"I will sing of well-founded Earth, mother of all, eldest of all beings. She feeds all creatures that are in the world, all that go upon the goodly land, and all that are in the paths of the seas, and all that fly: all these are fed of her store. Through you, O queen, men are blessed in their children and blessed in their harvests, and to you it belongs to give means of life to mortal men and to take it away."*
>
> (Homeric Hymns, see Anonymous 1914).

Abstract The Gaia hypothesis has been heavily discussed ever since its first formulation in the early 1970s. While parts of the hypothesis can be accepted right away, the optimizing view of Gaia as stabilizing conditions on Earth to support life on the planet has raised several doubts. In particular, the evolutionary aspect of conditions evolving in such an optimizing way is hardly in line with classical evolutionary theory. Considering Gaia as a holobiont-like system allows to clarify these issues. Accepting that every form of life interacts in manifold ways with its biotic and abiotic environments and that these interactions form a multilevel network from which higher-level properties can emerge, the self-stabilizing effect of the interactions of the biosphere with the geosphere and the atmosphere can be explained. With viewing evolution as adaptation of interaction and fitness as a shift in probability distribution of the observable, even the optimizing interpretation of Gaia does make sense and leads to valuable insight.

W. zu Castell (✉)
Scientific Computing Research Unit, Helmholtz Zentrum München, Ingolstädter Landstraße 1, Neuherberg, Germany85764
e-mail: castell@helmholtz-muenchen.de

W. zu Castell
Department of Mathematics, Technische Universität München, Munich, Germany

U. Lüttge
Department of Biology, Technical University of Darmstadt, Darmstadt, Germany

R. Matyssek
Ecophysiology of Plants, Center of Life and Food Sciences Weihenstephan, Technische Universität München, Freising, Germany

L. H. Wegner and U. Lüttge (eds.), *Emergence and Modularity in Life Sciences*,
https://doi.org/10.1007/978-3-030-06128-9_12

Nourishing Mother as a Provocative Scientific Metaphor

Gaia (Γαῖα), mother Earth has ever inspired mankind. Gaia feeds all creatures on land and in the sea, providing the basis of life for all living creatures on the planet. The metaphor of the mother comprises our interpretation of Gaia providing conditions for life on Earth in a supporting and comforting manner. As such, Lovelock and Margulis (1974) strikingly used the Greek notion of Gaia to coin a name for their hypothesis, known today as the *Gaia hypothesis* (Lovelock 1972). In its simplest form, the hypothesis states that the interactions of the biosphere with the atmosphere and the geosphere constitute a self-organizing system which optimizes conditions on Earth to support life on the planet. Taken in this form, the hypothesis comprises several concepts to be made explicit to serve a sound scientific discussion. The notion of the **biosphere**, as a first issue, relies on a suitable definition of life (see Lovelock 1988). Since it includes all forms of life on Earth, we need to focus on generic properties of life rather than explicit forms as being exemplified by specific organisms. Second, the hypothesis makes a statement about the **interactions** of the biosphere with other 'spheres'. Thus, we need to consider the set of interactions and focus on the potential of the biosphere to interact rather than any other property one might be interested in. Third, Gaia addresses the self-organizational nature of the system. We, therefore, need to consider **self-organization** contrasting any other, extrinsically controlled form of organization. Finally, the hypothesis characterizes the set of interactions as optimizing toward certain conditions. Clearly, speaking about **optimization**, we need to clarify in what sense optimization is to be understood. Before doing so, let us first recapture the history of the hypothesis and comment on some recent developments.

Soon after Lovelock's book (Lovelock 1979) was published, an intense debate started within the scientific community, splitting into those who consider Gaia as being an inspiring concept to think about Earth as basis for life, and others, quickly opposing by questioning the scientific nature of the hypothesis as being untestable, imprecise and not scientific in its core (see Doolittle 1981 for a summary). Kirchner (2002) opened-up grounds for an informed debate on the hypothesis by working out various variants showing-up in the discussion.

Undoubtedly, Earth's environment has been shaped by biotic interactions (Kasting and Siefert 2002), as well as it itself shapes conditions for all forms of life on Earth (see e.g., Beerling and Berner 2005; Berner et al. 2007) (see Fig. 1). In this sense, Earth as environment of life has coevolved with all forms of life, and both cannot be thought of without each other (*coevolutionary hypothesis*) (Kirchner 2002). The question remains to what extent this coevolutionary interaction can be understood as an optimization.

Matyssek and Lüttge (2013) provided several arguments for considering Gaia as an entity of interacting organisms being subject to selection in evolution. Their approach is based on the notion of a *holobiont*, which in its original form has been defined as "*the host organism and all of its associated symbiotic microorganisms*" (Rosenberg et al. 2007, p. 356). For illustrating this, Matyssek and Lüttge (2013) go through examples at increasing levels of scaling or spatial dimensions of biological

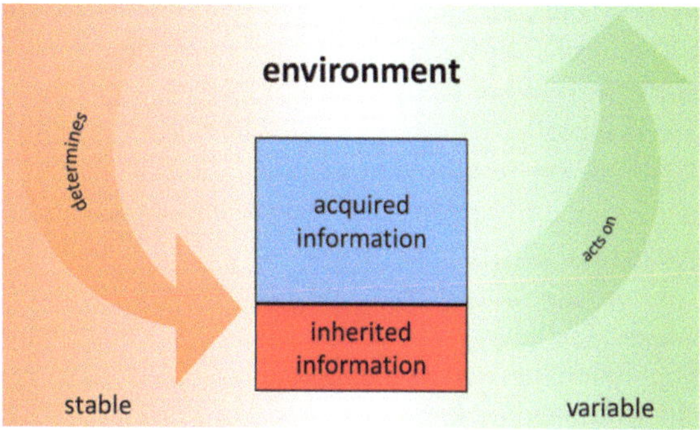

Fig. 1 The internal state of an agent reflects its environment. As there are (locally) stable factors in the environment and variable ones, the state consists of stable parts (inherited information) and dynamic parts (acquired information)

organization with the emergence of networks through integration of modules. This ladder starts with symbioses on cellular levels, including eukaryotic cells of animals and plants, where the organelles mitochondria and chloroplasts in the theory of endosymbiosis are considered as symbionts. It moves on to whole multicellular organisms, including man carrying a good kilogram of innumerable bacteria with him with 100–150-fold the number of genes of his own genome. It then arrives at habitats, biotopes, ecosystems, biomes, and eventually at the biosphere/Gaia as a supraorganism or the ultimate holobiont.

The central argument of the Gaia hypothesis can then be comprised in considering the biosphere as a dynamic system which self-organizes through feedback mechanisms keeping the system within a (pseudo-)stable state (Kirchner 2002; Lenton 2004). These feedback loops stabilize the system and thus provide the basis for the claim that at least within certain limits, the system maintains stability in a self-organizing manner and as such 'optimizes' conditions for life as such. The concept of the holobiont introduces the idea of evolution being the driving force of the observed 'optimization'.

Tyrrell (2013) objects to this point of view by pointing toward an apparent asymmetry in the argumentation. Without any doubt, the agents of the biosphere are subjected to natural selection, while neither atmosphere nor geosphere seem to adapt in an evolutionary sense. Taking a closer look, the difficulty in recognizing evolutionary forces acting on the abiotic components of Gaia is rooted in the gene-centric perspective of neo-Darwinism. In order to see evolutionary forces acting here, we need to broaden our view, looking for units of selection on all scales of biological organization. The idea of the holobiont already reflects the concept of multilevel selection (Theis et al. 2016). In this wider sense, geological formations retain historical information of the geosphere in the same way as, e.g., ice shelves preserve the history of past atmospheric events. Thus, abstracting from the gene as carrier of information, both, geosphere and atmosphere in their present form result from a developmental process and still provide informational content documenting their historic contingency. Note that it is not the presence of a specific mechanism of deciphering information which characterizes evolutionary development, but capability of utilizing stored information toward gaining 'knowledge' about the environment. This is indeed the case for geological formations being 'utilized' by mechanisms of Gaian self-control. For example, consider amplification of rock weathering (Schwartzman and Volk 1989) or the impact of arctic sea ice on stabilizing the global environment (Wadhams 2016). In its core, the theory of evolution describes the interacting forces of generation of information through random variation (mutation) and transmission of information (heredity) and their relation with respect to supporting some persistent organizational structure (selection).

Originally being introduced as the central unit of organization for the *hologenome theory of evolution* (Zilber-Rosenberg and Rosenberg 2008), the concept of the holobiont inherently carries a temporal and spatial dimension. Indeed, Zilber-Rosenberg and Rosenberg acknowledged the network of coenobiotic relations every animal and plant establishes, as well as the fact that associations with microorganisms can be transmitted to the next generation of the host organism. Nevertheless, through accentuating the role of a central host, the concept of the holobiont fixes the spatial dimension at the level of an organism. Likewise, temporal stability of the network of associations is located at the scale derived from generations in terms of the host. Note that these might comprise several 'generations' of microorganisms interacting with the same host individual.

On the other side, the concept of the holobiont emphasizes two important observations concerning living systems: (1) It is the network of interactions which is subject to natural selection rather than the individual organism on its own. Living systems exhibit themselves through their interaction with their environment (see Fig. 1) and, in particular, associated living systems. (2) The network of interactions as such comprises a level of information which can persist beyond the life span of a single agent participating in the network. Therefore, the network contributes to survival, i.e., stability of a level of organization beyond what can be expected from pure randomness.

The properties of systems leading to an observable characteristic pattern in both, spatial and temporal terms, has lead zu Castell et al. (2016) to generalize the concept of the holobiont, introducing a *holobiont-like system*. The central observation laying the foundation for the latter concept is the observation that life in its various forms on all scales of biological organization essentially deals with information, its generation, accumulation and transmission. A holobiont-like system is a conceptual model acknowledging these properties. It has been defined as a system of interacting biological agents developing through adaptation, the latter being a process generated by generic mutation, generic heredity and generic selection (cf. zu Castell et al. 2016). While utilizing notions from the classical Darwinian theory of evolution, the generalizing addition of the word 'generic' indicates that mutation, heredity and selection are understood in a broader sense, beyond their rather narrow definition derived from genetics. The concept of a holobiont-like system thus shifts the focus away from the *carrier of information*, as materialized in the structural bases of genes and other components, to the *process of information transmission*. As transformation of information can be termed *computation* (cf. Gershenson 2012), the shift therefore turns our focus away from the physical/chemical realization of an informational code in terms of DNA toward the computational processing of the code.

The second, important aspect being captured in the definition of a holobiont-like system is a definite choice of the type of algorithm being considered for developing the system of interacting agents. Adaptation describes a change in the informational representation of an entity in response to information being sensed from the environment of this entity. Gershenson (2012) calls such an entity an *agent*. The *environment* of an agent then consists of all the information an agent interacts with (Gershenson 2012).

Evolutionary Motivation for Holobiont-like Systems

Let us now take a closer look at evolution, i.e., the computational process driving adaptation. As mentioned, an organism's DNA is one way of coding information about the organism as a whole, its structure, its characteristics and its potential. Organisms interact with other organisms. At the level of habitats, in the sense of *biocoenosis*, this describes the network of interacting organisms within a certain habitat. Thus, organisms are agents, sensing their environment and reacting upon the sensed information. Thereby they alter their local environment. At large planetary scale this occurs to the extent that life has globally shaped the entire image of Gaia (Lüttge 2016). Life characterizes the large zonobiomes sensu Walter and Breckle (1984) from the equator to the poles of the planet with (1) the equatorial zonobiome with evergreen wet tropical forests, (2) the tropical zonobiome with summer rain and various types of tropical forests and savannas, with evergreen to deciduous vegetation, (3) the subtropical arid zonobiome with desert climate, (4) the mediterranian zonobiome with winter rain and evergreen laurel-like hard-leaf and deciduous forests, (5) the warm temperate zonobiome with temperate rainforests, (6) the temperate nemoral zonobiome (from latin *nemus* = grove) with summer-green deciduous broad-leaf forests, (7) the arid temperate continental zonobiome with steppes and prairies, i.e., grasslands subject to frost, (8) the cold-temperate boreal zonobiome (from latin *borealis* = northern) with the taiga of evergreen coniferous forests, and (9) the arctic and ant-arctic zonobiome with the tundra dominated by evergreen and deciduous shrub species. At smaller scales life actively changes the environment by niche construction (see below).

Taken together, a biocoenosis and its physical environment form an *ecosystem* in the sense of Tansley (1935). Being identifiable as a system, such a network of interactions carries characteristics which appear to persist longer than what we as an observer would expect from pure random choice. At the end, this is what makes us recognize a certain set of interactions as a system.

All the intrinsic properties being attributed to the set of interactions forming an ecosystem, such as waves of succession, trophic structure, or cycles of resources form an emerging characteristic, which we—again as being an observer—term an ecosystem. Organisms, being the central agents within the system, evolve from, adapt to and shape this ecosystem such that both are heavily affected by each other in an intertwined way (see Fig. 1). At the same time, an individual's life span within this system typically is much shorter than the period during which we can observe the level of organization characterizing the ecosystem. On the other side, the genetic information coded in DNA and as such, the blueprint for the interaction potential an individual is carrying, is bound to the one living representative. Thus, information has to be transmitted over time, being handed over to other individuals. The sum of these processes is what we call **heredity**.

The next observation which is crucial for the concept of a holobiont-like system is the fact that activity of an individual leaves an impact on the environment the individual is acting in. In this way, the network of interacting agents shapes its environment, leaving traces, which alter environmental conditions for the time to follow. These dynamic changes in the environment challenge all agents acting within it, being one factor that promotes adaptation in maintaining the agents' intrinsic level of organization. From an agent's point of view, these environmental changes are unforeseen as they integrate actions of all agents in the local environment. On top of that, there is a second source of change resulting from random fluctuations in the parameters of the environment. Summarizing, in order to persist, an agent has to deal with unforeseen changes in its environment and maintain its own stability. This is what we call *acclimation* when it is based on short-term plastic performance but **adaptation** when it is fixed long term through evolution.

The process of natural evolution uses random fluctuations as a source of creativity (in the sense of maintaining variability), from which ever new potential for reacting to this changing environment can evolve. The random seed being brought into the system results from incomplete transmission of genetic information. In combination with other sources of variation such as recombination, stability of interaction patterns being conserved by the genetic code (heredity) is paired with a source of change (**mutation**), creating variation in the gene pool of a biocoenosis. Both processes together provide a basis for adaptability. In order for these processes to work together, random changes in the informational representation of an agent (e.g., mutations) must be transmissible. Otherwise, they would just being wiped out at the end of life of the individual, leaving no persistent trace.

Darwin's theory of evolution now implies the presence of a third process, which acts above the level of the individual on the population as a whole. As such, **selection** is considered to be an autonomous process, i.e., not being directly driven by the agents in the system. The role of selection is to enhance subsets of genetic code while sorting out other sets and thus reducing variation. In a certain sense, mutation and heredity can be understood as active processes, being realized by the action of agents in their environment. In contrast, selection emerges in an indirect way. Selection as a process can be observed over time, although no agent can be identified which on its own selects in the sense of holding control over the system. We will dwell deeper on its characteristics in the following. For the moment being, let us assume that such an autonomous, variation reducing process exists.

Interestingly, mutation and heredity on the one side and selection on the other side do not act in parallel. Evolution proceeds through phases of development followed by phases of reproduction. Generations mark each such step, discretizing the continuous flow of time. Another view of the same alternation is that of phases of information transmission being followed by information encoding. The 'try-and-error' strategy of evolution might have a deeper reason we will come back to later on. Let us first return to our rather general point of viewing individual organisms as one form of representation of structured information.

Agents as being carriers of a certain informational representation, as well as the processes of mutation and heredity perfectly make sense when replacing the genome by any other means of coding information. **Generic mutation** thus means any form of randomly altering the informational state of an agent which is capable of persisting over longer periods of time. The boundary between alterations which prevent us from recognizing the agent as an entity and those which we consider as maintaining its essential characteristics is fuzzy. It reflects the role of the observer as being immanent in every definition of a system (Gershenson and Heylighen 2005). In a similar sense, **generic heredity** is understood as a process allowing for transmission of information such that emerging patterns, resulting from interactions of agents, can be observed longer than the individual life span of a single agent within the system. Adding **generic selection** in an analogous sense completes the triple generating adaptation of the holobiont-like system.

The advantage of using an abstract concept rather than a concrete biological representation, i.e., a given instantiation of these processes within the living world, lies in the possibility to transfer the concept to any level of biological organization. As such the concept of a holobiont-like system matches Bedau's *theory of supple adaptation* (Bedau 1998) which provides the basis for his perspective on a fundamental theory for life. Bedau postulates that the primary form of life is "*an automatic and continually creative evolutionary process of adapting to changing environments*" (Bedau 1998, p. 3). What distinguishes life from non-life "*is the way in which evolution automatically fashions and refashions appropriate strategies for coping as local contexts change*" (Bedau 1998, p. 3). In the same spirit, the concept of a holobiont-like system emphasizes the evolutionary process driving the system to continually adapt to an ever-changing environment.

Human experience clearly knows of several mechanisms which can be used to realize an adaptive process. Obviously, *learning from experience* provides one example. Nevertheless, our anthropocentric way of perception tends to be biased toward assuming teleological principles underlying suitable mechanisms of adaptation. For example, our common concept of learning assumes an intrinsic goal such that learning can be understood as a process of optimizing behavior toward achieving this goal. Scientifically, in epistemology, we clearly do not permit ourselves to assume the presence of any teleological principle underlying organization of nature. Although this might sound self-evident, it is a delicate issue when speaking about evolution in natural systems. Bedau, for example, purposely attributes a teleological notion of propriety as being part of supple adaptation (Bedau 1996). In the same way, Darwin's principle of 'survival of the fittest' carries an implicit goal of being 'fit'. In our understanding, such a teleological glimpse again reflects the role of the observer. On the long run, among the set of all possible networks of interactions, some managed to persist longer than what would be expected from random interaction. Observing these patterns, we tend to imply directedness in the evolution of such patterns, allowing for rational explanation. This *break of symmetry* as physicists call such phenomena (Hopfield 1994) can be understood by dismantling the steps leading to such a pattern and arranging them in a logically appealing order. As an ex post way of explaining the presence of breaks of symmetry, such a logically ordered chain of events is perfectly reasonable. On the other side, the existence of such a chain of events does not imply the influence of any directedness a priori during the evolution of the pattern. Therefore, natural selection is a statistical phenomenon capturing our ex post analysis of a set of natural patterns being observed. As a consequence, we cannot predict ex ante how selection will act in the future. We thus need to restrict ourselves to focusing on mechanisms providing the basis for future adaptability without relying on a concise model of the future environment.

The statistical nature of natural selection is well reflected in the commonly used way to measure selection in terms of an increase in allele frequencies within a given population. Doolittle (2014) is laying out the idea that increases in allele frequency can likewise result from an adoption of survival-enhancing, extinction-avoiding mutations. Obviously, mutants which are more likely to survive will last longer and therefore be more likely to be observed within a population. Second, acquiring survival-enhancing capabilities will, on the long run, effectively lead to more persistent individuals. At the end, differential selection of non-competing and non-reproducing individuals will also result in increasing allele frequencies (cf. Doolittle 2014). Thus, our commonly used practice to think of natural selection in terms of reproductive success should not blur the core of Darwin's theory explaining the observed likelihood of a certain stable configuration which results from the process of selection. In this sense, Darwin's theory of natural selection is indeed a special case of the more general principle of "*survival of the stable*" (cf. Dawkins 1976, p. 12). Indeed, the mere formation of stable patterns of atoms and molecules and their persistence beyond the duration of unstable configurations can be seen as the first steps of natural selection (cf. Dawkins 1976).

The question arises, whether such a form of persistence is the outcome of some lucky chain of events, only, or indeed more likely of being observed. Ulanowicz (1998, 2014) builds his theory of *autocatalysis* on the principle of self-enhancing reactions as given in many chemical reactions. Such autocatalytic configurations act in a growth enhancing way and thereby exhibit a tendency of *centripetality* since over time more resources will be brought into the autocatalytic cycle. Similarly, survival-enhancing adaptations will more likely be found in evolutionary lineages being observed later in their development compared to such being observed at an earlier state (Doolittle 2014). Therefore, the mere existence of a stable configuration over long periods of time already provides an indication that there are underlying mechanisms enhancing stability rather than observing a phenomenon purely resulting from chance.

Self-organization as Mechanism Driving Adaptation

The process of generating complex patterns from simple interactions with each of them not being able of capturing the pattern as a whole, is what is known as *self-organization*. Our way of understanding self-organization rests on two pillars. First, studying processes of self-organization in nature allows identifying common principles and underlying mechanisms. In particular, the study of self-organization in eusocial colonies with group members of different qualifications and specializations for social contributions leads to the unraveling of general principles resulting in emergent organizational patterns on the level of the group. Among those are *decentralization*, *stigmergy*,[1] *positive* and *negative feedback*, and *fluctuations* (cf. Garnier et al. 2007). The realization of such principles allows insect colonies to solve the problem of adaptation of the group to environmental changes. Interestingly, none of these general principles rests on altering the genetic state of the individual agent rather than utilizing modulation of an individual's behavior (e.g., via the hormone system). Thus, the risk of lethal consequences from random alterations of an individual's fundamental (genetic) code is being avoided by providing actions which can be fine-tuned upon sensing environmental signals. While this is not exclusively true (e.g., consider somatic hypermutation, e.g., Martin et al. 2015), this observation emphasizes the need for an adaptive system to find ways for transmission of such environmental modulation. Otherwise, adaptations upon environmental signals would not persist, since they cannot be passed on to the next generation. A growing body of knowledge sheds light on epigenetic mechanisms, which solve the problem even on the level of the genome (see, e.g., Allis and Jenuwein 2016) because by methylation/acetylation of nucleic acids and histone proteins they determine and thus modify the reading of the information inherent in a given structure (sequence of nucleotides) of the DNA (Grunstein 1997; Zhang and Reinberg 2001; Chinnusami and Zhu 2009; Yaish et al. 2011).

Consequently, in order to systematically understand the principles of self-organization, it is not sufficient to analyze the potential of action of all individual agents in the system (i.e., the genome). The emergent phenomena we observe on the level of the system of agents rather directly result from the interactions themselves and therefore are a matter of communication and information transfer among the members of the group (e.g., the *interactome*, see Aitchison and Rout (2015) on the cellular level).

The second pillar being used to study self-organization can be seen as the bottom-up approach complementing the study of self-organizing biological systems. Through actively modeling swarm intelligence in artificial systems, we have a tool at hand which allows analyzing the concepts underlying self-organization in a constructive way. Most importantly, in silico simulations allow us to experimentally test our hypotheses on interaction principles leading to emergent patterns of interaction. For example, in the first half of the last century, ornithologists still were look-

[1] *Stigmergy* denotes the principle of "*indirect communication mediated by modifications of the environment*" (Marsh and Onof 2008, p. 137).

ing for the 'common soul' of a flock of birds, suggesting telepathy or yet unobserved forms of electromagnetic communication (Selous 1931). Detailed video analysis then led toward emergent properties such as self-generated synchrony (Davis 1980). Through computer simulations, Reynolds (1987) finally showed that flocking emerges from the local processes of cohesion, separation and alignment, all of which being explained on the level of a single bird.

Both pillars meet in the recent research field of *artifical life* (Gershenson et al. 2018). For our purpose here, it suffices to note that the concept of a holobiont-like system extracts the core of information processing from natural systems such as insect colonies.

Considering the above-mentioned principles underlying self-organization in swarm intelligence, the holobiont-like system provides a suitable conceptual framework. For example, decentralization matches the motivation for moving from the holobiont sensu stricto toward a holobiont-like system. There is no central host organism, providing 'control' on an overall level of organization for the system. Stigmergy inherently is a concept of information transfer. Adapting to actions of agents within the local environment reflects an individual's way to capture information using a try-and-error strategy and thereby collecting information concerning its environment. Note that the concept of stigmergy builds on existing systems being used to sense the environment, rather than relying on specialized communication as for example given by human language. A similar concept can be found in *quorum sensing* as a principle of bacterial communication (Whiteley et al. 2017). Finally, positive and negative feedbacks in combination with fluctuations represent the core principles of the evolutionary process. Hereby, feedback acts as a mechanism of amplification, allowing small fluctuations to enfold impact on the level of the system as a whole. Thus, through feedback via transmission of hereditary information, fluctuations such as mutations can constitute phenotypic variation. The advantage of using the concept of a holobiont-like system in contrast to other, more genetically oriented concepts, again lies in the ability to study systems systematically at all levels of biological organization. For example, it can readily be used on the level of a collection of individuals.

Noble (2017) stresses the importance of interactions at all levels of biological organization as important factors for both, development and evolution of biological systems. "*Such interactions are also analogue in nature because they depend on constraints of lower (e.g., molecular) levels by higher-level processes that are formed as dynamic patterns. Those patterns represent continuous variation in expression levels of genes and many other factors.*" (Noble 2017, p. 127)

Self-organization of Control

For the purpose of this chapter, we are particularly interested in self-organization leading to self-stabilization. Reacting to criticism concerning the possible self-stabilizing interaction between the biosphere and the atmosphere, Watson and Lovelock (1983)

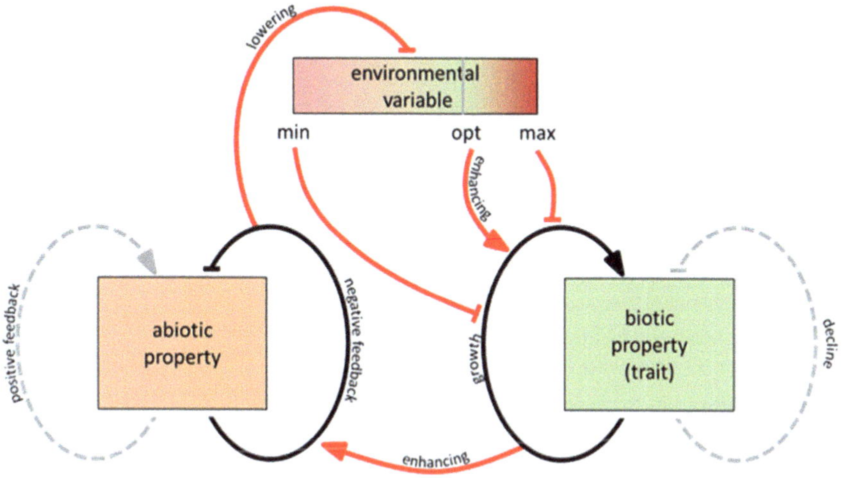

Fig. 2 Control of environmental variables can be achieved through the coupling of positive and negative feedback in both, the biotic and the abiotic 'sphere'

presented Daisyworld, a model of a planet covered with two species of daisy exhibiting different reflection properties due to different colors. Daisyworld provides an example of *rein control* (Dyke and Weaver 2013), the property of a system to regulate a set of variables in a homeostatic way (see Fig. 2). Daisyworld and successors of the model solve the problem of regulation through direct coupling between atmospheric variables and the biotic sphere reflecting the outcome of selection. In the original example, the plants reflect light and thus contribute to the thermal regulation of the atmosphere. While showing that self-regulation is possible in principle, the model nevertheless relies on very particular types of interaction and their immediate influence on the variables being controlled. Therefore, doubts have not been fully clarified since the presented coupling of life with environment appeared to be rather special.

Abstracting from the concrete example, Daisyworld exemplifies just one particular way of self-regulation. The process of evolution has led to a wealth of examples of biological systems stabilizing themselves (Lenton and Wilkinson 2003). Comparing stabilizing regulatory systems to destabilizing ones, the pure property of being stabilizing makes such systems more likely of being observed compared to destabilizing ones. Allowing for self-stabilizing interactions to emerge, the statistical nature of natural selection will make it more likely to observe more of them on the long run, given the potential that such systems manage to adapt to changing environmental conditions. Betts and Lenton (2008) argue that biotic regulatory feedback, only, can lead to the emergence of such self-stabilizing systems. The reason lies in the discrete progression of evolution already mentioned above.

Looking at evolution as a form of trial-and-error process, ever new strategies to adapt to environmental changes can be exploited as long as the creative potential of the pool of possible interactions has not been exhausted. With stabilizing mechanisms being as likely as destabilizing ones, just by chance, destabilizing biotic interactions tend to quickly lead to the extinction of the realizing biotic agents. Eventually, such biotic agents will deteriorate their conditions of existence. Thus, it is the capacity for death which allows evolution to advance, with less favorable trials diminishing themselves and the more favorable ones emerging and creating novelty. Combined with a capacity for change, the process of evolution on the long run will increase the likelihood of stabilizing interactions to be observed (see Betts and Lenton 2008).

Abiotic regulation does not have the same potential of supporting stabilizing systems since there is no way of stopping destabilizing interactions apart from them coming to a natural end due to physical or chemical constraints. Thus, the property of death on the one hand allows for unfavorable variants to deteriorate, while at the same time imposing the need for hereditary processes to help favorable ones to persist. Summarizing, self-regulation can indeed emerge through a process of evolution which is not optimizing for self-regulation per se, but effectively leading to a preference of self-stabilizing interactions as being a statistical outcome of a trial-and-error strategy.

Note that beside the possibility to create new trials as a result of chance, the process relies on some kind of memory. Without memory, a trial-and-error strategy would not much differ from pure random effects. Again, this is where generic heredity comes into play. The weakest form of information conservation can already be seen in the traces being left on the environment by an acting system. The process of fitness-enhancing environmental transformation is known as *niche construction* in ecological theory (cf. Krakauer et al. 2009). Thus, the existence of a niche can, if resulting from active transformation, already be interpreted as information transfer, since the niche lasts beyond the life span of a single occupant. Occupation of new niches can result from genetic mutations as well as from a subgroup of the population choosing the new niche (e.g., Skinner et al. 2014). Niche construction varies among genotypes. It rather establishes a correlation of the genotype with environmental conditions (Saltz and Nuzhdin 2014), although it depends on the functionality of the phenotypes emerging from the expression of information of genotypes. Similarly, there are many other ways of leaving traces of biotic action on the environment. Clearly, not all of them have the capacity of being used by further agents but undoubtedly, some of them have. For example, the sand-stabilizing effects of *Leymus arenarius* are assumed to contribute to the plants decline in dune habitats due to competition with other species (Greipsson and Davy 1994).

Longitudinal memory, i.e., preservation of information over longer periods of time and being derived from experience, can also actively be created by transfer of information from one generation to another. Such transfers need not necessarily be done in the sense of active teaching. To give an example, sodium derived from mud puddles by the moth *Gluphisia septentrionis* is transferred to the offspring, thus providing a fitness-enhancing advantage (Smedley and Eisner 1996). Multiple examples of such *parental effects* exist (Badyaev and Uller 2009), all of which can be seen as passing of information concerning the environment from the parents to their offspring. Epigenetic modifications in response to environmental cues can be stored and recalled and be also handed over to sexually produced subsequent generations (Molinier et al. 2006; Bond and Finnegan 2007; Adams 2010; Daxinger and Whitelaw 2010; Lüttge and Thellier 2016). This we name epigenetic memory (Thellier and Lüttge 2013; Lüttge and Thellier 2016).

Making use of information in an explicit way is given by *social learning* (e.g., Giurfa 2012). Social learning also produces information about the environment and thus, depending on the specific mechanism, provides a selective advantage being passed on to successive generations (van der Post et al. 2016). Furthermore, the actual process of learning itself can be supported by parental interaction with their offspring. Active teaching reveals fitness-enhancing effects since it speeds up individual learning and at the same time prevents erosion of traditions (see Thornton and Clutton-Brock 2011).

A third type of memory consists in community compositions. Albeit traditionally considered as originating from random circumstances, there is growing evidence, that certain aspects of community composition are the outcome of coevolutionary processes, leading to increased stability (Rosenberg and Zilber-Rosenberg 2018). The mere pool of possible associations might indeed be seen as random, while the results originating from interactions being created out of the pool typically show stable patterns. Random associations will originate during niche occupation with facilitation and competition between organisms establishing certain types of (bio-)diversity. Diversity establishes complexity, and this confers stability (Souza and Lüttge 2015).

For example, colonization of the volcanic island of Surtsey by vascular plants is the outcome of plant communities populating neighboring islands (Magnússon et al. 2014). From the potential brought in by wind and sea dispersal and, in addition, dispersal via birds, a widely observed association between the sea sandwort *Honckenya peploides* and costal grass *Leymus arenarius* started the formation of the plant ecosystem on the volcanic island (Fridreiksson 2005). This association between the deep-rooting dune grass and a salt-tolerant Caryophyllaceae has developed via coevolution in coastal regions of the northern hemisphere (Greipsson and Davy 1994). The information concerning this association has been carried over to the newly generated island via seed dispersal.

The amount of evidence showing transfer of microbial associations to the next generation in the holobiont sensu stricto is constantly growing. The well-known example of the maternal gut microbiome being transferred to the offspring passing through the birth canal endows the child with a first set of microbes, several of them being considered as beneficial for the development of the child (Kovatcheva-Datchary et al. 2013; van de Guchte et al. 2018). Similar examples can be found in plant systems (e.g., Vannier et al. 2018). The advantage of the host-microbe associations, providing a stable environment by the host on the one hand and a suite of quickly adaptable genomes by the high functional and species bio-diversity of the microorganisms, has been found to provide an efficient way to adapt to changes of the environment and thus providing stability of the system (Rosenberg and Zilber-Rosenberg 2018).

Return to Gaia as a Holobiont-like System

Being equipped with a solidly laid-out theoretical basis, we may now confidently ask to what extent Kirchner's form of coevolutionary Gaia carries properties of supple adaptation and thus, allows to consider Gaia as a holobiont-like system.

Koshland (2002) suggests the following **Seven Pillars of Life** being summarized in the acronym **PICERAS** to define the characteristics of life.
P—program. All forms of life follow some kind of internal program (e.g., the DNA).
I—improvisation. Living forms react to changes (unforeseen) in their environment.
C—compartmentalization. Living systems exhibit a boundary between the interior and the environment. At the same time, they make use of modularity, with modules also being compartmentalized.
E—energy. Living systems as ordered structure must be open systems and thus use energy and absorb entropy to the environment.
R—regeneration. Making use of catalysts and other chemical substances to transform energy, living systems have to cope with thermodynamic losses.
A—adaptability (which we would call acclimation). Improvisation alone is too slow. Therefore, living systems also reveal plasticity to react to environmental triggers on a short timescale.
S—seclusion. Living systems shield themselves to some extent from losing resources. For example, critical chemicals for one pathway may be prevented from being metabolized within another pathway.

As mentioned above, in order to discuss the hypothesis, we need to provide a proper basis of the concepts implicitly underlying the formulation of the hypothesis. First, there is the concept of the biosphere. It is commonly understood as the summary of all forms of life on Earth. From a broad perspective, life can be understood in terms of supple adaptation. But being an essential part of the concept of a holobiont-like system, adaptation has to be seen as a generic property. Do properties of the Earth allow us to identify adaptability of the system planet Earth? The fact that Earth adapts in a certain way seems to be widely accepted even among the most severe critics of Gaia (Kirchner 2002). The point of discussion is to what extent the biosphere and thus life per se is needed to drive adaptability of the system. We have already seen that life increases the probability of observing self-stabilizing systems, if not being an essential prerequisite (Betts and Lenton 2008).

There is an ongoing debate on the definition of life. We cannot arrive at a comprehensive definition (Lüttge and Scarano 2019, this book). Life appears more like a philosophical category than a natural-scientific one (Nachtigall 2010). We can only comprehend life by listing its essential characteristic properties. However, the discussion on a minimal set of characteristic properties is far from being settled. On the other hand, several suggestions are available for sets of characteristics, which, if being all present in a given system, allow the conclusion that we consider the system as being living. One of those sets is PICERAS (Koshland 2002) (see Textbox). Checking this list against Gaia, some characteristics appear to be obvious and commonly accepted. For example, no one would doubt the Earth being a compartmentalized system, i.e., one planet (C), transforming solar energy (E) into various forms of entropy being released to space (R), and improvising in the sense that the system adapts to environmental changes such as the increase of solar energy, or cosmic impact (I). Furthermore, Lovelock (1979) presents a careful analysis of global processes, maintaining a quite unlikely state, considered form the point of view of thermodynamic stability. Yet, we must remember that all living systems are open systems through which a continuous flow of matter and energy is occurring (Netter 1959) and this includes Gaia composed of living subsystems. These systems are thermodynamically not in steady state but in a kind of dynamic pseudo-steady state in terms of irreversible thermodynamics. Thus, with stability, we mean maintenance and sustainment of the characteristics of the dynamics as will be dealt with in more detail below in relation to homeostasis. Also seclusion is somewhat of a void condition in the context of Gaia with the Earth being the only system carrying life within a reasonable vicinity of the planet (S).

The essential argument raised by opponents of Lovelock's theory is that Gaian and anti-Gaian feedback mechanisms are likewise likely to evolve by natural selection and thus, there is no process of selection driving the evolution of Earth (Kirchner 2002). The core of this criticism is tightly linked to what essentially is the unit of selection. Opposing Gaia, Doolittle (1981) argues that natural selection always has to operate in a way, such that the unit of selection is more capable of maintaining long-term persistence than its competitors. Thus, selection is operating on a set of representatives of a certain genetic type, i.e., the population. While Charles Darwin considered selection to act on individual organisms, Stephen Jay Gould argued that selection is also acting on entire species (Gould 2002). Following Noble (2017), we need to consider multilevel interactions reflecting the life-environment relationship. We, therefore, have to search for persistence of information being stored in some way within a system. Koshland demands the presence of a program which can be adjusted through mechanisms of adaptability (Koshland 2002). In order to facilitate discussion, let us call the program—in the sense of Koshland—giving rise to the performance of a system within its environment the *infotype*. The classical genotype can therefore be seen as a subset of the infotype of a given organism. We thus have to ask: What is Gaia's infotype (P) and how does it relate to Gaia's 'fitness' (A)?

Broadening our view, selection should be interpreted as a statistical property, leading to deviations from an initial distribution of infotypes, which are perceived as being non-random (i.e., some kind of bias). *Fitness* thus is an ex post property, being attributed to a system after observing that the system outcompeted its competitors. As such, fitness is intrinsically bound to the realization of a certain infotype (e.g., the phenotype) within the context of given environmental conditions. Considering fitness as being the property optimized through the process of natural selection, selection consequently combines the phenotype and its specific environmental conditions in an intrinsic way. Being laid-out symmetrically, the concept allows to be adjusted on either side. Fitness of a system can thus change by either changing the properties of the system, or altering its environmental conditions. The gene-centric interpretation of population genetics captures fitness through the adaptation of the genotype by measuring the number of offspring. However, we need a more general view on fitness. For example, clonal growth of plants in habitats may cover large areas and thus demonstrates obvious fitness in the absence of any sexual reproduction (Kempermann and Barnes 1976; Grant 1993; Brulheide et al. 2004). Fitness might likewise arise from the ability of the phenotype to change its environmental conditions in a favorable way, e.g., as revealed by successful niche construction (see above). Thinking of anthropogenic traits, our ability to build houses guarantees *Homo sapiens* a higher level of survivability providing locally stable climatic conditions. The basis for this specific advantage leads to higher fitness which is not being stored primarily in the genotype of the organism, rather than within its cultural heritage. Summarizing, we need to uncover processes of information storage, i.e., memory, and change of information, i.e., novelty, characterizing the homeostatic processes of planet Earth. The characterizing property of Gaia is its set of interactions among the various 'spheres'. Thus, we need to focus on infotypes representing these interactions.

Hopfield (1994) described biology as the study of meaningfulness of information in natural systems contrasting the passive role information plays in physical systems. Following von Uexküll information is "*anything that an agent can sense, perceive or observe*" (von Uexküll 1957; cited in Gershenson 2012). Being a generic concept, information allows to consider interaction among agents as transfer of information. Clearly, this is a very broad definition. From a cybernetics perspective, we aim at studying those interactions which are capable of maintaining the characteristics of the system. Note that the latter is an essentially observer-dependent property (Gershenson and Heylighen 2005). Through observing a system, we as an observer are implicitly defining the system's characteristics. For example, we can consider an enterprise from an economical point of view, studying its action as player in an economic environment. We might also consider the same enterprise as a social system, aiming at its inner functioning, its management system and its way to communicate with customers. Neither one of these perspectives is right or wrong. It is the perspective of the observer giving a system its identity, i.e., the set of properties which make the system appear as an entity in the view of the observer.

In the same way, Gaia comprises a certain view of planet Earth as being a cybernetic system controlling certain global physical, chemical, geological, and biological properties. A central theme in Lovelock's work is the network of bio-/geo-chemical processes, maintaining control of the system. His hypothesis is based on the observation that conditions on Earth are highly unlikely from the point of view of thermodynamics, and at the same time surprisingly constant over the course of time. These two properties lead to the question of (1) how such conditions emerge from interactions among the various 'spheres', and (2) how homeostasis in a dynamic sense of stable open systems (see above) has been maintained over eons. Thus, we are being faced with an instantiation of emergence of control via self-organization and the question of the extent to which adaptation plays a role in maintaining such a system of control.

Coming back to the interactions, Lovelock (1979) considers three major factors being controlled by Gaia: surface temperature, chemical composition of the atmosphere, and salinity of oceans. He then discusses several processes which contribute to homeostasis of these factors. In what sense can information, needed to maintain control loops for these three factors, be found on Earth, and how is this information being stored and transmitted? These sound like unfeasible questions from a neo-Darwinian perspective, since Gaia is not captured in a single genotype. To answer the questions, let us start with the observation that living systems are open systems, taking up energy from their environment in exchange of entropy. Persistent life has to sustain homeostatic control of the energy-entropy cycle. Thus, living systems have developed processes maintaining the internal stability of the organisms in the presence of environmental fluctuations. Homeostatic control is realized through the coordinated interaction of positive and negative feedback loops. In accordance with Kirchner's objection (Kirchner 2002), initially, stabilizing and destabilizing feedbacks are equally likely to evolve. Nevertheless, in an a posteriori analysis stabilizing feedbacks have a greater chance of being observed since destabilizing components will lead to obliteration of the systems carrying them (see Betts and Lenton 2008). Torday suggests that "*moving in a prograde direction, by thinking about the evolutionary adaptations in the context of the ever-changing environment, the causal relationships become clear*" (Torday 2015, p. 580). Homeostasis can be seen as one characteristic of evolutionary adaptation. It "*can act simultaneously as both a stabilizing agent and as the determining mechanisms for evolutionary change*" (Torday 2015, p. 575). A convincing example is given by the evolution of the lung, which shows how ontogeny and phylogeny are playing together in order to master major transitions resulting from ocean-salinity and land-transition (Lane and Martin 2012).

Initially, random interactions lead to feedback cycles which in some cases might have enhanced a system's capability of maintaining internal homeostasis, while in others deterioration was the consequence. Once being in place, feedback cycles will lead to changes of the environment through action of the system. These changes will then be reflected by adaptations of other systems, also altering environmental conditions. Progressing, fluctuations occur, which the initially considered feedback system will have to compensate. From this perspective, *homeostatic control* is a mechanism to reduce noise in perception of information from environmental sensors (cf. Woods and Wilson 2013). Under changing environmental conditions, systems utilizing mechanisms to adjust their homeostatic control will have a better chance to persist. Interestingly, the latter development seems to depend on predictability of noise. While predictable noise tends to be integrated within the system's function, unpredictable noise leads to emergence of mechanisms of noise exclusion (see Stern 1999).

Evolution through homeostatic control also provides an explanation for major transitions in evolution, in particular via 'coupling of forces'. Through combining control systems which might have evolved separately, new features can emerge and higher-order control levels can be reached. The importance of pairing as driving evolutionary change has been highlighted by Lynn Margulis (then L. Sagan) supporting the theory of endosymbiosis (Sagan 1967). Clear parallel developments can be seen in Torday's study of embryogenesis and physiology of the lung. "*Perhaps the reason why we go through the life cycle from zygote to zygote is to acquire epigenetically-heritable information from the environment and selectively integrate it into our genome. The 'filtering' mechanisms are those of ontogeny and phylogeny, providing both the short-term and long-term 'histories' of the organism as a means of monitoring the homeostatic relevance of the acquired mutations.*" (Torday 2015, p. 579).

Summarizing, Gaia is represented by the set of interactions of the communities of living organisms, the ecosystems they create and the modulation of environmental processes they generate, as well as the buffers they maintain. All of them comprise an infotype of interactions which develops and adapts. Such an *integrative view* of evolution (Noble 2017) has no privileged level of causation (Noble 2012). But as an observer, we perceive the set as a self-regulatory system leading to unexpected stability over eons.

Gaia as Optimizing for Interaction

In its strongest form, the hypothesis concludes that self-organization of Gaia is optimized toward maintaining conditions for life on the planet. As mentioned before, we are far from assuming the existence of any extrinsic principle providing the basis for the criterion according to which optimization might be driven. Thus, we have to explain, how optimization can occur in absence of an a priori given functional to be optimized. We will built up the argumentation by the following thought experiment. At the end, it will follow that optimization can be seen as an ex post property being generated by a shift in probability.

Let us begin with assuming that an agent has a finite set of options for interactions. At first, the probability for choosing any one out of this set is uniform, i.e., every interaction has the same probability of being chosen. Picking one, the agent experiences the outcome of the chosen action evaluating it by mapping it onto its internal state. For the sake of simplicity, assume the result of this evaluation being either positive or negative. Hereby, *positive* is understood in the sense of helping the agent to maintain its current state of organization. In the opposite case, the chosen interaction will lead to a decrease of the agent's potential for self-maintenance (*negative*). Recall that the constant flow of entropy in the universe forces the agent to invest energy in order to perform work being needed to maintain its state of low entropy.

On the long run, negative actions will lead to a decreasing likelihood of observing such an agent, due to its increasing need of energy to work against the current break of symmetry. Recall that every form of organization does need to invest in work to resist the tendency of increasing entropy. On an average basis, over time, agents having predominantly chosen negative interactions will vanish. In the other case, the probability of observing agents which (randomly) chose several positive interactions accumulating in a net increase of their capability for self-maintenance

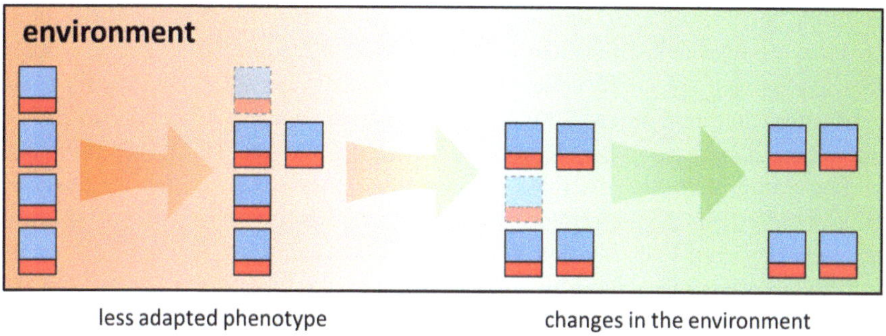

Fig. 3 Negative selection of less adapted phenotypes eventually leads to a distribution of phenotypes which retrospectively is interpreted as adaptation

will display a higher likelihood of being observed over longer time periods. Summarizing, starting with a population of agents being equal with respect to their ability for self-maintenance, successive interactions will shift the probability distribution toward agents exhibiting higher potential. This is what is called *sequential selection* (Fig. 3, Betts and Lenton 2008).

Our thought experiment rests on a critical assumption. In order to be able to accumulate an increase in the potential for self-maintenance, the agents need to have the capability to preserve the outcome of the previously chosen interaction. In a wider sense, we can call such a capability '*being able to learn from experience*'. If the population misses this capability, the results of positive and negative interactions would average out on the long run. Thus, we have found a first crucial ingredient for being able to observe organization being generated through interactions. The system must have the possibility to allow for *positive feedback*.

As mentioned before, the increasing need of energy to maintain the organizational state in the case of unfavorable interaction creates a *negative feedback* eventually leading to the situation where loss of energy exceeds energy uptake. Being less effective in using its resources, such configurations will have less likelihood of being observed over longer periods of time.

Having identified positive feedback, the question arises why evolution does not stop, i.e., converge to the ultimate, i.e., an optimal agent. The answer lies right at the start of our thought experiment. Recall that the seed for producing organization was a random selection from an initial set of options. Being crucial for the start of the process, these random seeds also play a crucial role in driving evolution. In practical terms, no system of storing and transforming information is perfect. There is always a chance of error. There is always a chance of random alteration of the 'chosen' action as there is a chance of error in the observation of its consequences. Both sources of error generate a source of novelty, producing new, not yet tested actions and not yet observed reactions. Taking into account that every action is reflected in a resulting change of the environment the latter also carries a source of random change. In consequence, the resulting source of variability leads to ever-changing conditions the agent has to cope with. As long as the agent manages to maintain its capability to react to such changes, i.e., as long as there also is an internal source of variability (e.g., plasticity), the agent will be able to persist. This property of an agent is termed *autopoiesis* by Maturana and Varela (see, e.g., Varela et al. 1974). Altogether, this creates what we observe as adaptation. Adaptation thus rephrases the observation that self-sustaining organization on the long run is more likely of being observed. In the same sense, selection means the long-term observation of change in probability distribution from a (theoretical) uniform distribution over the set of all possible interactions toward higher likelihood of interactions maintaining self-organization.

Ulanowicz' theory of autocatalysis (Ulanowicz 1998) leading to increasing ecosystem *ascendency* argues along these lines, postulating that positive interactions tend to create a self-supporting feedback. Can these arguments also be lifted to the global scale? Indeed, there is no theoretical reason for stopping us from doing so. In the same way as we have argued earlier that the integrative view of evolution relies on multilevel interactions, the biosphere–geosphere–atmosphere interactions driven by actions of living agents accumulate to Gaian interactions at the global scale. Modeling approaches have shown that Gaian self-organization, including principles of rein control, can indeed emerge from individual-based interaction (see Downing and Zvirinsky 1999). Thus, at least the *lucky version of Gaia* can be brought in line with evolution and natural selection. In their assessment of arguments, Free and Barton (2007) come to the conclusion that biotic feedback on the environment should tend to stabilize the interaction system leading to greater stabilization of the biosphere. This exactly matches our thought experiment. Hereby, stability must clearly not depend on the presence/absence of a particular species or a particular ecosystem. In this sense, we can think of **selection for stabilization** rather than selection of traits or particular genotypes.

Betts and Lenton, go further, arguing that it is the property of life itself which makes the trial-and-error strategy of evolution work. With regulatory and anti-regulatory feedback being likewise probable, the temporal boundedness of life allows anti-regulatory mechanisms to destroy themselves leading to a 'reset of the system' (cf. Betts and Lenton 2008). The difficulty of missing a population of examples to act on for selection is solved by considering advantages vs. disadvantages in a sequential order. Thus, in terms of sequential selection, Gaia is a valid perspective coming to terms with the principles of Darwin's theory. In this sense, Gaia's optimizing property results from the self-stabilizing feedback of the biotic–abiotic interaction systems. Clearly, Gaia is not competing with other comparable systems for limited resources. But as we have pointed out, competition for space and resources is not sufficient to be claimed as being the sole driving force of evolution. As has already been shown by the Baldwin effect (Simpson 1953), the outcome of natural selection also reflects choices individuals have taken. Popper (2013) has been led to formulate a concept of *active Darwinism*. In this context, the current version of the system Gaia as we observe it today is the one which evolved out of several other versions before. Obviously, none of the precursors 'survived' (e.g., the Hadean, the Archean, the Proterozoic, etc.), since the atmosphere–geosphere–biosphere interactions constantly forced Gaia to adapt (see, e.g., Wassenburg et al. 2016).

One point of criticism remains to be eliminated. While Downing and Zvirinsiky's guild model (1999) showed that self-regulation might emerge from individual-based natural selection, the actual nature of the interaction between life and its environment remains predefined in their model. This condition has been released by the Flask model (Williams and Lenton 2007). The model shows nutrient recycling and stable ranges of abiotic factors emerging from a microbial-interaction model without referring back to an explicit assumption on the nature of the life–environment interaction.

A purely interaction-based approach not even using interpretations such as nutrients or biotic/abiotic factors is given by the Tangled Nature Model (Christenssen et al. 2002). Arthur and Nicholson (2017) used a modified version of the Tangled Nature Model to allow for effects of the individuals on the physical environment of another individual. While the original version of the Tangled Nature Model only considered direct interaction among individuals, this generalization reflects both, direct and indirect effects of action by an individual on neighboring individuals. The essential features of the Tangled Nature Model remain, i.e., quasi-stable evolutionary states emerge, in which a network of species forms a *stable core* which eventually breaks down if a new species evolves, which destabilizes the existing core via *parasitic* or *symbiotic quakes*. Over time, it gets more and more difficult for a newly generated species to induce such quakes, leading to increasing, so-called *entropic barriers*. The alternation of periods determined by self-stabilizing interactions with shorter periods of destabilizing interactions confirms the pattern of sequential selection. Through the coevolution of the species with their environment, stability again increases as a consequence of increasing entropic barriers. The life–environment interaction in this model does not necessarily improve conditions for life, but makes them more probable (cf. Arthur and Nicholson 2017). Thus, at the end, it is the more likely we observe, leading us to accept Gaia even in its optimizing version.

Conclusions

The Gaia system being defined as the sum of all interactions of life with its environment on planet Earth thus forms a holobiont-like system with many processes of processing information in both, the biotic and the abiotic 'spheres'. These processes undoubtedly have led to conditions which support life on Earth through a network of positive and negative feedbacks. Self-stabilization of this network can indeed arise from action of individual agents. In this sense, Gaia has evolved over eons of time. The outcome of a sequential process of selection for stabilization has led to the interpretation of Gaia as optimizing conditions for life on Earth.

Gaia cannot be understood from one level of biological organization alone. It is a system of multilevel interaction. And we must keep in mind that no single organism, no single ecosystem and no single interaction can be considered as a must. The history of planet Earth has shown several quakes, both, parasitic and symbiotic. Superplume eruptions leading to changes in the atmosphere with mass extinctions being the consequence (Matyssek and Lüttge 2013) might be considered as example of the first, the development of vascular plants as an example for the second type. Either way, Lovelock has been misinterpreted early on as risking that based on his hypothesis men would resist from taking action to stop threatening the global ecosystem due to human perturbation (Doolittle 1981). But this conclusion does not hold. The optimizing view of Gaia is acceptable, indeed, but with stabilizing Gaia being more probable, it does not follow that the improbable is not possible. Gaia does not hold up as an excuse not to carefully study the sensible interactions of feedbacks stabilizing every complex system and base ones decisions to interact on the fact that every action in such a system leads to counter-action.

References

Adams KL (2010) Dandelions 'remember' stress: heritable stress-induced methylation patterns. New Phytol 185:867–868

Aitchison JD, Rout MP (2015) The interactome challenge. J Cell Biol 211(4):729

Allis CD, Jenuwein T (2016) The molecular hallmarks of epigenetic control. Nat Rev Gen 17:487–500

Anonymous (1914) Homeric hymns. In: Evelyn-White HG (ed) The Homeric Hymns and Homerica (Evelyn-White HG, With an English Translation). Harvard University Press, London, William Heinemann Ltd., Cambridge, MA

Arthur R, Nicholson A (2017) An entropic model of Gaia. J Theor Biol 430:177–184

Badyaev AV, Uller T (2009) Parental effects in ecology and evolution: mechanisms, processes and implications. Phil Trans R Soc B 364(1520):1169–1177

Bedau MA (1996) The nature of life. In: Boden M (ed) The philosophy of artificial life. Oxford Univ Press, New York, pp 332–357

Bedau MA (1998) Four puzzles about life. Artif Life 4:125–140

Beerling DJ, Berner RA (2005) Feedbacks and the coevolution of plants and atmospheric CO_2. Proc Natl Acad Sci USA 102:1302–1305

Berner RA, VandenBrooks JM, Ward PD (2007) Oxygen and evolution. Science 316:557–558

Betts RA, LentonTM (2008) Second chances for lucky Gaia: a hypothesis of sequential selection. Hadley Centre, Technical Note 77 (6 Oct 2008)

Bond DM, Finnegan EJ (2007) Passing the message on: inheritance of epigenetic traits. Trends Plant Sci 12:211–216

Brulheide H, Manegold M, Jandt U (2004) The genetical structure of *Populus euphratica* and *Alhagi sparsifolia* stands in the Taklimakan desert. In: Runge M, Zhang X (eds) Ecophysiology and habitat requirement of perennial plant species in the Taklimakan desert. Shaker, Aachen, pp 153–160

zu Castell W, Fleischmann F, Heger T, Matyssek R (2016) Shaping theoretic foundations of holobiont-like systems. Prog Bot 77:219–244

Chinnusami V, Zhu J-K (2009) Epigenetic regulation of stress responses in plants. Current Opinion Plant Biol 12:133–139

Christensen K, Di Collobioano SA, Hall M, Jensen HJ (2002) Tangled nature: a model of evolutionary ecology. J Theor Biol 216:73–84

Davis JM (1980) The coordinated aerobatics of dunlin flocks. Anim Behav 28:668–673

Dawkins R (1976) The selfish gene. Oxford University Press, Oxford

Daxinger L, Whitelaw E (2010) Transgenerational epigenetic inheritance: more questions than answers. Genome Res 20:1623–1628
Doolittle WF (1981) Is nature really motherly? Coevol Quart 29:58–63
Doolittle WF (2014) Natural selection through survival alone, and the possibility of Gaia. Biol Philos 29:415–423
Downing K, Zvirinsky P (1999) The simulated evolution of biochemical guilds: reconciling Gaia theory and natural selection. Artif Life 5:291–318
Dyke JG, Weaver IS (2013) The emergence of environmental homeostasis in complex ecosystems. PLoS Comput Biol 9(5):e1003050
Free A, Barton NH (2007) Do evolution and ecology need the Gaia hypothesis? Trends Ecol Evol 22(11):611–619
Fridriksson S (2005) Surtsey. Ecosystems formed. The Surtsey Research Society, Reykjavik
Garnier S, Gautrais J, Theraulaz G (2007) The biological principles of swarm intelligence. Swarm Intell 1:3–31
Gershenson C (2012) The world as evolving information. In: Minai AA, Braha D, Bar-Yam Y (eds) Unifying themes in complex systems VII. Springer, Berlin, Heidelberg
Gershenson C, Heylighen F (2005) How can we think the complex? In: Richardson K (ed) Managing organizational complexity: philosophy, theory and application. Information Age Publishing, Greenwich, pp 47–61
Gershenson C, Trianni V, Werfel J, Sayama H (2018) Self-organization and artificial life: a review. arXiv:1804.01144 (3 Apr 2018)
Giurfa M (2012) Social learning in insects: a higher-order capacity? Front Behav Neurosci 6:57
Gould SJ (2002) The structure of evolutionary theory. Harvard University Press, Cambridge
Grant MC (1993) The trembling plant. Discover. (Los Angeles) 84:82–89
Greipsson S, Davy A (1994) *Leymus arenarius*. Characteristics and uses of a dune-building grass. Icel Agr Sci 8:41–50
Grunstein M (1997) Histone acetylation in chromatin structure and transcription. Nature 389:349–352
van de Guchte M, Blottière HM, Doré J (2018) Humans as holobionts: implications for prevention and therapy. Microbiome 6:81
Hopfield JJ (1994) Physics, computation, and why biology looks so different. J Theor Biol 171:53–60
Kasting JF, Siefert JL (2002) Life and the evolution of Earth's atmosphere. Science 296:1066–1068
Kemperman JA, Barnes BV (1976) Clone size in American aspens. Can J Bot 54:2603–2607
Kirchner JW (2002) The Gaia hypothesis: fact, theory, and wishful thinking. Clim Change 52:391–408
Koshland DE (2002) The seven pillars of life. Science 295:2215–2216
Kovatcheva-Datchary P, Tremaroli V, Bäckhed F (2013) The gut microbiota. In: Rosenberg E, DeLong EF, Lory S, Stackebrandt E, Thompson F (eds) The Prokaryotes. Human microbiology, 4th ed. Springer, Berlin, pp 3–24
Krakauer DC, Page KM, Erwin DH (2009) Diversity, dilemmas, and monopolies of niche construction. Am Nat 173(1):26–40
Lane N, Martin WF (2012) The origin of membrane bioenergetics. Cell 151:1406–1416
Lenton TM (2004) Clarifying Gaia: regulation with or without natural selection. In: Schneider SH, Miller JR, Crist E, Boston PJ (eds) Scientists debate Gaia. The next century. MIT Press, Cambridge, MA, pp 15–25
Lenton TM, Wilkinson DM (2003) Developing the Gaia theory. Clim Change 58:1–12
Lovelock JE (1972) Gaia as seen through the atmosphere. Atmos Envir 6:579–580
Lovelock JE (1979) Gaia: a new look at life on earth. Oxford University Press, Oxford
Lovelock JE (1988) The ages of Gaia. W.W. Norton & Company, New York
Lovelock JE, Margulis L (1974) Atmospheric homeostasis by and for the biosphere: the Gaia hypothesis. Tellus 24:2–9

Lüttge U (2016) Plants shape the terrestrial environment on earth: challenges of management for sustainability. Progr Bot 77:187–217

Lüttge U, Scarano FR (2019) Emergence and Sustainment of Humankind on Earth: the Categorical Imperative. In: Wegner LH, Lüttge U (eds), Emergence and modularity in life science, Springer, Heidelberg, pp 235–254

Magnússon B, Magnússon SH, Ólafsson E, Sigurdsson BD (2014) Plant colonization, succession and ecosystem development on Surtsey with reference to neighbouring islands. Biogeosciences 11:5521–5537

Marsh L, Onof C (2008) Stigmergic epistemology, stigmergic cognition. Cogn Syst Res 9(1–2):136–149

Martin A, Chahwan R, Parsa JY, Scharff MD (2015) Somatic hypermutation: the molecular mechanisms underlying the production of effective high-affinity antibodies. In: Honjo T, Reth M, Radbruch A, Alt F (eds) Molecuar biology of B cells. Academic Press, London, pp 363–388

Matyssek R, Lüttge U (2013) Gaia: The planet Holobiont. In: Matyssek R, Lüttge U, Rennenberg H (eds), The alternatives growth and defense: resource allocation at multiple scales in plants. Nova Acta Leopoldina, vol 114/391, pp 325–344

Molinier J, Ries G, Zipfel C, Hohn B (2006) Transgeneration memory of stress in plants. Nature 442:1046–1049

Nachtigall W (2010) Bionik als Wissenschaft. Erkennen – Abstrahieren – Umsetzen. Springer, Heidelberg

Netter H (1959) Theoretische Biochemie. Springer, Heidelberg

Noble D (2012) A theory of biological relativity: no privileged level of causation. Interface Focus 2(1):55–64

Noble D (2017) Digital and analogue information in organisms. In: Walker SI, Davies PCW, Ellis GFR (eds) From matter to life. Information and causality. Cambridge University Press, Cambridge, pp 114–129

Popper K (2013) Eine Neuinterpretation des Darwinismus. Die erste Medawar-Vorlesung 1986 (German). Aufklärung und Kritik 1:7–20

van der Post DJ, Franz M, Laland KN (2016) Skill learning and the evolution of social learning mechanisms. BMC Evol Biol 16(1):166

Reynolds CW (1987) Flocks, herds and schools: a distributed behavioral model. ACM SIGGRAPH Comp Graph 21:25–34

Rosenberg E, Koren O, Reshef L, Efrony R, Zilber-Rosenberg I (2007) The role of microorganisms in coral health, disease and evolution. Nat Rev Microbiol 5:355–362

Rosenberg E, Zilber-Rosenberg I (2018) The hologenome concept of evolution after 10 years. Microbiome 6:78

Sagan L (1967) On the origin of mitosing cells. J Theor Biol 14:225–274

Saltz JB, Nuzhdin SV (2014) Genetic variation in niche construction: implications for development and evolutionary genetics. Trends Ecol Evol 29(1):8–14

Schwartzman DW, Volk T (1989) Biotic enhancement of weathering and the habitability of Earth. Nature 340:457

Selous E (1931) Thought-transference (or what?) in birds. Constable, London

Simpson GG (1953) The Baldwin effect. Evolution 7(2):110–117

Skinner MK, Gurerrero-Bosagna C, Haque MM, Nilsson EE, Koop JAH, Knutie SA, Clayton DH (2014) Epgenetics and the evolution of Darwin's finches. Genome Biol Evol 6(8):1972–1989

Smedley SR, Eisner T (1996) Sodium: a male moth's gift to its offspring. Proc Natl Acad Sci USA 93:809–813

Souza GM, Lüttge U (2015) Stability as a phenomenon emergent from plasticity—complexity—-diversity in eco-physiology. Progr Bot 76:211–239

Stern M (1999) Emergence of homeostasis and 'noise imprinting' in an evolution model. Proc Natl Acad Sci USA 96:10746–10751

Tansley AG (1935) The use and abuse of vegetational concepts and terms. Ecology 16(3):284–307

Theis KR, Dheilly NM, Klassen JL, Brucker RM, Baines JF, Bosch TCG, Cryan JF, Gilbert SF, Goodnight CJ, Lloyd EA, Sapp J, Wandenkoornhuyse P, Zilber-Rosenberg I, Rosenberg E, Bordenstein SR (2016) Getting the hologenome concept right: an eco-evolutionary framework for hosts and their microorganisms. MSystems 1(2):e00028–16
Thellier M, Lüttge U (2013) Plant memory: a tentative model. Plant Biol 15:1–12
Thornton A, Clutton-Brock T (2011) Social learning and the development of individual and group behaviour in mammal society. Phil Trans Roy Soc B Biol Sci 366:978–987
Torday JS (2015) Homeostasis as the mechanism of evolution. Biology 4:573–590
Tyrrell T (2013) On Gaia: a critical investigation of the relationship between life and earth. Princeton University Press, Princeton
von Uexküll J (1957) A stroll through the worlds of animals and men. In: Schiller CH (ed) Instinctive behavior: the development of a modern concept. Internships University Press, New York, pp 5–80
Ulanowicz RE (1998) Network orientors: theoretical and philosophical considerations why ecosystems may exhibit a propensity to increase ascendency. In: Müller F, Leupels M (eds) Eco targets, goal functions, and orientors. Springer, Berlin, pp 177–192
Ulanowicz RE (2014) Reckoning the nonexistent: putting the science right. Ecol Modell 293:22–30
Vannier N, Mony C, Bittebiere A-K, Michon-Coudouel S, Biget M, Wandenkoornhuyse P (2018) A microorganisms' journey between plant generations. Microbiome 6:79
Varela FG, Maturana HR, Uribe R (1974) Autopoisis: the organization of living systems, its characterization and a model. BioSystems 5:187–196
Wadhams P (2016) The global impacts of rapidly disappearing arctic sea ice. YaleEnvironment360, 26 Sept 2016
Walter H, Breckle SW (1984) Ökologie der Erde. 2. Spezielle Ökologie der tropischen und subtropischen Zonen. Gustav Fischer, Stuttgart
Wassenburg JA, Dietrich S, Fietzke J, Fohlmeister J, Jochum KP, Scholz D, Richter DK, Sabaoui A, Spötl C, Lohmann G, Andreae MO, Immenhauser (2016) Reorganization of the north atlantic oscillation during early Holocene deglaciation. Nat Geosci 9:602–605
Watson AJ, Lovelock JE (1983) Biological homeostasis of the global environment: the parable of Daisyworld. Tellus 35B:284–289
Whiteley M, Diggle SP, Greenberg EP (2017) Progress in and promise of bacterial quorum sensing research. Nature 551:313–320
Williams HTP, Lenton TM (2007) The Flask model: emergence of nutrient-recycling microbial ecosystems and their disruption by environment-altering rebel organisms. Oikos 116:1087–1105
Woods HA, Wilson JK (2013) An information hypothesis for the evolution of homeostasis. Trends Ecol Evol 28(5):283–289
Yaish MW, Colasanti J, Rothstein SJ (2011) The role of epigenetic processes in controlling flowering time in plants exposed to stress. J Exp Bot 62:3727–3735
Zhang Y, Reinberg D (2001) Transcription regulation by histone methylation: interplay between different covalent modifications of the core histone tails. Genes Dev 15:2343–2360
Zilber-Rosenberg I, Rosenberg E (2008) Role of microorganisms in the evolution of animals and plants: the hologenome theory of evolution. FEMS Microbiol 32:723–735

MIX
Papier aus verantwortungsvollen Quellen
Paper from responsible sources
FSC® C105338

If you have any concerns about our products,
you can contact us on
ProductSafety@springernature.com

In case Publisher is established outside the EU,
the EU authorized representative is:
Springer Nature Customer Service Center GmbH
Europaplatz 3, 69115 Heidelberg, Germany

Printed by Libri Plureos GmbH
in Hamburg, Germany